Y0-AEU-375

Macromolecular Symposia

Symposium Editor: A.S. Abd-El-Aziz

Editor: I. Meisel
Deputy Editor: C.S. Kniep
Senior Associate Editor: S. Spiegel
Associate Editor: A. Carrick

Executive Advisory Board: M. Antonietti, M. Ballauff, S. Kobayashi, K. Kremer, T.P. Lodge, H.E.H. Meijer, R. Mülhaupt, A.D. Schlüter, H.W. Spiess, G. Wegner

196

pp. 1–353

June 2003

Macromolecular Symposia publishes lectures given at international symposia and is issued irregularly, with normally 14 volumes published per year. For each symposium volume, an Editor is appointed. The articles are peer-reviewed. The journal is produced by photo-offset lithography directly from the authors' typescripts.
Further information for authors can be obtained from:
Editorial office "Macromolecular Symposia"
Wiley-VCH Verlag GmbH & Co. KGaA,
Boschstrasse 12, 69469 Weinheim,
Germany
Tel. +49 (0) 62 01/6 06-2 38 or -5 81; Fax +49 (0) 62 01/6 06-3 09 or 5 10;
E-mail: macro-symp@wiley-vch.de
http://www.ms-journal.de
Suggestions or proposals for conferences or symposia to be covered in this series should also be sent to the Editorial office at the address above.

Macromolecular Symposia:
Annual subscription rates 2003 (print only or online only)*
Germany, Austria € 1318; Switzerland SFr 2168; other Europe € 1318; outside Europe US $ 1568.
Macromolecular Package, including Macromolecular Chemistry & Physics (18 issues), Macromolecular Bioscience (12 issues), Macromolecular Rapid Communications (18 issues), Macromolecular Theory & Simulations (9 issues) is also available. Details on request.
* For a 5 % premium in addition to **Print Only** or **Online Only**, Institutions can also choose both print and online access.
Packages including Macromolecular Symposia and Macromolecular Materials & Engineering are also available. Details on request.
Single issues and back copies are available. Please inquire for prices.

Orders may be placed through your bookseller or directly at the publishers:
WILEY-VCH Verlag GmbH & Co. KGaA, P. O. Box 10 11 61, 69451 Weinheim, Germany, Tel. +49 (0) 62 01/6 06-400, Fax +49 (0) 62 01/60 61 84. E-mail: service@wiley-vch.de

Macromolecular Symposia (ISSN 1022-1360) is published with 14 volumes per year by WILEY-VCH Verlag GmbH & Co. KGaA, P. O. Box 10 11 61, 69451 Weinheim, Germany. Air freight and mailing in the USA by Publications Expediting Inc., 200 Meacham Ave., Elmont, NY 11003, USA. Application to mail at Periodicals Postage rate is paid at Jamaica, NY 11431, USA. US POSTMASTER please send address changes to: Macromolecular Symposia, c/o Wiley-VCH, III River Street, Hoboken, NJ 07030, USA.

WILEY-VCH Verlag GmbH & Co. KGaA grants libraries and other users registered with the Copyright Clearance Center (CCC) Transactional Reporting Service the right to photocopy items for internal or personal use, or the internal or personal use of individual clients, subject to a base fee of $ 17.50 per copy, per article, payable to CCC, 21 Congress St., Salem, MA 01970. 1022–1360/99/$17.50+.50.

© WILEY-VCH Verlag GmbH & Co. KGaA, Weinheim, Germany, 2003
Printing: Strauss Offsetdruck, Mörlenbach. Binding: J. Schäffer, Grünstadt

QD380
.M328
v.196

Invited lectures from

Metal- and Metalloid-Containing Macromolecules

The 39th IUPAC Congress and 86th Conference of The Canadian Society for Chemistry,

Ottawa, Canada
August 10–15, 2003

Symposium Editor

Alaa S. Abd-El-Aziz

Department of Chemistry
The University of Winnipeg
Winnipeg, Manitoba R3B 2E9
Canada

Symposium Organizing Committee

Alaa S. Abd-El-Aziz, Ian Manners, Hiroshi Nishihara, Martel Zeldin

---

Copyright © 2003 WILEY-VCH Verlag GmbH & Co. KGaA
ISBN 3-527-30700-1

The **tables of contents** of the published issues are displayed on the WWW.

This service as well as further information on our journals can be found at the following WWW address:

http://www.ms-journal.de

Contents of Macromolecular Symposia 196

Metal- and Metalloid-Containing Macromolecules Conference
Ottawa (Canada), 2003

* The asterisk indicates the name of the author to whom inquiries should be addressed.

Author Index

Preface

The rapid pace at which the field of metal- and metalloid-containing macromolecules is expanding is a result of the applications that these materials may find in the electronic, biomedical and other fields. While the properties of organic macromolecules are based on the chemistry of carbon and a few other elements such as hydrogen, nitrogen, oxygen, sulfur and the halogens, the properties of metal- and metalloid-containing macromolecules are dependent on a large number of main group and transition metals as well as other elements with metal-like character.
This volume is a result of the Metal- and Metalloid-Containing Macromolecules Symposium to be held in Ottawa, Canada at the 39th IUPAC Congress and 86th Conference of the Canadian Society for Chemistry in August, 2003. It contains 30 articles that provide an overview of this field and covers recent developments in the synthesis of monomers and polymers as well as examining the properties and applications of these new materials. These articles were written by some of the leading researchers in this field from Japan, China, France, United Kingdom, The Netherlands, Germany, Italy, United States and Canada. The purpose of this symposium was to generate dialogue between these researchers and to allow them to present some of their most recent findings.
This symposium was supported by the American Chemical Society, Petroleum Research Funds, as well as the Inorganic and Macromolecular Chemistry and Engineering Divisions of the Canadian Society for Chemistry. The organizers of this symposium would like to extend their appreciation to all of the contributors to this volume for their willingness to prepare their manuscripts and for their participation at the conference.

Alaa S. Abd-El-Aziz

Macromol. Symp. **196**, 1–25 (2003)

Nano-Scale Metallodendritic Complexes in Electron-Transfer Processes and Catalysis

Didier Astruc,[*a] Jean-Claude Blais,[b] Marie-Christine Daniel,[a] Victor Martinez,[a] Sylvain Nlate,[a] Jaime Ruiz*[a]

[a]Groupe Nanosciences Moléculaires et Catalyse, LCOO, UMR CNRS N° 5802, Université Bordeaux I, 33405 Talence Cedex., France
E-mail : d.astruc@lcoo.u-bordeaux1.fr
[b]LCSOB, UMR CNRS N° 7613, Université Paris VI, 75252 Paris, France

Summary: Nano-sized metallodendrimers in which the equivalent metal fragments are located at the periphery can be assembled covalently, by H-bonding (supramolecular) or onto dendronized nanoparticles. They can be used as electron-reservoirs, i.e. molecular batteries, redox catalysts and sensors for the recognition of biologically relevant anions. They can also be deposited on metal surfaces or electrodes, which optimizes their use as recoverable sensors.

Keywords: catalysis; dendrimers, nanocomposites; sensors; transition metal chemistry

Content

Introduction

Metallodendrimers [1-10] with redox stability are electron-reservoir systems that should prove useful as molecular batteries, catalysts and sensors. With a low polydispersity, their molecular definition is much more precise than that of organometallic polymers, yet they can be very large. Moreover, their use in molecular electronics is promising in the context of nano-technology.

© 2003 WILEY-VCH Verlag GmbH & KGaA, Weinheim CCC 1022-1360/00/$ 17,50+.50/0

Ferrocenyl dendrimers are now a well-spread field of dendrimer research to which we have contributed.[11-14] In this chapter, we review our strategy as well as electron-transfer processes in other metallodendrimers involving monomeric electron-reservoir iron-sandwich complexes.

Organo-Iron Syntheses of Dendritic Cores, Dendrons and Large Dendrimers

In the robust, very easily accessible cationic complexes $[FeCp(arene)][PF_6)]$, the benzylic protons are more acidic than in the free arene because of the electron-withdrawing character of the 12-electron $CpFe^+$ moiety. For instance, $[FeCp(C_6Me_6)][PF_6)]$ is more acidic by 15 p*K*a units (p*K*a = 28 in DMSO) than in the corresponding free arene (p*K*a = 43 in DMSO). As a result, these complexes are much more easily deprotonated than the free arene. [15,16] This key proton-reservoir property led us to synthesize stars and dendrimers in an easy way. [15] Indeed, reaction of $[FeCp(C_6Me_6)][PF_6)]$,[17-19] with excess KOH (or *t*-BuOK) in THF or DME and excess methyl iodide, alkyl iodide, allyl bromide or benzylbromide result in the one-pot hexasubstitution (Scheme 1a).[20-22] With allyl bromide (or iodide) in DME, the hexaallylated complex has also been easily isolated and its X-ray crystal structure determined. With alkyliodides, the reaction using *t*-BuOK only leads to dehalogenation of the alkyl iodide giving the terminal olefin. Thus, one must use KOH, and the reactions with various alkyl iodides (even long-chain ones) were shown to work very well with this reagent to give the hexaalkylated Fe^{II}-centered complexes. The hexa-alkylation was also performed with alkyl iodides containing functional groups at the alkyl chain termini. [23] For instance, 1-ferrocenylbutyliodide reacts nicely to give the hexaferrocene star containing the $CpFe^+$ center.[24] The reaction with excess benzylbromide, [20,24] *p*-alkoxybenzylbromide [24] or *p*-bromobenzylbromide [24] only gives the hexabenzylated, hexa-*p*-alkoxybenzylated or hexa *p*-bromobenzylated complex as the ultimate reaction product. Cleavage of the methyl group in the *p*.methoxybenzyl derivatives synthesized in this way yields the hexaphenolate stars that could be combined with halogen containing organometallic compounds.[24b,c]

t-BuOK THF, RT — RX THF, RT

$RX = CH_3I, PhCH_2Br, SiMe_3Cl, PPh_2Cl, FeCp(CO)_2Br$ (also CO_2 and metal carbonyls)

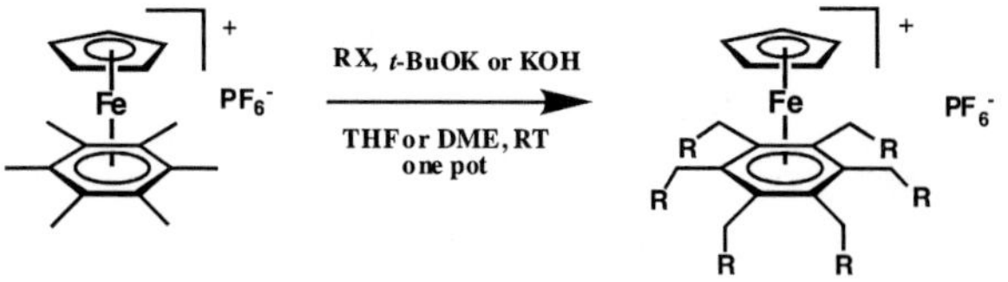

RX, *t*-BuOK or KOH — THF or DME, RT one pot

$R = CH_3, (CH_2)_nCH_3, (CH_2)_4Fc$ (X = I), CH_2Ph, *p*-CH_2PhOR', $CH_2CH{=}CH_2$, (X = Br)

Scheme 1: Deprotonation of $[FeCp(\eta^6\text{-}C_6Me_6)][PF_6]$ followed by reactions with electrophiles (top) and one-pot hexafunctionalization of this complex under ambient conditions (bottom). The top reaction illustrates the mechanism of the bottom one.

It is remarkable that the allyl group (as allyl bromide or iodide) is the only one leading to complete double branching of the C_6Me_6 complex. $CpFe^+$-induced dodecaallylation of C_6Me_6 indeed gives the extremely bulky dodeca-allylation product [26] that can be reached when the reaction is prolonged for two weeks at 40°C. The chains are blocked in a directionality that cannot convert into its enantiomer and makes the metal complex chiral. Both the hexa- and dodeca-allylation reactions are well controlled.

Alkynyl halides cannot be used in the $CpFe^+$-induced hexafunctionalization reaction, but alkynyl substituents can be introduced from the hexa-alkene derivative by bromination followed by dehydrohalogenation of the dodecabromo compound.[27] The hexa-alkene is also an excellent starting point for further syntheses, especially using hydroelementation reactions. Hydrosilylation reactions catalyzed by Speir's reagent led to long-chain hexasilanes [28] and hydrometallations were also achieved using $[ZrCp_2(H)(Cl)]$.[29] The hexa-zirconium compound obtained is an intermediate for the synthesis of the hexa-iodo derivative.[29] One of most useful

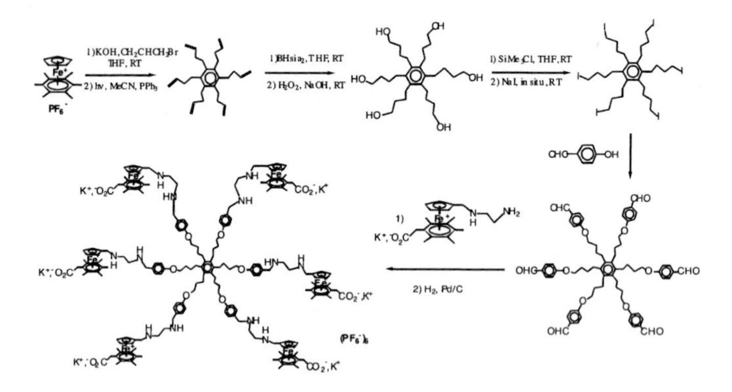

Scheme 2: $CpFe^+$-induced hexa-allylation of C_6Me_6 and subsequent hexafunctionalization of the aromatic stars with the heterodifunctional, water soluble organometallic redox catalyst (bottom) for the cathodic reduction of nitrates and nitrites to ammonia in water).

hydroelementation reactions of the hexabutenyl derivatives is the hydroboration leading to the hexaborane. The latter is oxidized to the hexa-ol using H_2O_2 under basic condition.[21] This chemistry can be carried out on the iron complex or alternatively on the free hexa-alkene which may be liberated from the metal by photolysis in CH_2Cl_2 or MeCN using visible light.[20] The polyol stars and dendrimers can be transformed into mesilates and iodo derivatives that are useful for further functionalization. The hexa-ol is indeed the best source of hexa-iodo derivative either using HI in acetic acid or even better by trimethylsilylation using $SiMe_3Cl$ followed by iodination using NaI.[30] Williamson coupling reactions between the hexa-ol and 4-bromomethylpyridine or -polypyridine led to hexa-pyridine and hexa-polypyridine and to their ruthenium complexes (31,32). This hexa-iodo star was condensed with *p*-hydroxybenzaldehyde to give an hexa-benzaldehyde star, which could further react with substrates bearing a primary amino group. Indeed, this reaction yielded a water-soluble hexametallic redox catalysts which was active in the electroreduction of nitrate and nitrite to ammonia in basic aqueous solution, *vide infra*.[33-35]

If the hexafunctionalization of hexamethylbenzene leads to stars, the octafunctionalization of durene leads to dendritic cores. The first of these octa-alkylation reactions was reported as early as 1982, and led to a primitive dendritic core containing a metal-sandwich unit.[20] Thus, as the hexafunctionalization, this reaction is very specific. Two hydrogen atoms of each methyl group are now replaced by two methyl, allyl or benzyl groups.[26] Applications to the synthesis of dendrimers containing 8 [36] or 24 redox-active groups has recently been reported. Double branching, *i.e.* replacement of two out of three hydrogen atoms by two groups on each methyl substituent of an aromatic ligand coordinated to an activating cationic group CpM^+ in an 18-electron complex is also easily obtained in the pentamethylcyclopentadienyl ligand (in pentamethyl cobaltocenium [37] and in penta- [38] and decamethylrhodocenium [39]). The interconversion of the two directionalities of decafunctionalized ligands coordinated to $CpCo^+$ or $CpRh^+$ whose could be observed by ^1H NMR for the decaisopropyl- and decaisopentyl cyclopentadienyl cobalt and rhodium complexes [37-39] (Scheme 3).

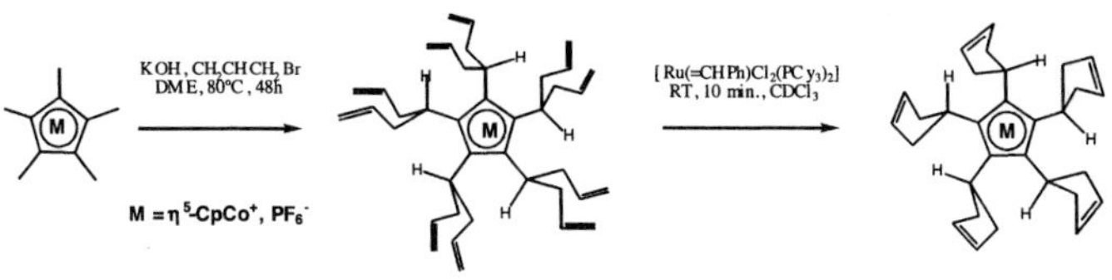

Scheme 3: Deca-allylation of 1,2,3,4,5-pentamethylcobaltocenium in a one-pot reaction consisting in 10 deprotonation-allylation sequences (steric constraints inhibit further reaction, and the 10 groups introduced are self-organized according to a single directionality) and follow-up RCM of the deca-allylated complex.

In all the above examples, the polybranching reaction of arene ligands was limited by the steric bulk. In the toluene and mesitylene ligands, the deprotonation-allylation reactions are no longer restricted by the neighborhood of other alkyl groups. All the benzylic protons, *i.e.* three per benzylic carbon, can be replaced by methyl or allyl groups in the one-pot iterative methylation or allylation reactions.[30] Thus, the toluene complex can be triallylated and the resulting tripod can be disymetrized by stoichiometric [40] or catalytic reaction [41] with transition metals shown in Scheme 18. The metathesis reaction, in particular, is complete in 5 min. at room temperature using the first-generation Grubb's catalyst $[Ru(=CHPh)Cl_2(PCy_2)_2]$ [2] with many polyallylated complexes [FeCp(arene)]$^+$ described above as well as to the decaallylated cobalt complex.[41] The reaction is very selective and terminal double bonds remain unreacted using this catalyst at room temperature.

The mesitylene complex can be nonaallylated, these reactions being carried out smoothly at room temperature in the presence of excess KOH and allyl bromide. The nonaallyl complex was photolyzed using visible light to remove the metal group CpFe$^+$, then hydroborated using 9-BBN, and the nonaborane was oxidized using H_2O_2/OH^- to the nona-ol.[30]

The triple branching reaction being very straightforward, we sought a more sophisticated version compatible with a functional group in the para position of the tripod in order to open the access to a functional dendron. Serendipitously, we found that KOH or *t*-BuOK easily cleaved the iron complexes of aromatic ethers under very mild conditions. The activating CpFe$^+$ group again induces this reaction, which is very general for a variety of aromatic ether complexes (42,43).

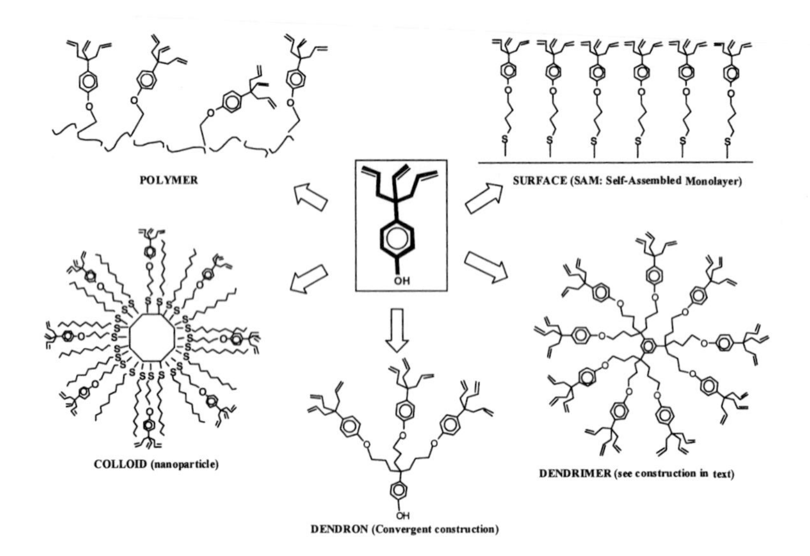

Scheme 4: Example of the linkage of the phenoltriallyl dendron to various nanostructures.

Since this cleavage reaction is carried out with the same reagent and solvent as the one used in the trialkylation reaction (ideally *t*-BuOK in THF), we have attempted to perform both reactions in a well-defined order (triallylation before ether cleavage) in a one-pot reaction. Indeed, this works out well and the $CpFe^{II}$ complex of the phenol tripod was made in 50% yield in this way. This complex can be photolyzed in the usual way using visible light, which yields the free phenol tripod. However, we have also further investigated the possibility to obtain the cleavage of the arene ligand *in situ* at the end of the phenol tripod construction; *t*-BuOK is a reductant when it cannot perform other reactions. Since the two important reactions are over, then comes the third role of *t*-BuOK: single-electron reductant. Reasoning in this way turned out to be correct: the cleavage of the arene intervenes rapidly at the 19-electron stage because 19-electron complexes of this kind are not stable with an heteroatom located in exocyclic position (most probably because the heteroatom coordinates to the metal from the labile 19-electron structure). After optimizing the reaction conditions, a 50%-yield of free phenol dendron from the ethoxytoluene complex could be reproducibly obtained,[44,45] and this reaction is now currently used in our laboratory to synthesize this very useful dendron as a starting material.

This phenoltriallyl dendron has been functionalized at both the phenolic and allylic positions. For instance, the dendron can be bound, after suitable molecular engineering, to the branches of a phenolic-protected dendron (convergent construction) onto stars and dendritic cores (divergent construction), nonaparticles, surfaces and polymers. An example is provided by the $CpFe^{+}$-induced hexafunctionalization by a phenol-nonaallyl dendron (prepared according to such a convergent synthesis) that was functionalized in phenolic position by a tail terminated by a benzylbromide group. This type of strategy allows direct access to large dendrimers by simply using the $CpFe^{+}$-induced hexafunctionalization reaction that gives hexa-branch stars with linear organic halides.

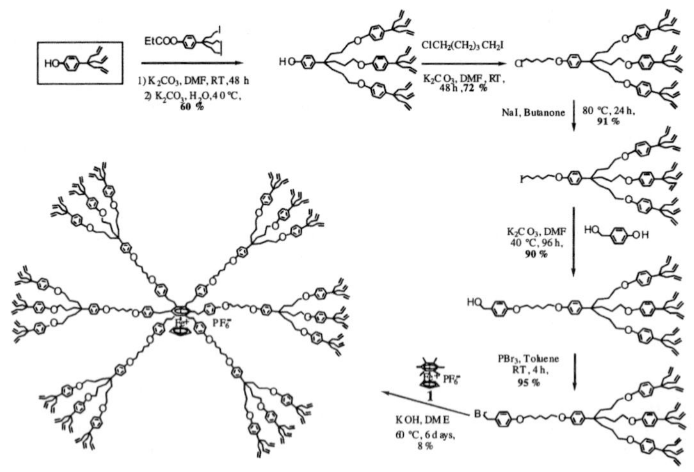

Scheme 5: $CpFe^+$-induced hexabenzylation of C_6Me_6 applied to direct convergent dendrimer synthesis of a 54-allyl dendrimer.

Decoration of Dendrimers with Redox-Active Groups: Towards Molecular Batteries

The functionalization of the three allyl chains of the phenol dendron could be achieved by hydrosilylation reaction catalyzed by the Karsted catalyst.[46] Indeed, it is very interesting that there is no need to protect the phenol group before performing these reactions. For instance, catalyzed hydrosilylation using ferrocenyldimethylsilane gives a high yield of the triferrocenyl dendron $HO\textit{p}\text{-}C_6H_4C(CH_2CH_2CH_2SiMe_2Fc)_3$ that is easily purified by column chromatography.[46,47] Protection of the phenol dendron using propionyliodide gave the phenolate ester which was hydroborated. Oxidation of the triborane using H_2O_2/OH^- gave the triol, then reaction with $SiMe_3Cl$ gave the tris-silyl derivative. Reaction with NaI yielded the tri-iodo compound, and reaction with the tri-ferrocenyl dendron provided the nona-ferrocenyl dendron that was deprotected using K_2CO_3 in DMF. The nona-ferrocenyl dendron was allowed to react with hexakis(bromomethyl)benzene, which gave the 54-ferrocenyl dendrimer. This convergent synthesis is clean and the 54-ferrocenyl dendrimer gave correct analytical data, although a mass spectrum could not be obtained (Scheme 6).This approach is somewhat limited, however, since

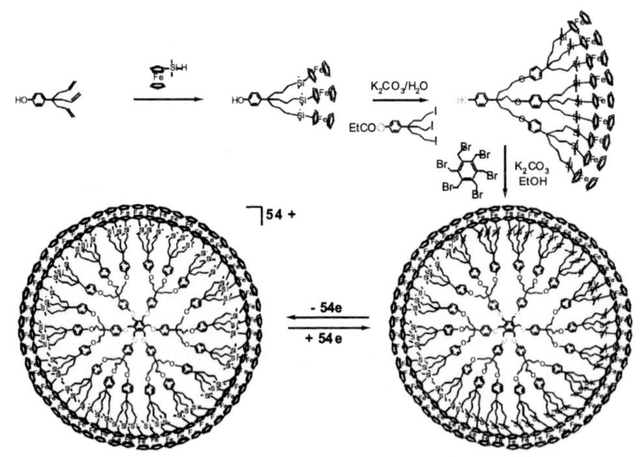

Scheme 6: Convergent synthesis of a redox-robust 54-silylferrocenyl dendrimer.

larger dendrons, which one would like to synthesize in this way, cannot be made because dehydrohalogenation becomes faster than nucleophilic substitution of the iodo by phenolate for bulkier higher generations of dendrons. Although this problem might be overcome by modifying the iodo branch in such a way that there would be no hydrogens in β positions, the condensation of higher dendrons onto a core would become tedious or impossible for steric reasons. This well known inconvenient is intrinsic to the convergent dendritic synthesis. On the other hand, divergent syntheses are not marred by such a problem since additional generations and terminal groups are added at the periphery of the dendrimer. The limit is that indicated by De Gennes, *i. e.* the steric congestion encountered at a generation where the peripheral branches can no longer by divided. Another obvious limit intervenes if the molecular objects added onto the termini of the branches are large and interfere with one another. We have developed a divergent synthesis of polyallyl dendrimers indicated on *Scheme 20* whereby each generation consists in hydroboration, oxidation of the borane to the alcohol, formation of the mesylate, and reaction of the phenol dendron with the mesylate. This strategy has allowed us to synthesize dendrimers of generation 0, 1, 2 and 3 with respectively 9 (G_0), 27 (G_1), 81 (G_2) and 243 branches (G_3) (Scheme 7 and Chart 1).

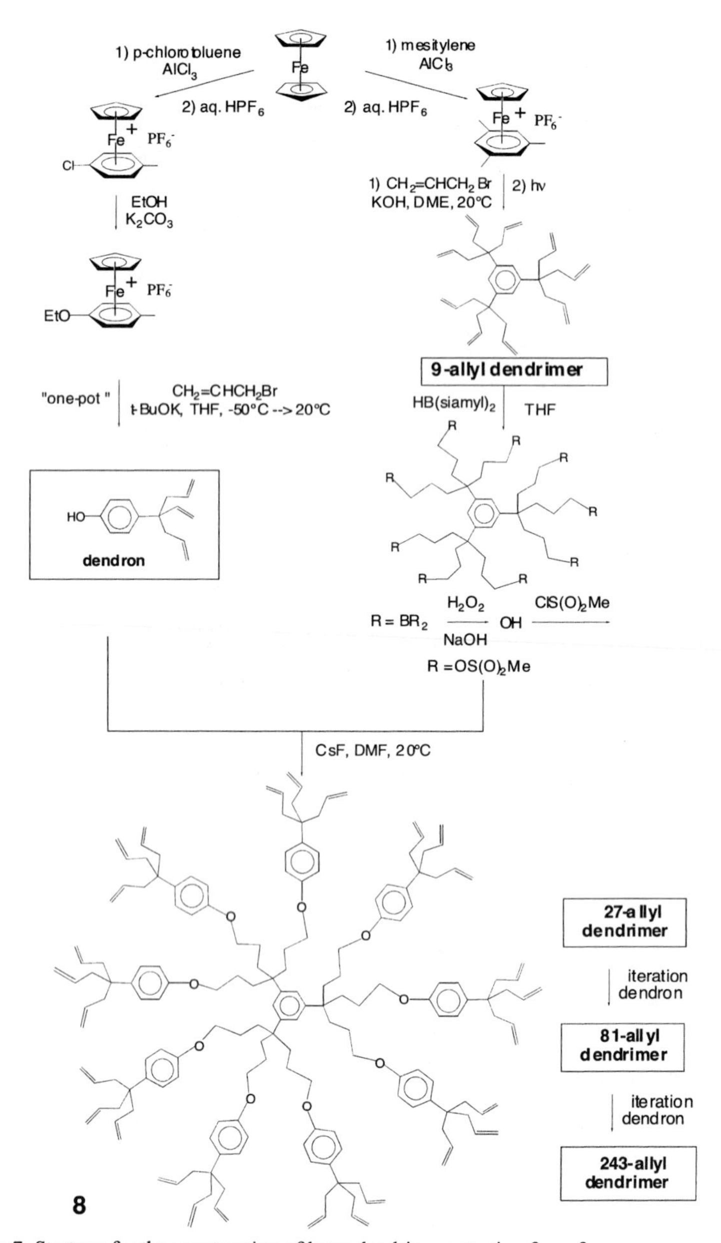

Scheme 7: Strategy for the construction of large dendrimers starting from ferrocene.

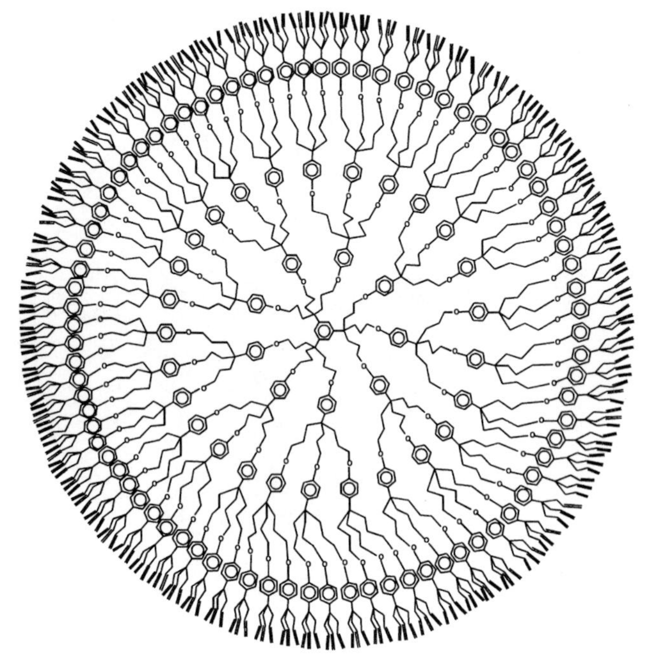

Chart 1: 243-allyl dendrimer (3rd generation, see the construction on Scheme 7).

The MALDI TOF mass spectrum of the 27-allyl dendrimer only shows the molecular peak with only traces of side product. That of the 81-allyl shows a dominant molecular peak, but also important side products resulting from incomplete branching. That of the 243-allyl could not be obtained possibly signifying that this dendrimer is polydisperse (correct ^{1}H and ^{13}C NMR spectra were obtained, however, indicating that the ultimate reactions had proceeded to completion). This dendrimer was soluble which indicated that this generation is not the last one, which might be reached. Larger dendrimers have recently been synthesized using a slightly diferent strategy. The ferrocenylsilylation of all these polyallyl dendrimers was carried out using ferrocenyldimethylsilane in ether or toluene and was catalyzed by the Karsted catalyst [48,49] at 40-45°C. The reactions were complete after two or three days except for the ferrocenylsilylation of 243-allyl that required a reaction time of one weeks indicating some degree of steric congestion (Scheme 8).

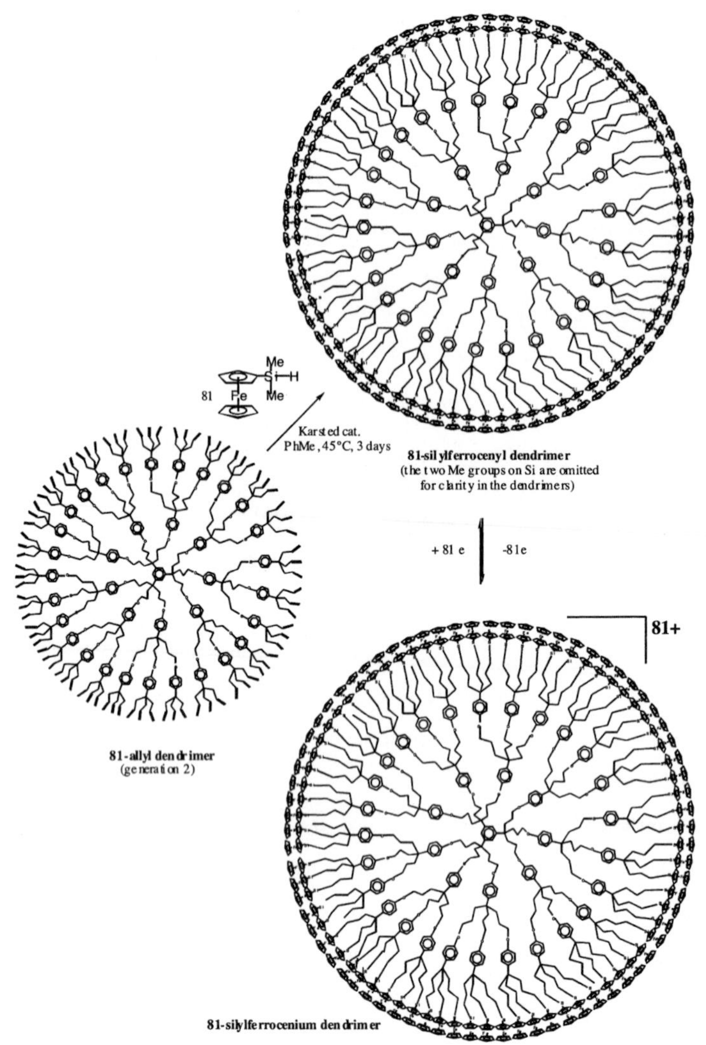

Scheme 8: Ferrocenylsilylation of the polyallyl dendrimers synthesized. Example of the 2nd generation 81-allyl dendrimer.

The ^{1}H and ^{13}C spectra indicated the absence of regioisomer. The solubility in pentane decreased from good for the 9-Fc dendrimer to low for the 27-Fc dendrimer and nil for the superior dendrimers, but the solubility in ether remained good for all the ferrocenyl dendrimers. Likewise, the retention times on plate or column chromatography increased with generation and no migration was observed for the "243-Fc" dendrimer. The silane use here, $HSi(Fc)Me_2Cl$, reported by Pannel and Sharma,[50] was already used by Jutzi [51] to synthesize the decaferrocenyl dendrimer $[Fe(CCH_2CH_2SiMe_2Fc)_{10}]$ (with Fc = ferrocenyl) from deca-allylferrocene.

The cyclic voltammetry of all the ferrocenyl dendrimers on Pt anode shows that all the ferrocenyl centers are equivalent and only one wave was observed. It was possible to avoid adsorption using even CH_2Cl_2 for the small ferrocenyl dendrimers, but it was required to use MeCN for the medium size ones (27-Fc, 54-Fc and 81-Fc). Finally, adsorption was not avoided even with MeCN for the "243-Fc" dendrimer. From the intensity of the wave, the number of ferrocenyl units could be estimated using the Anson-Bard equation,[52] and the number found were within 5% of the branch numbers except for the "243-Fc"dendrimer, for which the experimental number was too high (250) because of the adsorption.

The first polyferrocenium dendrimers reported by our group in 1994 and characterized *inter alia* by Mössbauer spectroscopy (a "quantitative" technique) were mixed valence Fe^{II}/Fe^{III} complexes.[23] Since then, we have been seeking to synthesize larger ferrocenyldendrimers, which could also withstand oxidation to their ferrocenium analogues. The syntheses of amidoferrocene dendrimers were reported five years ago simultaneously by our group [11,53] and the Madrid group using different cores.[54,55] In our reports, we were able to show the use of these metallodendrimers as redox sensors for the recognition of oxo-anions, with remarkable positive dendritic effects when the generation increased. The amidoferrocenyl dendrimers are not the best candidates for a stable redox activity on the synthetic scale, however, and thus even less so for molecular batteries. Indeed, although they give fully reversible cyclic voltammetry waves, it is known that ferrocenium derivatives bearing an electron-withdrawing substituent are at least fragile, if stable at all. This inconvenient is probably enhanced in the dendritic structures because of the steric effect which forces ferrocenium groups to encounter one another more easily than as monomers. Thus, we have oxidized our silylferrocenyl dendrimers using $[NO][PF_6]$ in CH_2Cl_2 and obtained stable polyferrocenium dendrimers as dark-blue precipitates, as expected from the known characteristic color of ferrocenium itself. These polyferrocenium dendrimers were reduced back to soluble orange polyferrocenyl dendrimers using decamethylferrocene as the reductant.[56] No decomposition was observed either in the oxidation or in the reduction reactions which were very clean, and this redox cycle could be achieved in quantitative yield even with the "243-ferrocenyl" dendrimer. The zero-field Mössbauer spectrum of the 243-ferrocenium dendrimer (Figure 1) showed a single line corresponding to the expected spectrum known for ferrocenium itself,[57] confirming its electronic structure. Thus, these polyferrocenyl dendrimers are molecular batteries, which could be used, in specific devices. Indeed, as large as they may be, they transfer a very large number of electrons rapidly and "simultaneously" with the electrode. By "simultaneously", we mean that, visually, the cyclic voltammogram looks as if it were that of a monoelectronic wave. One must question the notion of the isopotential for the many ferrocenyl units at the periphery of a dendrimer. In theory, all the standard potentials of the n ferrocenyl units of a single dendrimer are distinct even if all of them are equivalent and independent. This situation arises since the charge of the overall dendrimer molecule increases by one unit of charge every time one of its ferrocenyl units is oxidized to ferrocenium. The next single-electron oxidation is more difficult than the preceding one since, the dendritic molecule having one more unit of positive charge, it is more difficult to oxidize because of the increased electrostatic factor. Thus, the potentials of the n redox units are statistically distributed around an average standard potential centered at the average potential (Gaussian distribution).[52] In practice, the situation is complicated by the fact that the dendritic molecule, as large as it may be, is rotating much more rapidly than the usual electrochemical time scales.[58,59] Under these conditions, all the potentials are probably averaged. The fast rotation is also responsible for the fact that all the ferrocenyl units come close to the electrode within the electrochemical time scale. Consequently, there is no slowing down of the electron transfer due to long distance from the electrode even in large dendrimers. Indeed, the waves of the ferrocenyl dendrimers always appear fully electrochemically reversible indicating fast electron transfer.

The ferrocenyl dendrimers also adsorb readily on electrodes, a phenomenon already well known with various kinds of polymers.[60] When polymers contain redox centers, the adsorbed polymer have long been shown to disclose a redox wave for which the cathodic and anodic waves are located at exactly the same potential and the intensity of each wave is proportional to scan rate. Continuous cycling shows the stability of the adsorption of the electrode modified in this way. The ferrocenyl dendrimers described show this phenomenon as expected. The stability of the electrode modified by soaking the Pt electrode in a CH_2Cl_2 solution containing the ferrocenyl dendrimer and cyclic scanning between the ferrocenyl and ferrocenium regions is all the better as the ferrocenyl dendrimer is larger. For instance, in the case of the 9-ferrocenyl dendrimer, scanning twenty times is necessary before obtaining a constant intensity, and this intensity is weak. With the 27-, 54-, 81-, and 243-ferrocenyl dendrimers, only approximately ten cyclic scans are necessary before obtaining a constant wave, and the intensity is much larger. When such derivatized electrodes are washed with CH_2Cl_2 and re-used with a fresh, dendrimer-free CH_2Cl_2 solution, the cyclic voltammogram is obtained with $\Delta E_p = 0$. Other characteristic features are the linear relationship between the intensity and scan rate and the constant stability after cycling many times with no sign of diminished intensity (Figure 1).

Under these conditions, one may note that the argument of the fast rotation of the dendritic molecule to bring all the redox centers in turn close to the electrode does not hold for modified electrodes. Some redox centers must be close to the electrode and some must be far. It is probable that a hoping mechanism in the solid state is responsible for fast electron transfer and for averaging all the potentials of the different ferrocenyl groups of a single dendritic molecule around a mean value. The proximity of the ferrocenyl groups at the periphery of the dendrimer is a key factor allowing this hoping to occur since it is known that electron transfer with redox sites which are remote or buried inside a molecular framework is slow, if at all observable.[61-67]

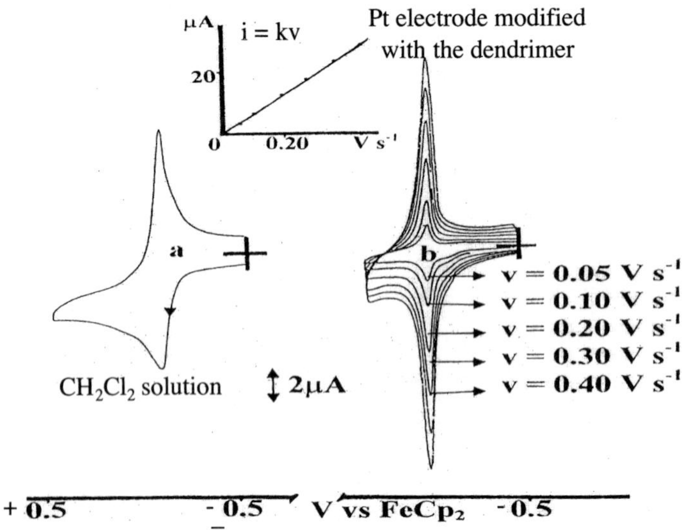

Figure 1: Cyclic voltammogram of the 243-ferrocenyl dendrimer ("243-Fc") in CH_2Cl_2 solution containing 0.1M $[n\text{-}Bu_4N][PF_6]$: a) in solution (10-4 M) at 100 mV. s-1 on Pt anode; b) Pt anode modified with "243-Fc" at various scan rates, dendrimer-free clear CH_2Cl_2 solution (inset: intensity as a function of scan rate: the linearity shows the expected behavior of a modified electrode with a fully adsorbed dendrimer).

Ferrocenes and ferrocenyl dendrimers are poor reductants. On the other hand, the complexes $[Fe(\eta^5\text{-}C_5R_5)(\eta^6\text{-}C_6Me_6)]^{2+/+/0}$ (R = H or Me) have been shown to be efficient for various stoichiometric and catalytic electron-transfer reactions.[16,68] The covalent linkage of this sandwich complex to the Cp ligand by means of a chlorocarbonyl substituent leads, upon reaction with dendritic polyamines, to soluble Fe^{II} metallodendrimers. Moreover, these Fe^{II} metallodendrimers can be reduced to Fe^{I} by $[Fe^{I}Cp(\eta^6\text{-}C_6Me_6)]$, **1**. Reduction of the monomeric model $[Fe^{II}(\eta^5\text{-}C_5H_4CONH\text{-}n\text{-}Pr)(\eta^6\text{-}C_6Me_6)][PF_6]$ by Na/Hg in THF (RT) gives the deep-blue-green, thermally stable 19-electron complex $[Fe^{I}(\eta^5\text{-}C_5H_4CONH\text{-}n\text{-}Pr)(\eta^6\text{-}C_6Me_6)]$ that shows the classic rhombic distorsion of the Fe^{I} sandwich family, observable by EPR in frozen THF at 77K (3 g values around 2).[69] Given this stability, we carried out the same reaction of $[Fe^{II}(\eta^5\text{-}C_5H_4COCl)(\eta^6\text{-}$

$C_6Me_6)][PF_6]$ with the commercial polypropyleneimine dendrimer of generation 5 (64 amino termini) in $MeCN/CH_2Cl_2$: 2/1 in the presence of NEt_3. The polycationic metallodendrimer DAB *dendr*-64--$NHCOCpFe^{II}(\eta^6\text{-}C_6Me_6)$, was obtained as the PF_6^- salt, soluble in MeCN and DMF (Equation 1).

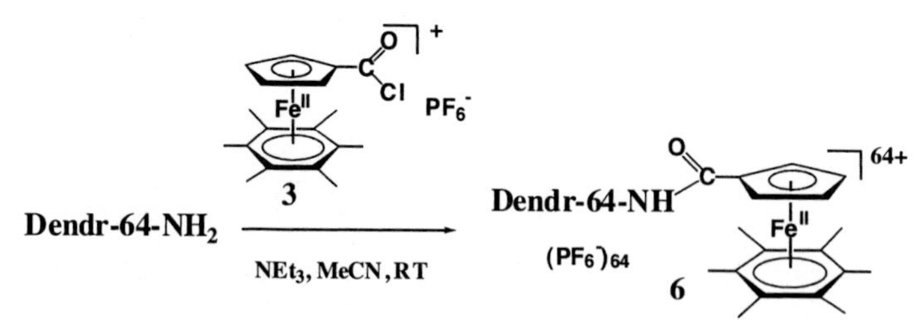

Equation 1: Covalent linkage of the complex $[FeCpCOCl(\eta^6\text{-}C_6Me_6)][PF_6]$ to the DSM polyamine dendrimer of generation 5 (64 branches). Example of the 64-NH_2 dendrimer (generation 5).

This dendritic complex was characterized by ^1H and ^{13}C NMR and IR spectroscopies and cyclovoltammetry (a single reversible wave in DMF, at $E_{1/2} = -1.84$ V *vs*. $FeCp_2^{0/+}$; $\Delta E = 70$ mV). Attempts to reduce it with the classic reductants that reduce monomeric complexes $[Fe^{II}(\eta^5\text{-}Cp)(\eta^6\text{-arene})][PF_6]$ such as Na sand, Na/Hg or $LiAlH_4$ in THF or DME failed due to the insolubility of both the metallodendrimer and the reductant in the required solvents. The only successful reductant was the parent 19-electron complex $[Fe^ICp(\eta^6\text{-}C_6Me_6)]$ [70,71] (in pentane or THF) that reduced the metallodendrimer in MeCN at –30°C to the neutral, deep-green-blue 19-electron Fe^I dendrimer **6** in a few minutes (Equation 2).[72]

The exoergonicity of this electron-transfer reaction is 0.16 V, which is due to the electron-withdrawing effect of the juxta-cyclic carbonyl group on the Cp ring that lowers the reduction potential of the metallodendrimer as compared to that $[Fe^ICp(\eta^6\text{-}C_6Me_6)]$. Although the Fe^I dendrimer decomposes a 0°C, it was also characterized by its EPR spectrum at 10 K confirming, as the deep-blue-green color, the Fe^I -sandwich structure analogous to that of the monomeric model.

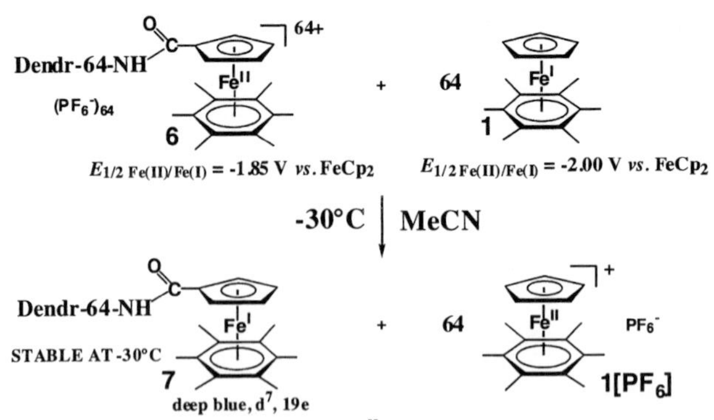

Equation 2: Exergonic reduction of the cationic Fe^{II} dendritic sandwich groups by the parent 19-electron complex $[FeICp(\eta^6\text{-}C_6Me_6)]$ to the Fe^I dendrimer complex.

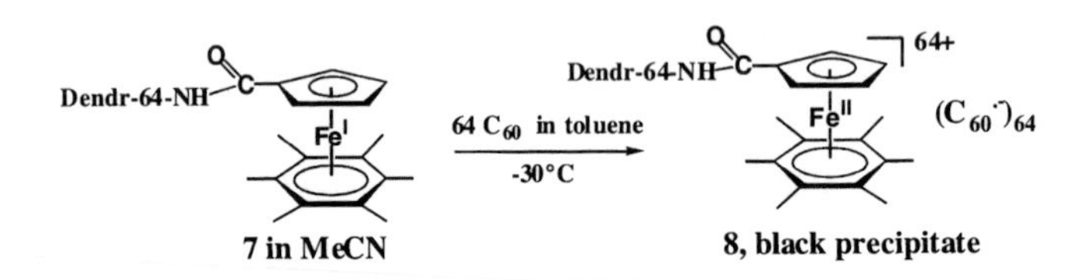

Equation 3: Dendr-64-$NHCOCpFe(C_6Me_6)^{64+}$, 64 $C_{60}^{\cdot-}$ resulting from the reaction of the 64-Fe^I dendrimer with C_{60} in MeCN/toluene at –30°C yielding the 64-Fe^{II}-C_{60} - dendrimer with EPR spectrum (bottom, right) in MeCN at 10K and Mössbauer spectrum at 77K (bottom, left) of the latter.

Contrary to the case of $[Fe^ICp(\eta^6\text{-}C_6Me_6)]$,[69] however, it was not possible to record the EPR spectrum of the solution of the Fe^I dendrimer above 10 K. This is presumably due to the intramolecular relaxation among the peripheral Fe^I sandwich units. The intermolecular version of this relaxation effect is known to preclude observation of the spectrum of monomeric Fe^I sandwich complexes in the solid state above 4K and in solution above 77K.[69] This acetonitrile solution of the 64-Fe^I dendrimer was used for the reaction with C_{60}, the stoichiometry being Fe^I/C_{60}: 1/1 (64 C_{60} *per* dendrimer). Upon reaction with a toluene solution of C_{60}, the deep-blue-green color of the Fe^{I} dendrimer disappeared, leaving a yellow solution that contained $[Fe^{II}Cp(\eta^6\text{-}C_6Me_6)][PF_6]$ and a black precipitate (Equation 3). Tentative extraction of this precipitate with toluene yielded a colorless solution, which indicated that no C_{60} was present. The Mössbauer spectra of this black solid at 298K discloses parameters that show the presence of an Fe^{II} sandwich complex of the same family as $[Fe^{II}Cp(\eta^6\text{-}C_6Me_6)]^+$.[69-71] Its EPR spectrum recorded at 77 K shows the same EPR spectrum as that of $[Fe^{II}Cp(\eta^6\text{-}C_6Me_6)]^+$ C_{60}.[73] It could thus be concluded that C_{60} had been reduced to its monoanion, as designed for a process that is exergonic by 0.9 V (74). The $[dendr\text{-}Fe^{II}]^+$ C_{60}^- units being very large, they must be located at the dendrimer periphery, presumably with rather tight ion pairs although the number of fullerene layers and

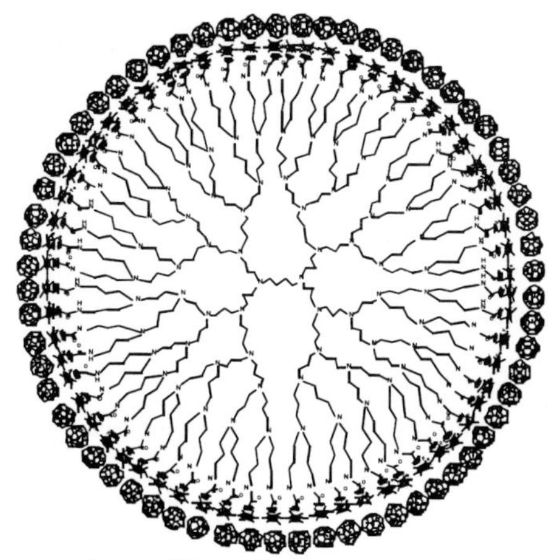

overall molecular size are unknown (Figure 2).

Figure 2 : Dendr-$[NHCOCp\text{-}Fe^{II}C_6Me_6]_{64}[C60]64$ (top) ; Mössbauer (left, bottom) and EPR (right, bottom) spectra.

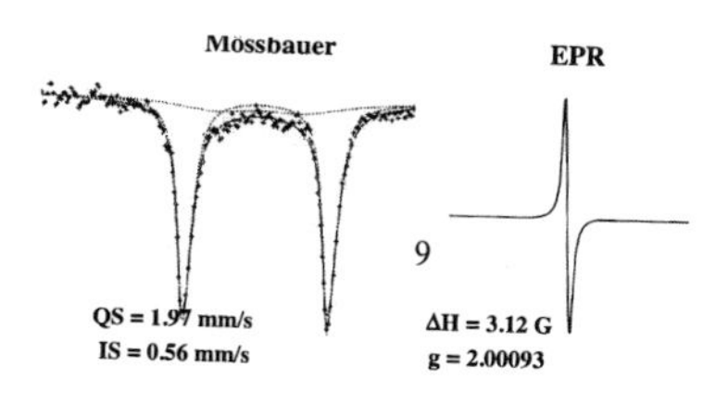

Decoration of Dendrimers with Ruthenium Clusters: Towards Dendritic Catalysts

The clean introduction of clusters onto the termini of polyphosphine dendrimers is a real challenge because of the current interest of dendritic clusters in catalysis and the mixtures usually obtained in thermal reactions of $[Ru_3(CO)_{12}]$ with phosphines. The diphosphine $CH_3(CH_2)_2N(CH_2PPh_2)_2$ (abbreviated P-P below) was used as a simple, model ligand. The reaction between P-P and $[Ru_3(CO)_{12}]$ [75] (molar ratio: 1/1.05) in the presence of 0.1 equiv. $[Fe^ICp(\eta^6\text{-}C_6Me_6)]$ in THF at 20°C led to the complete disappearance of $[Ru_3(CO)_{12}]$ in a few minutes and the appearance of a mixture of chelate [P-P. $Ru_3(CO)_{10}$], monodentate [P-P. $Ru_3(CO)_{11}$] and bis-cluster [P-P. $\{Ru_3(CO)_{11}\}_2$]. These reactions were reported by Bruce with simple diphosphines.[76] On the other hand, the reaction of P-P with $[Ru_3(CO)_{12}]$ in excess (1/4) and only 0.01 equiv. $[Fe^ICp(\eta^6\text{-}C_6Me_6)]$ in THF at 20°C led, in 20 minutes, to the formation of the air-stable, light-sensitive bis-cluster [P-P. $\{Ru_3(CO)_{11}\}_2$] as the only reaction product. Given the simplicity of the above characterization of the reaction product by ^{31}P NMR and the excellent selectivity of this model reaction when excess $[Ru_3(CO)_{12}]$ was used, the same reaction between Reetz's dendritic phosphines,[77a] derived from DSM's dendritic amines,[77b] and $[Ru_3(CO)_{12}]$ could be more confidently envisaged. This reaction, catalyzed by 1% equiv. $[Fe^ICp(\eta^6\text{-}C_6Me_6)]$ was carried out in THF at 20°C. The dendrimer-cluster assembly was obtained in 50% yield. This shows the selectivity and completion of the coordination of each of the 32 phosphino

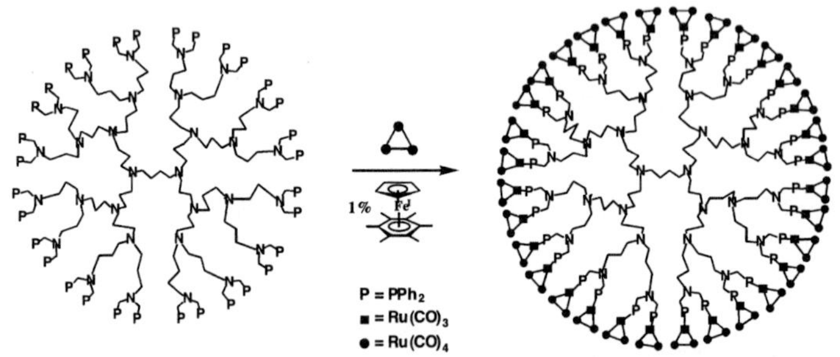

Scheme 9: Electron-Transfer-Chain catalyzed ligand substitution of one Ru-coordinated CO by a dendritic phosphine termini in Reetz's 32-phosphine dendrimer under ambiant conditions leading to the 32-$Ru_3(CO)_{11}$ dendrimer-cluster.

ETC mechanism [78-80] proceeds for the introduction of the 32 cluster fragments in the dendrimer for ligation of the first $Ru_3(CO)_{11}$ fragment to the dendritic phosphine. Then, this first complex [dendriphosphine.$Ru_3(CO)_{11}$] would undergo the same ETC cycle as [$Ru_3(CO)_{12}$] initially does to generate the bis-cluster complex [dendriphosphine.$\{Ru_3(CO)_{11}\}_2$], and so on (Scheme 10).

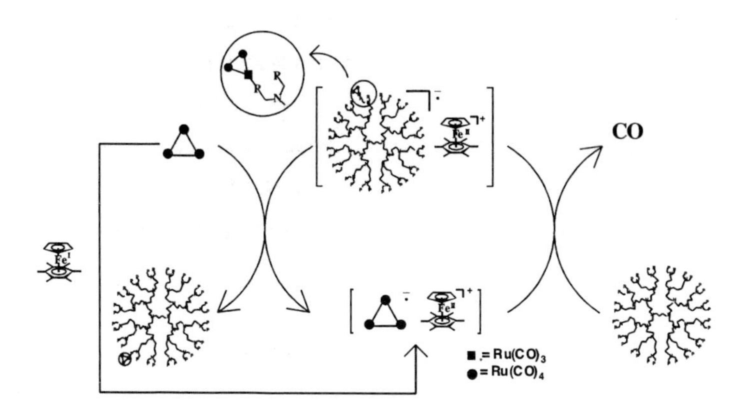

Scheme 10: Electron-Transfer-Chain mechanism for the synthesis of the 96-Ru dendrimer-cluster complex.

Finally, the 64-branch phosphine DAB-*dendr*-G4-$[N(CH_2PPh_2)_2]_{32}$ analogously reacts with [$Ru_3(CO)_{12}$] and 1% [$Fe^{I}Cp(\eta^6-C_6Me_6)$] (20°C, THF, 20 min.) to give the dark-red 192-Ru dendrimer. Characterization of the purity of these dendrimer-cluster assemblies is conveniently monitored by ^{31}P NMR. This application should find extension to other metal-carbonyl clusters and other families of phosphine dendrimers.

Dendritic Catalysts and Sensors

It is of interest to compare the efficiency of homologous metallo-stars and metallo-dendrimers in catalysis and sensing. Recently, we were able to compare the rate of the redox catalyzed cathodic reduction of nitrate and nitrite to ammonia using metallo-stars and cyclic voltammetry. It was found that the metallo-stars in which the catalytic metal centers were located at the periphery of the metallo-stars showed about the same rate of reduction of the nitrogen oxides NO_2^- and NO_3^- by the 19-electron form of the catalysts as mononuclear complexes with the same driving force. On the other hand, those containing the redox-active site at the center of the dendritic- or star core showed rates that were at least an order of magnitude lower.[81]

In another example using dendritic ruthenium carbene dendrimers for the living ROMP of norbornene, it was found that the most efficient dendrimer was that of first generation (4 arms) whereas the efficiency decreased as the generation increased. This finding is due to the increasing steric bulk around the ruthenium-carbene centers that increases with the dendrimer generation.[82] This type of steric congestion is not found is metallostars. Thus, first-generation dendrimers (which are in fact stars rather than dendrimers) are best for catalytic activity as long as they are large enough to be removed by ultra-filtration or centrifugation.[8-10]

For dendritic amido- or silylferrocene nano-sensors, stars show some activity, but dendrimers are far better, and their efficiency increases as the dendritic generation increases. In this case, the steric bulk at the dendrimer periphery provides narrow channels that are desired and ideal with an optimized topology in view of efficient recognition.[83-85]

Outlook

Very large dendrimers [86] have been synthesized and functionalized with transition metal fragments including ferrocenyl and other metal-sandwich type complexes that include a supramolecular function (amide, silicon center).[87]. Their large size is a very important parameter for improved adsorption on metal surfaces and electrodes. It is also essentiel to have in hands nano-sized, well defined macromolecules for integration in devices as nano-wires. For instance, comparison of surfaces and electrodes modified with several layers and average-size metallodendrimers with those modified with giant metallodendrimers should provide insight into the mechanism of electron-hoping in the solid (compare intramolecular vs. intermolecular hoping). It is also essential to control reactions such as those involving metal-ligand bond formation at the nanoscale. The behavior of nano-catalysts is highlighted in this review and different topologies have been compared in term of efficiency. These studies are needed in order to make progress in green catalysis.[88] Applications of these fundamental studies are awaited in the fields of catalysts, sensors and nano-devices.

Acknowledgement

We are grateful to the colleagues, students and post-docs cited in the references who have contributed to the ideas and efforts of this research program. We are especially indebted in this respect to Dr Sylvain Lazarre (AFM: LCPT, University Bordeaux I), Dr Eric Cloutet (SEC: LCPO, University Bordeax I) and Prof. François Varret (Mössbauer spectroscopy: University of Versailles). Financial support from the Institut Universitaire de France (IUF), the University Bordeaux I, the Centre National de la Recherche Scientifique (CNRS), the Alexander von Humboldt Fundation, Iberdrola, the Picasso and Erasmus-Socrates programs, the European Community and the Region Aquitaine is gratefully acknowledged.

[1] G. R. Newkome, C. N. Moorefield, F. Vögtle, *"Dendrimers and Dendrons. Concept, Syntheses, Applications"*, Wiley-VCH, Weinheim, 2001.

[2] Dendrimers and other Dendritic Polymers, (Eds.: Tomalia, D.; Fréchet, J. M. J.), Wiley-VCH, New York, 2002.

[3] V. Balzani, S. Campagna, G. Denti, A. Juris, S. Serrini, M. Venturi, *Acc. Chem. Res.* **1998,** 31, 26.

[4] G. R. Newkome, E. He, C. N. Moorefield, *Chem. Rev.* **1999**, 99, 1689.

[5] A. W. Bosman, E. W. Jansen, E. W. Meijers, *Chem. Rev.* **1999**, 99, 1665.

[6] I. Cuadrado, M. Morán, C. M. Casado, B. Alonso, J. Losada, *Coord. Chem. Rev.* **1999**, 193-195, 395-445.

[7] M. A. Hearshaw, J. R. Moss, *Chem. Commun.* **1999,** 1.

[8] G. E. Oosterom, J. N. H. Reek, P. C. J. Kamer, P. W. N. M. van Leeuwen, *Angew. Chem. Intern. Ed. Engl.* **2001**, 40, 1828.

[9] R. Kreiter, A. W. Kleij, R. J. M. Klein Gebbink, G. van Koten In Dendrimers IV: Metal Coordination, Self Assembly, Catalysis, (Eds.: F. Vögtle, C.A. Schalley) Top. Curr. Chem. Springer-Verlag, Berlin, 2001, 217, 163.

[10] D. Astruc, F. Chardac, *Chem. Rev.* **2001**, 101, 2991; N. Ardoin, D. Astruc, *Bull. Soc. Chim. Fr.* **1995**, 132, 875 (review).

[11] C. Valério, J-L. Fillaut, J. Ruiz, J.-C. Guittard, J.-C. Blais, D. Astruc, *J. Am. Chem. Soc.* **1997**, 117, 2588.

[12] C. Valério, E. Alonzo, J. Ruiz, J.-C. Blais, D. Astruc, *Angew. Chem. Int. Ed. Engl.* **1999**, 38, 1747.

[13] a) A. Labande, J. Ruiz, D. Astruc, *J. Am. Chem. Soc.* **2002**, 124, 1782; b) A. Labande, D. Astruc, *Chem.Commun.* **2000**, 1007.

[14] E. Alonso, A. Labande, L. Raehm, J.-M. Kern, D. Astruc, *C. R. Acad. Sci. Ser. IIc* **1999**, 2, 209.

[15] a) H. Trujillo, C. Casado, J. Ruiz, D. Astruc, *J. Am. Chem. Soc.* **1999**, 121, 5674; b) D. Astruc *Acc. Chem. Res.* **2000**, 33, 287.

[16] C. C. Lee, B. R. Steele, K. J. Demchuk, R. G. Sutherland, *Can. J. Chem.* **1979**, 57, 946.

[17] For the synthesis of $[Fe^{II}Cp(\eta^6\text{-}C_6Me_6)][PF_6]$, see references 18,19 and 35.

[18] P. L. Pauson, W. E. Watts, *J. Chem. Soc.* **1963**, 2990.

[19] D. Astruc, J.-R. Hamon, M. Lacoste, M.-H. Desbois, E. Román, *Organometallic Synthesis* (Ed.: R. B. King), 1988, Vol. IV, p. 172.

[20] J.-R. Hamon, J.-Y. Saillard, A. Le Beuze, M. McGlinchey, D. Astruc, *J. Am. Chem. Soc.* **1982**, 104, 3755.
[21] F. Moulines, D. Astruc, *Angew. Chem. Int. Ed. Engl.* **1988**, 27, 1347.
[22] F. Moulines, D. Astruc, *J. Chem. Soc. Chem. Commun.* **1989**, 614.
[23] See also: B. R. Steele, C. G. Screttas, *J. Am. Chem. Soc.* **2000**, 122, 2391.
[24] a) J.-L. Fillaut, J. Linares, D. Astruc, *Angew. Chem. Int. Ed. Engl.* **1994**, 33, 2460; b) J.-L. Fillaut, R. Boese, D. Astruc, *Synlett* **1992**, 55; c) J.-L. Fillaut, D. Astruc, *New J. Chem.* **1996**, 20, 945.
[25] B. Alonso, J.-C. Blais, D. Astruc, *Organometallics* **2000**, 21, 1001.
[26] F. Moulines, B. Gloaguen, D. Astruc, *Angew. Chem. Int. Ed. Engl.* **1992**, 28, 458.
[27] H. W. Marx, F. Moulines, T. Wagner, D. Astruc, *Angew. Chem. Int. Engl.* **1996**, 35, 1701.
[28] J. Ruiz, E. Alonso, J. Guittard, J.-C. Blais, D. Astruc, *J. Organomet. Chem.* **1999**, 582/1, 139 (issue dedicated to Alan H. Cowley).
[29] F. Moulines, L. Djakovitch, J.-L. Fillaut, D. Astruc, *Synlett* **1992**, 57.
[30] F. Moulines, L. Djakovitch, R. Boese, B. Gloaguen, W. Thiel, J.-L. Fillaut, M.-H. Delville, D. Astruc, *Angew. Chem. Int. Ed. Engl.* **1993**, 105, 1132.
[31] V. Marvaud, D. Astruc, *Chem. Commun.* **1997**, 773.
[32] V. Marvaud, D. Astruc, *New J. Chem.* **1997**, 21, 1309.
[33] S. Rigaut, M.-H. Delville, D. Astruc, *J. Am. Chem. Soc.* **1997**, 119, 1132.
[34] S. Rigaut, M.-H. Delville, J. Losada, D. Astruc, *Inorg. Chim. Acta* **2002**, *334*, 225.
[35] D. Astruc In *Electron Transfer in Chemistry* (Ed.: V. Balzani), Vol II (J. Matay, D. Astruc Vol. Eds.), Wiley, Weinheim, 2001, pp 714-803.
[36] C. Valério, F. Moulines, J. Ruiz, J.-C. Blais, D. Astruc, *J. Org. Chem.* **2000**, 65, 1996.
[37] B. Gloaguen, D. Astruc, *J. Am. Chem. Soc.* **1990**, 112, 4607.
[38] D. Buchholz, B. Gloaguen, J.-L. Fillaut, M. Cotrait, D. Astruc, *Chem. Eur. J.* **1995**, 1, 374.
[39] D. Buchholz, D. Astruc, *Angew. Chem. Int. Ed. Engl.* **1994**, *33*, 1637.
[40] S. Marcen, S. Jimenez, M. V. Dobrinovitch, F. Lahoz, L. Oro, J. Ruiz, D. Astruc *Organometallics* **2002**, 21, 326.
[41] V. Martinez, J.-C. Blais, D. Astruc, *Org. Lett.* **2002**, 4, 651.
[42] F. Moulines, L. Djakovitch, M.-H. Delville, F. Robert, P. Gouzerh, D. Astruc, *J. Chem. Soc., Chem. Commun.* **1995**, 463.
[43] F. Moulines, L. Djakovitch, D. Astruc, *New J. Chem.* **1996**, 20, 1071.
[44] a)V. Sartor, L. Djakovitch, J.-L. Fillaut, F. Moulines, F. Neveu, V. Marvaud, J. Guittard, J.-C. Blais, D. Astruc, *J. Am. Chem. Soc.* **1999**, 121, 2929; b) V. Sartor, S. Nlate, J.-L. Fillaut, F. Djakovitch, F. Moulines, V. Marvaud, F. Neveu, J.-C. Blais, *New J. Chem.* **2000**, 24, 351.
[45] S. Nlate, Y. Neto, J.-C. Blais, J. Ruiz, D. Astruc *Chemistry Eur. J.*, **2002**, 8, 171.
[46] S. Nlate, J. Ruiz, D. Astruc, *Chem. Commun.* **2000**, 417.
[47] S. Nlate, J. Ruiz, V. Sartor, R. Navarro, J.-C. Blais, D. Astruc *Chemistry Eur. J.* **2000**, 6, 2544.
[48] B. Marciniec, In "*Applied Homogeneous Catalysis with Organometallic Compounds*" (Eds.: B. Cornils, W. A. Herrmann), VCH, Weinheim, 1996, Vol. 1, Chap. 2.6.
[49] L. N. Lewis, J. Stein, K. A. Smith, In *Progress in Organosilicon Chemistry* (Eds.: B. Marciniec, J Chojnowski) Gordon and Breach, Langhorne, USA, 1995, p. 263.
[50] K. H. Pannel, H. Sharma, *Organometallics* **1991**, 10, 954.
[51] P. Jutzi, C. Batz, B. Neumann, H. G. Stammler, *Angew. Chem. Int. Engl.* **1996**, 35, 2118.
[52] J. B. Flanagan, S. Margel, A. J. Bard, F. C. Anson, *J. Am. Chem. Soc.* **1978**, 100, 4248.
[53] a) D. Astruc, C. Valério, J.-L. Fillaut, J.-R. Hamon, F. Varret, In *Magnetism, a Supramolecular Function* (Ed.: O. Kahn), NATO ASAI Series, Kluver, Dordrecht, 1996, p. 1107; b) C. Valério, *PhD Thesis*, Université Bordeaux I, 1996.
[54] a) I. Cuadrado, M. Morán, C. M. Casado, B. Alonso, F. Lobete, B. Garcia, J. Losada, *Organometallics* **1996**, 15, 5278; b) K. Takada, D. J. Diaz, H. Abruña, I. Cuadrado, C. M. Casado, B. Alonso, M. Morán, J. Losada, *J. Am. Chem. Soc.* **1997**, 119, 10763.
[55] Review: C. M. Casado, I. Cuadrado, M. Moran, B. Alonso, B. Garcia, B. Gonzales, J Losada, *Coord. Chem. Rev.* **1999**, 185-6, 53.
[56] J. Ruiz, D. Astruc, *C. R. Acad. Sci.* Paris, t. 1, Série II *c*, **1998**, 21.
[57] R. L. Collins, *J. Chem. Phys.* **1965**, 42, 1072.
[58] S. J. Green, J. J. Pietron, J. J. Stokes, M. J. Hostetler, H. Vu, W. P. Wuelfing, R. W. Murray, *Langmuir* **1998**, 14, 5612.
[59] C. B. Gorman, J. C. Smith, M. W. Hager, B. L. Parhurst, H. Sierzputowska-Gracz, C. A. Haney *J. Am. Chem. Soc.* **1999**, 121, 9958.
[60] R. Murray In *Molecular Design of Electrode Surfaces* (Ed.: R. Murray), Wiley, New York, 1992, p. 1.
[61] P. J. Dandliker, F. Diederich, M. Gross, B. Knobler, A. Louati, E. M. Stanford, *Angew. Chem. Int. Ed. Engl.*

1994, 33, 1739.
[62] G. R. Newkome, R. Güther, C. N. Moorefield, F. Cardullo, L. Echegoyen, F. Pérez-Cordero, H. Luftmann, *Angew. Chem. Int. Ed. Engl.* **1995**, 34, 2023.
[63] H.-F. Chow, I. Y.-K. Chan, D. T. W. Chan, R. W. M. Kwok, *Chem. Eur. J.* **1996**, 2, 1085.
[64] P. J. Dandliker, F. Diederich, H.-F. Chow, I. Y.-K. Chan, R. W. M. Kwok *Chem. Eur. J.* **1996**, 2, 1085.
[65] J. Issberner, F. Vögtle, L. De Cola, V. Balzani, *Chem. Eur. J.* **1997**, 3, 706.
[66] C. B. Gorman, B. L. Parkhurst, W. Y. Su, K. Y. Chen, *J. Am. Chem. Soc.* **1997**, 119, 1141.
[67] D. K. Smith, F. Diederich, *Chem. Eur. J.* **1998**, 4, 2353.
[68] J. Ruiz, F. Ogliaro, J.-Y. Saillard, J.-F. Halet, F. Varret, D. Astruc, *J. Am. Chem. Soc.* **1998**, 120, 11693.
[69] M. V. Rajasekharan, S. Giesynski, J. H. Ammeter, N. Oswald, J.-R. Hamon, P. Michaud, D. Astruc *J. Am. Chem. Soc.* **1982**, 104, 129.
[70] D. Astruc, J.-R. Hamon, G. Althoff, E. Roman, P. Batail, P. Michaud, J;_P. Mariot, F. Varret, D. Cozak, *J. Am. Chem. Soc.* **1979**, 101, 5445. This paper also reports the first $CpFe^{+}$-induced iterative starburst hexa-alkylation of C_6Me_6.
[71] J.-R. Hamon, D. Astruc, P. Michaud, *J. Am. Chem. Soc.* **1981**, 103, 758.
[72] J. Ruiz, C. Pradet, F. Varret, D. Astruc, *Chem. Commun.* **2002**, 1108.
[73] C. Bossard, S. Rigaut, D. Astruc, M.-H. Delville, G. Félix, A. Février-Bouvier, J. Amiell, S. Flandrois, P. Delhaès, *J. Chem. Soc., Chem. Commun.* **1993**, 333.
[74] Redox potentials of the six cathodic monoelectronic reductions of C_{60}: A. Xie, E. Pérez-Cordero, L. Echegoyen, *J. Am. Chem. Soc.* **1992**, 114, 3978.
[75] E. Alonso, D. Astruc, *J. Am. Chem. Soc.* **2000**, 122, 3222.
[76] a) M. I. Bruce, D. C. Kehoe, J. G. Matisons, B. K. Nicholson, P. H. Rieger, M. L. J. Williams, *J. Chem. Soc. Chem. Commun.* **1982**, 442; b) M. I. Bruce, J. G. Mattisons, B. K. Nicholson, *J. Organomet. Chem.* **1983**, 247, 321.
[77] a) M. T. Reetz, G. Lohmer, R. Scwickardi, *Angew. Chem. Int. Ed. Engl.* **1997**, 36, 1526. b) E. M. M. de Brabander-van den Berg, E. W. Meijers, *Angew. Chem. Int. Ed. Engl.* **1993**, 32, 1308.
[78] For pioneering work in the field of ETC catalysis, see: R. Rich, H. Taube, *J. Am. Chem. Soc.* **1954**, 76, 2608.
[79] Review on ETC catalysis: D. Astruc, *Angew. Chem. Int. Ed. Engl.* **1988**, 27, 643. See also ref. 80.
[80] a) D. Astruc, *"Electron Transfer and Radical Processes in Transition-Metal Chemistry"*, VCH, New York, 1995; b) ref. 80 a), chapter 6: Chain Reactions, pp. 413-478.
[81] S. Rigaut, M.-H. Delville, J. Losada, D. Astruc, *Inorg. Chim. Acta*, **2000**, 334, 225 (issue dedicated to Andrew Wojcicki).
[82] S.Gatard , S. Nlate, E. Cloutet , G. Bravic, J.-C. Blais, D. Astruc, *Angew. Chem. Int. Ed.* **2003**, 42, 452.
[83] A. Labande, J. Ruiz, D. Astruc, *J. Am. Chem. Soc.* **2002**, 124, 1782.
[84] M.-C. Daniel, J. Ruiz, D. Astruc, *J. Am. Chem. Soc.* **2003**, 125, 1150.
[85] M.-C. Daniel, J. Ruiz, S. Nlate, J.-C. Blais, D. Astruc, *J. Am. Chem. Soc.* **2003**, 125, 2617.
[86] J. Ruiz, G. Lafuente, S. Marcen, C. Ornela, S. Lazarre, J.-C. Blais, E. Cloutet, D. Astruc, *J. Am. Chem. Soc.* **2003**, 125, ASAP.
[87] S. Nlate, J. Ruiz, V. Sartor, R. Navarro, J.-C. Blais, D. Astruc, *Chem. Eur. J.* **2000**, 6, 2544.
[88] "Green Chemistry" P. T. Anastas, T. C. Williamson Eds., *ACS Symp. Ser.* 626, ACS. Washington DC, 1996.

Organometallic Conducting Polymers Synthesized by Metallacycling Polymerization

Hiroshi Nishihara, Masashi Kurashina, Masaki Murata*

Department of Chemistry, School of Science, The University of Tokyo, 7-3-1 Hongo, Bunkyo-ku, Tokyo 113-0033, Japan

Summary: The study on the synthesis of π-conjugated polymers using cobaltacyclopentadiene formation reaction of $CpCo(PPh_3)_2$ and conjugated diacetylene and the structure and physical properties of the polymers is overviewed. The substituents on the diacetylene affect crucially the solubility, the degree of polymerization, redox properties, and electronic structures and so on. Recent synthesis of a ruthenacyclopentatriene polymers by metallacyling polymerization is also described.

Keywords: cobalt; conducting polymers; metallacycle; organometallic polymers; redox properties; ruthenium

Introduction

Discovery of so-called "conducting polymers" comprising organic π-conjugated polymer chains such as polyacetylene is one of the epoch-making scientific events in the last century.[1] This is because they exhibit various curious physical and chemical characteristics that exploded the concept of "durable" plastics. Main characteristics such as facile oxidation and reduction leading to charge storage, high electronic conductivity in the doped states, electrochromism, photo- and electro-luminescence are derived from π-electrons which can be delocalized along the π-conjugated chain.[2] One of the extended studies of such organic conducting polymers is to incorporate other functional molecular units into the polymer chain. For example, synthesis of ferromagnetic materials by attaching organic radicals as pendant groups of the π-conjugated main chain has been investigated extensively in recent years.[2] Another effective way to functionalize conducting polymers is to combine them with functional metal complex units, which gives a category of "organometallic conducting polymers".[3] Examples include polymetallocenylene,[4,5] poly(metalyne),[6-10] poly(metallo-phthalocyanine),[11,12] polydecker sandwich compounds,[13]

© 2003 WILEY-VCH Verlag GmbH & KGaA, Weinheim CCC 1022-1360/00/$ 17,50+.50/0

thiolate complex polymers,[14-16] cyclobutadienecobalt complex polymers,[17] and others.[18] A decade ago, we have developed a new class of organometallic conducting polymers wherein the framework of the polymer is π-conjugated and in part composed of cobaltacyclopentadienes.[19] In these polymers, the cobaltacyclopentadiene ring with its *d*-block heteroatom is structurally analogous to the rings found in other representative π-conjugated polymers, such as poly(pyrrole) and poly(thiophene), which contain *p*-block heteroatoms. Synthesis of the cobaltacyclopentadiene polymers has been accomplished by a new type of polymerization scheme, MetallaCycling Polymerization (MCP), that is based on successive metallacyclization. As for the similar kinds of metallacycle polymers, Endo *et al.* have reported the MCP reactions for cobalt complex polymers, independently,[20] and Tilley *et al.* have reported the MCP reactions for zirconacyclopentadiene polymers.[21] In these metallacyclopentadiene polymers, it is interesting to probe the effect that a *d*-transition metal heteroatom has on the π-system and electronic properties of a conducting polymer.

In the present article, we review the history of our study on the cobaltacyclopentadiene-based organometallic polymers and recent results on the ruthenacyclopentatriene-based organometallic polymers.

Synthesis of Cobaltacyclopentadiene Polymers

1. Polymers obtained from HC≡C-Ar-C≡CH

It is well established that the addition of two acetylenes to the metal center affords mtallacyclopentadiene, which is the precursor of the formation of six-membered aromatic rings.[22-25] This reaction accompanies the C-C bond formation, indicating that it can be used for MCP when diacetylene, which cannot cause intramolecular C-C coupling reaction, is employed. In the case of cobaltacyclopentadiene, the most common starting compound is $[CpCo(PPh_3)_2]$ (Cp = η^5-C_5H_5), which reacts with two acetylenes, $CR^1{\equiv}CR^2$ and $CR^3{\equiv}CR^4$ stepwise, affording $[CpCo(CR^1{\equiv}CR^2)(PPh_3)]$ and then $[CpCo(CR^1{=}CR^2\text{-}CR^3{=}CR^4)(PPh_3)]$ (Eq (1)). A further addition reaction with unsaturated compounds such as acetylene and nitrile gives aromatic six-membered rings. It is also known that the thermal reaction of cobaltacyclopentadiene $[CpCo(CR^1{=}CR^2\text{-}CR^3{=}CR^4)(PPh_3)]$ affords cyclobutadiene complex, $[CpCo(\eta^4\text{-}C_4R^1R^2R^3R^4)]$ (Eq (1)).[26] This kind of unique chemical reactivity of metallacyclopentadiene units can be utilized to obtain interesting polymeric substances from metallaycyclopentadiene polymers.

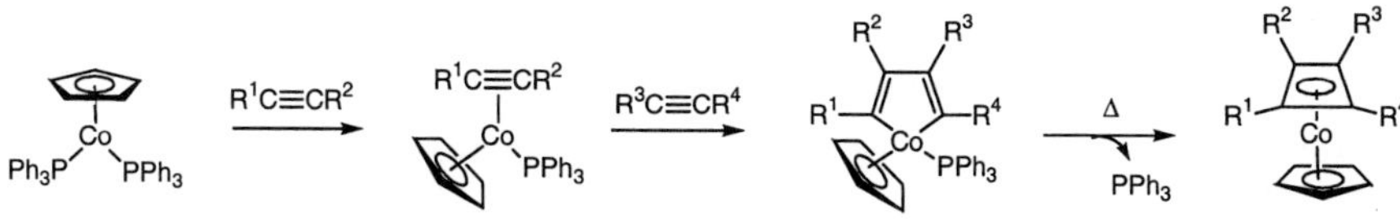

(1)

Application of the metallacyclization to the polymer synthesis has been successfully made using conjugated diacetylene with the formula, HC≡C-Ar-C≡CH (Ar = 1,4-phenylene, 2-fluoro-1,4-phenylene, 2,5-difluoro-1,4-phenylene, and 4,4'-biphenylene), and the polymers, **1** – **4**, respectively, were obtained as insoluble powders.[27] The insertion of the arylene moieties in the diacetylene is because the diacetylene without a spacer, RC≡C-C≡CR, could not yield polymers because of the high steric hindrance due to Cp and PPh_3 groups around the cobalt center.[28,29]

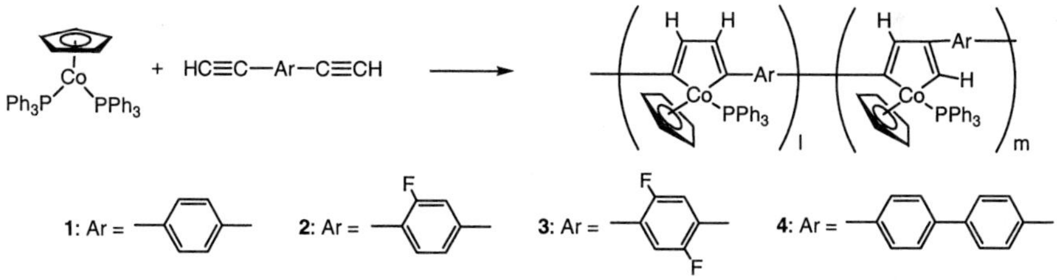

The regioselectivity of the metallacyclization is important to obtain highly π-conjugated polymers by MCP. When the cobaltacyclopentadiene unit is formed from two CR≡CR' molecules, there are three possible isomers, 2,4-R_2-3,5-R'_2, 2,5-R_2-3,4-R'_2 and 3,4-R_2-2,5-R'_2 forms. Wakatsuki *et al.* have shown the rule that the acetylenic carbon bearing a bulky group becomes the α-carbon of the metallacyclopentadiene.[30] The diacetylenes HC≡C-Ar-C≡CH of the first attempt have been chosen because they would give 2,5-Ar_2 structures according to the rule above. The polymers obtained are insoluble and could give films when the glass plates are immersed in the solution of the MCP synthesis. Electronic spectra of the polymers show the red shift of the absorption edge compared with the monomeric complex, indicating the extension of π-conjugation by the involvement of 2,5-Ar_2 structures in the polymers. However, the results on the soluble polymers suggest incomplete π-conjugation by the involvement of 2,4-Ar_2 units (*vide infra*).

2. Soluble polymers

The insolubility of the polymers often prevents their characterization and processability. An improvement of the solubility has carried out using hexylcyclopentadienyl (HexCp) ligand instead of Cp in the starting cobalt complex[31, 32] and/or using alkyl-terminated diacetylenes, RC≡C-Ar-C≡CR (R = Me, Bu).[33, 34] The purpose of the attempt to synthesize soluble cobaltacyclopentadiene polymers using (HexCp)Co(PPh_3)$_2$ was not only to increase the polymerization degree but also to characterize the polymers in detail including the regioselectivity in MCP reactions. An MCP reaction between (HexCp)Co(PPh_3)$_2$ and *p*-diethynylbenzene at 4 °C for 4 days gave no insoluble product, and the GPC spectrum of the reaction product indicated several oligomeric and polymeric components, **5**, the highest molecular weight of which is more than 7×10^4 based on the polystyrene standard.[32] The oligomeric components HC≡C-C_6H_4[C_4H_2{Co(HexCp)(PPh_3)}-C_6H_4]$_n$C≡CH were separated up to nonamer, **5**$_2$-**5**$_9$. The ^{1}H NMR analysis indicated that the products were mixtures of these two isomeric structures, 2,5- *vs.* 2,4-substituted cobaltacyclopentadienes with the ratio of 4 : 6. As mentioned above, the regioselectivity is principally dominated by the steric effect of the substituents on the acetylenic carbon. However, this result indicates another factor, probably the dipole-dipole interaction of two acetylenes ligated to cobalt, is concerted in dete rmining the conformation at the intermediate state. Reinvestigation of the stereochemistry of the metallacyclization for monomeric complexes actually support this consideration; a reaction of (HexCp)Co(PPh)$_2$ with ethynylbenzene afforded 2,5- and 2,4-diphenylcobaltacyclopentadienes with the ratio of 4 : 6.

Hex
Co
Ph_3P PPh_3
+ RC≡C—A—C≡CR →
R R Co PPh_3 Hex l R R Co PPh_3 Hex m

5: R = H, A = $-C_6H_4-$ **6**: R = Me, A = $-C_6H_4-$

R—≡—C_6H_4[H H Co PPh_3 Hex C_6H_4]$_n$≡—R

5$_2$ - **5**$_9$: n = 2 - 9

Our next attempt to increase the yield of 2,5-diarylcobaltacyclopentadiene using 1-propynylbenzene instead of ethynylbenzene, expecting the large steric repulsion between phenyl and methyl groups in the 2,4-diphenyl-3,5-dimethylcobaltacycllpentadiene, resulted in the yield of 80% of the desirable isomer.[32] This was applied to the polymer synthesis; MCP reactions between (HexCp)Co(PPh_3)$_2$ and *p*-di-1-propynylbenzene performed at room temperature and 40 °C affords a soluble product, **6**, of which GPC spectra shows a drastic enhancement of polymerization by the temperature increase.

In the course of the study using RC≡C-Ar-C≡CR, we recognized the increase in solubility in organic solvents compared with the polymers of HC≡C-Ar-C≡CH, even if CpCo(PPh_3)$_2$ was used as the starting material. The reaction of CpCo(PPh_3)$_2$ with MeC≡C-*p*-C_6H_4-C_6H_4-*p*-C≡CMe,[33] with MeOCC≡C-*p*-C_6H_4-C_6H_4-*p*-C≡CCOMe,[33] with BuC≡C$C_6H_4C_6H_4$C≡CBu,[34] with MeC≡C-*p*-C_6H_4-C≡CMe,[35] and with MeC≡C-2,5-C_4H_2S-C≡CMe[35] afforded soluble polymers, **7** with $M_w/M_n = 4.0 \times 10^5$ ($M_w/M_n = 4.5$), **8** with 3.0×10^4 ($M_w/M_n = 3.7$), **9** with $M_n = 2.7 \times 10^5$ ($M_w/M_n = 5.2$), **10** with $M_n = 3.8 \times 10^5$ ($M_w/M_n = 4.5$), and **11** with $M_n = 4.4 \times 10^3$ ($M_w/M_n = 1.3$), respectively.

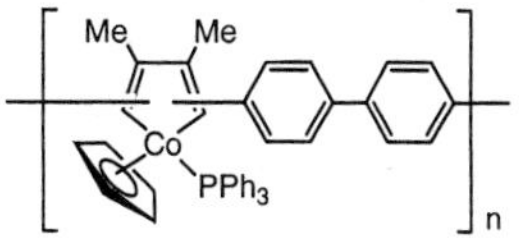

Me Me Co PPh3 n **7** MeOC COMe Co PPh3 n **8** Bu Bu Co PPh3 n **9**

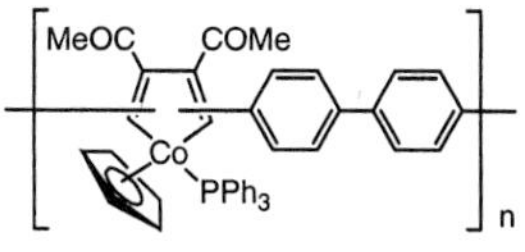

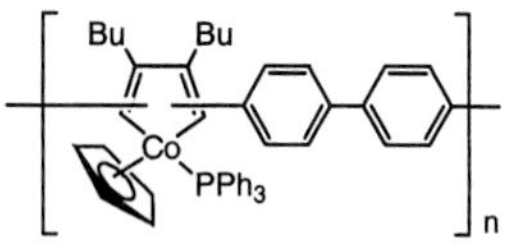

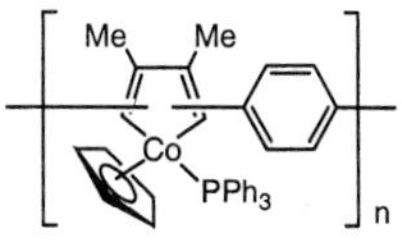

Me Me Co PPh3 n **10** Me Me S Co PPh3 n **11**

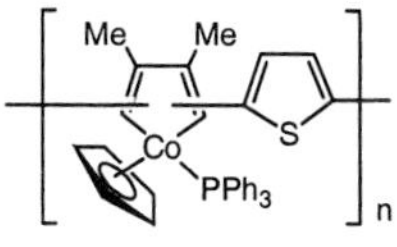

3. Perfectly conjugated polymers synthesized by polycondensation of a dihalogenated cobaltacyclopentadiene complex

As the regioselective polymerization had not been achieved by the metallacycling polymerization, another method utilizing a polycondensation of a dihalogenated cobaltacyclopentadiene complex, Cp(PPh_3)[Co-C(4-C_6H_4I)=CBu-CBu=C(4-C_6H_4I)] (**12**) , with [Ni(cod)$_2$] (cod = cycloocta-1,5-

diene), was applied for the synthesis of perfectly π-conjugated polymer, **13**.[34] When the reaction of **12** was carried out with an excess amount of $Ni(cod)_2$ (2.0 eq) at 50 °C, the molecular weight, M_n, reached to 2.0×10^5 ($M_w/M_n = 2.8$) after 12 h. When the reaction of **12** was carried out with an equimolar of $Ni(cod)_2$ at a room temperature, the oligomers up to a hexamer were obtained. The polymer and oligomer could be purified with a recycling preparative GPC method. Especially, a dimer $\mathbf{13_2}$ and a trimer $\mathbf{13_3}$ were isolated and used as samples for investigating physical properties in detail.

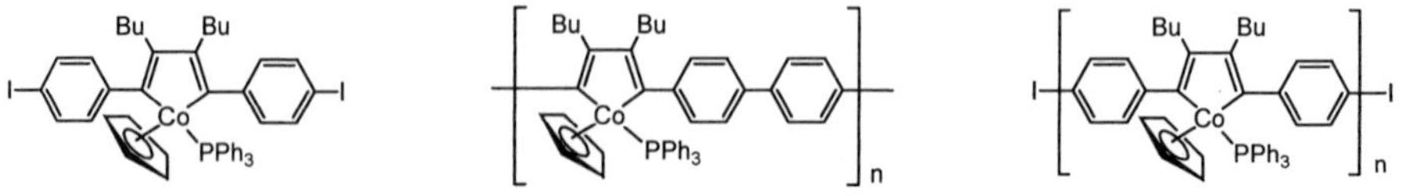

12 **13** $\mathbf{13_2}$: n = 2, $\mathbf{13_3}$: n = 3

Physical Properties of Cobaltacyclopentadiene Polymers

1. Electronic spectra

The color of all the cobaltacyclopentadiene polymers noted above is dark brown and the UV-Vis absorption spectra of the films of **1-4** coated on quartz glass show strong bands that have an edge at 500 to 600 nm.[32] As for the soluble polymer, the band edge shifts to the higher wavelength according to an increase in polymerization degree for **9** in CH_2Cl_2 solutions, indicating the formation of π-conjugated structure.[32] The perfectly π-conjugated polymer **13** prepared by polycondensation on dihalogenated complex exhibit a shift of the peak edge to the longer wavelength compared with the corresponding polymer, **9** prepared by metallacycling polymerization.[34]

Band gap energies, E_g, for the cobaltacyclopentadiene polymers were evaluated from the absorption edge based on the semiconductor theory.[36] The E_g values thus evaluated were 2.1 to 2.3 eV,[32] which correspond roughly to the value observed in poly(thiophene) (2.0 eV) and are relatively small when compared to the band gaps among previously known π-conjugated organic polymers.[2]

2. Redox properties

Cobaltacylopentadiene complexes undergo one-electron oxidation and their potential and chemical reversibility depend strongly on the substituents on the metallacycle.[37-40] This directly reflects the redox properties of the cobaltacylopentadiene polymers. The oxidation potential of the polymers becomes more positive when the electron-withdrawing substituents such as –COMe was bound to the metallacycle,[33] and the chemical reversibility increases in the order of the substituents, -COMe < -H < alkyl (-Me, Bu).[33] The polymers with alkyl groups as substituents show cyclic voltammograms indicating high chemical reversibility at a scan rate of 0.1 Vs^{-1} in NBu_4ClO_4-CH_2Cl_2. These results support our consideration that the HOMO based on *d*-orbital of the metal atoms in the polymer exists between the valence band (VB) and the conduction band (CB) derived from π-conjugation, and the oxidation occurs at metal sites.

In the cyclic voltammograms of a dimer $\mathbf{13}_2$ and a trimer $\mathbf{13}_3$, the waves are broader compared with that of the monomer, although only a single oxidation peak is observed for the dimer and the trimer.[34] This discrepancy suggests that Co(III) and Co(IV) sites are weakly interacted and the mixed-valence states, [Co(III), Co(IV)], [Co(III), Co(IV), Co(III)] and [Co(IV), Co(III), Co(IV)], are generated within a narrow potential range. On the basis of computer simulation, the oxidation potentials are calculated to be $E^{0'}{}_1$ = -0.285 V and $E^{0'}{}_2$ = -0.212 V *vs.* ferrocenium/ferrocene (Fc^+/Fc) for the dimer, and $E^{0'}{}_1$ = -0.291 V, $E^{0'}{}_2$ = -0.248 V, and $E^{0'}{}_3$ = -0.189 V *vs.* Fc^+/Fc for the trimer.

Aoki and Chen have reported a theoretical insight on the redox properties of a linearly combined multi-redox system based on the electronic interaction energy between the neighbouring redox sites.[41] The parameters for evaluating the stability of the mixed-valence states, u_1 and u_2 can be evaluated from the redox potentials. When $u_1 = (u_{RR} + u_{OO})/2 - u_{OR}$ and $u_2 = (u_{OO} - u_{RR})/2$, the difference in redox potentials $E^{0'}{}_2 - E^{0'}{}_1$ for dimer corresponds to $2u_1$, and the differences $E^{0'}{}_2 - E^{0'}{}_1$ and $E^{0'}{}_3 - E^{0'}{}_2$ for trimer correspond to $2u_1 - 2u_2$ and $2u_1 + 2u_2$, respectively. From the experimental results for the dimer, u_1 is calculated to be 3.5 kJ mol^{-1}.[34] Similarly for the trimer, u_1 and u_2 are estimated at 3.0 and 0.96 kJ mol^{-1}, respectively. These values are reasonable because the difference in u_1 between the dimer and the trimer is small. If the value of u_{RR} is assumed to be zero since R is a neutral form and the interaction between neutral forms should be weak, u_{OR} is estimated to be -2 kJ mol^{-1}. and is about one-fifths compared with that for oligo(1,1'-dihexylferrocenylene)s.[42,43]

The oxidation wave in the cyclic voltammograms of thenylene-bridged cobaltacyclopentadiene polymer is fairly broader than that of the phenylene-bridged one.[35] This is because the energy level for the highest occupied π-orbital of thiophene is closer than that of phenylene to *d*-orbital level of the cobalt site, so that the internuclear electronic interaction through the thiophene ring is considered to be stronger. In the oxidation process, therefore, more than one oxidation waves due to the formation of mixed-valence states overlap, resulting in a broad wave in the cyclic voltammogram.

3. Electrical conductivity and photoconductivity

The cobaltacyclopentadiene polymers **4** and **7** show an electrical conductivity of 10^{-12} - 10^{-6} Scm^{-1} in the neutral form at room temperature.[33,44] When **7** was treated with I_2, the conductivity increased up to 10^{-4} Scm^{-1}.[33] This result could be interpreted by the consideration that I_2-doping generates Co(III)/Co(IV) mixed-valence states in the polymer chain; this state is stable to some extent as suggested by the electrochemical properties; and consequently, Co(III) and Co(IV) sites are interacted through a p-conjugated chain, causing the mixed-valence conductivity.

The intrinsic photoconductive property was found for a cobaltacyclopentadine polymer **6**.[32] Photo-response of i-V characteristics for ITO/**6**/ITO indicates that the polymer has a low conductivity in dark, and photocurrent is four times larger than the dark current. This kind of remarkable photoconductivity does not appear for common π-conjugated organic polymers in the undoped state but is caused by forming charge-transfer complexes with donor or acceptor molecules such as fullerene.[45,46] We propose that the metal d-character orbitals localized at cobalt sites and their energy level lying between valence and conduction bands act as the trapping sites of holes generated by photo-activation of electrons from the valence band to the conduction band.

Cyclobutadienecobalt Polymer Containing Ferrocenyl Groups

The reaction of $CpCo(PPh_3)_2$ with a π-conjugated diacetylene, $FcC{\equiv}C$-*o*-C_6H_4-$C{\equiv}CFc$, in which Fc = ferrocenyl, was found to give a cyclobutadienecobalt mononuclear complex, $\{\eta^4$-C_4Fc_2(*o*-$FcC{\equiv}CC_6H_4)_2\}CoCp$ (**14**), the crystal structure of which was determined by X-ray crystallography.[47] The complex **14** shows reversible $1e^-$ and $3e^-$ redox waves at $E^{0'}$ = 0.067 and 0.000 V *vs.* Fc^+/Fc according to the strong electronic interaction between ferrocenyl groups on the cyclobutadiene ring. In contrast, the reaction of $CpCo(PPh_3)_2$ with $FcC{\equiv}C$-*p*-C_6H_4-$C{\equiv}CFc$

affords a polymer.[47] The reaction at 50 °C for 22 h, followed by reprecipitation from toluene-hexane afforded an orange powdery solid. Its ^1H NMR spectrum showed a signal of Cp coordinated to Co at δ5.00 and no signals due to PPh_3. This result and the broadness of the signals suggest the formation of a cyclobutadiene complex polymer, $[p\text{-}C_6H_4\{(\eta^4\text{-}C_4Fc_2)CoCp\}]_n$ (**15**). GPC analysis of **15** showed that M_n = 5500 and M_w = 9600, based on the polystyrene standard. The molecular weight of 5500 indicates that the degree of polymerization is ca. 9. There are two possibilities for the geometric structure of the cyclobutadiene unit, either a 1,2-diferrocenyl or 1,3-diferrocenyl conformation. The ratio of these two structures has not been determined because of the broadness of the ^1H NMR signals. However, the redox properties of **15** showing two chemically reversible redox waves at $E^{0\prime}$ = 0.149 and 0.210 V *vs*. Fc^+/Fc, indicating a strong electronic interaction between ferrocenyl groups suggest that 1,2-diferrocenylcyclobutadiene is the primary structure in the polymer.

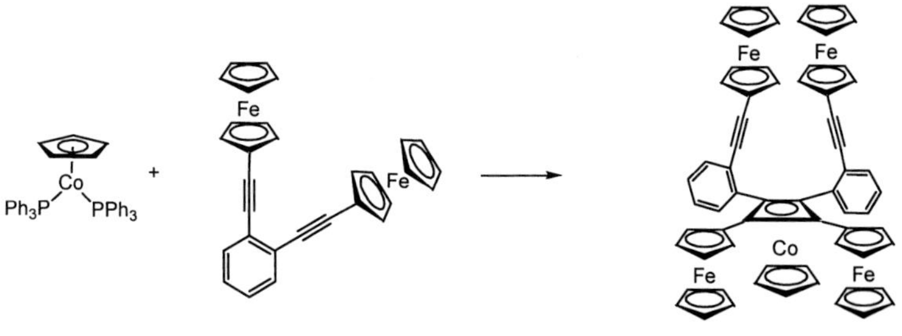

Co
Ph3P PPh3
+
Fe
Fe
Fe Fe
Co
Fe Fe
14

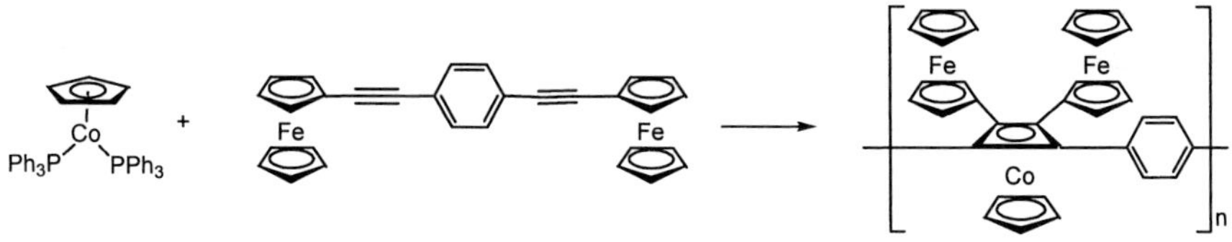

Co
Ph3P PPh3
+
Fe Fe
Fe Fe
Co
n
15

Ruthenacyclopentatriene Polymer

Ruthenacyclopentatriene is formed by the metallacycling reaction of two RC≡CH molecules with CpRuBr(cod) or $(\eta^5\text{-}C_5Me_5)RuCl(cod)$.[48,49] The difference in the cyclization reaction compared with the cobalt system described above is the regioselectivity of this reaction. It was reported that only 2,5-R_2 isomer is formed in the case R = Ph. Our study for the reaction using ethynylferrocene also afford one isomer, 2,5-bis(ferrocenyl)ruthenacyclopentatriene, **16**.[50] Cyclic voltammetry of this complex showed a reversible one-electron reduction due to the ruthenacycle and a two-step one-electron oxidation of the ferrocenyl moieties. Separation of the redox potentials of the ferrocenyl moieties was 0.24 V, and the electronic spectrum of one-electron oxidized species exhibited an intervalence-transfer band at 1180 nm. These results indicate the existence of significant electronic interactions between two ferrocenyl moieties through the ruthenacyclopentatriene ring.

Utilization of the ruthenacycle formation reaction for the polymer synthesis has been recently achieved by the reaction of (HexCp)RuBr(cod) with HC≡C-*p*-C_6H_4-C_6H_4-*p*-C≡CH.[51] The reaction carried out at 0 °C for 90 h gave a polymer **17** with M_n = 3400 (M_w/M_n = 1.7). UV-vis spectra of the polymer shows a shift of the π–π* band to a longer wavelength due to the enlargement of the π-conjugation. The polymer undergoes reversible reduction due to the ruthenacylcle moiety at –1.01 V vs. Fc^+/Fc, which is similar to that of the corresponding monomer, **18**.

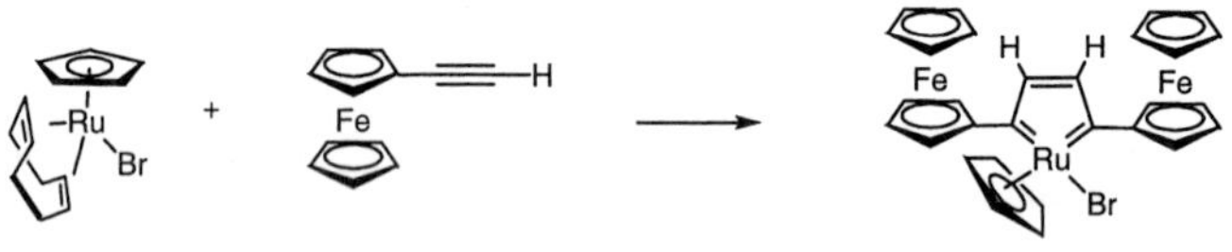

Ru Br + Fe H → H H Fe Fe Ru Br

16

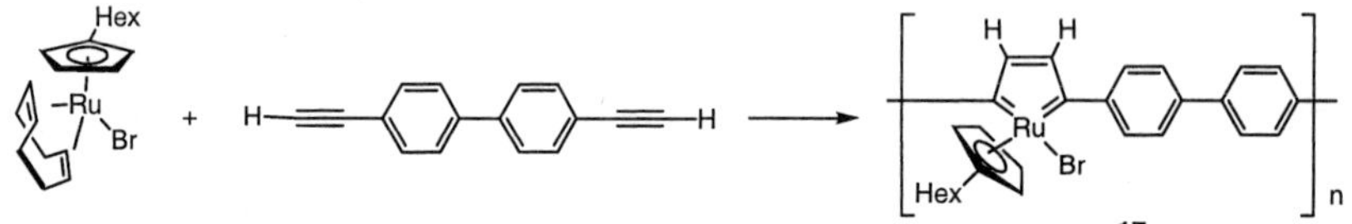

Hex Ru Br + H H → H H Ru Br Hex n

17

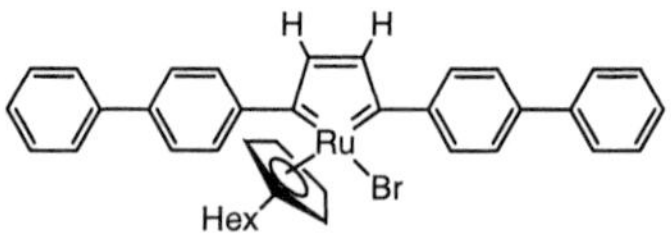

H H Ru Br Hex

18

Conclusion

The cobaltacyclopentadiene formation reaction using conjugated dieacetylnes affords a π-conjugated conducting organometallic polymers. Regioselectivity of the metallacyclization is important for the π-conjugation, and up to present, 80% of the regioselectivity has been achieved for the cobaltacylopentadiene polymers. Perfectly π-conjugated cobaltacyclopentadinene polymer was synthesized by a polycondensation of a dihalogenated cobaltacyclopentadiene complex. The oxidation of the cobalt center occurs facilely and the location of the Co *d*-orbital between the conduction and valence bands brings intrinsic photoconductivity. Ferrocenyl A π-conjugated ruthenacylcopentatriene polymer was also synthesized by the metallacyling polymerization.

Acknowledgments

This work was supported in part by Grants-in-Aid for Scientific Research (Nos. 14044021 (area 412) ,14204066) from the Ministry of Culture, Education, Science, Sports, and Technology, Japan, and by The 21st Century COE Program for Frontiers in Fundamental Chemistry.

[1] H. Shirakawa, *Angew. Chem. Int. Ed. Engl.* **2001,** 40, 2575, and the references therein.
[2] H. S. Nalwa (Ed.), "Handbook of Organic Conductive Molecules and Polymers", Wiley-VCH, Weinheim 1997.
[3] H. Nishihara, in "Handbook of Organic Conductive Molecules and Polymers", H. S. Nalwa, Ed., Wiley, Weinheim 1997, Vol. 2, Chapter 19, pp. 799-832.
[4] E. W. Neuse, *J. Macromol. Sci.-Chem.* **1981,** *A16*, 3.
[5] T. Yamamoto, K. Sanechika, A. Yamamoto, *Inorg. Chim. Acta* **1983**, *73*, 75.
[6] Y. Okamoto, M. C. Wang, *J. Polym. Sci., Polym. Let. Ed.* **1980**, *18*, 249.
[7] K. Krikor, M. Rotti, P. Nagles, *Synth. Met.* **1987**, *21*, 353.
[8] H. Matsuda, H. Nakanishi, M. Kato, *J. Polym. Sci., Polym. Lett. Ed.* **1984**, *22*, 107.
[9] N. Hagihara, K. Sonogashira, S. Takahashi, *Adv. Polym. Sci.* **1980**, *41*, 159.
[10] M. S. Khan, S. J. Davies, A. K. Kakkar, D. Schwartz, B. Lin, B. F. D. Johnson, J. Lewis, *J. Organomet. Chem.* **1992**, *87*, 424.
[11] J. W. P. Lin, L. P. Dudek, *J. Polym. Sci., Polym. Chem. Ed.* **1985**, *23*, 1579.
[12] S. Venkatachalam, K. V. C. Rao, P. T. Manoharan, *Synth. Met.* **1988**, *26*, 237.
[13] T. Kuhlmann, S. Roth, J. Rozière, W. Siebert, *Angew. Chem. Int. Ed. Engl.* **1986**, *25*, 105.
[14] R. A. Clark, K. S. Varma, A. E. Underhil, J. Becher, H. Toftlund, *Synth. Met.* **1988**, *25*, 227.
[15] J. R. Reynolds, F. E. Karasz, C. P. Lillya, J. C. W. Chien, *J. Chem. Soc., Chem. Commun.* **1985**, 268.
[16] R. Vincete, J. Ribas, P. Cassoux, L. Valade, *Synth. Met.* **1986**, *13*, 265.
[17] M. Altmann, U. H. F. Bunz, *Angew. Chem. Int. Ed. Engl.* **1995**, *34*, 569.
[18] H. S. Nalwa, *Appl. Organometal. Chem.* **1990**, *4*, 91.
[19] A. Ohkubo, K. Aramaki, H. Nishihara, *Chem. Lett.* **1993**, 271.
[20] I. Tomita, A. Nishio, T. Igarashi, T. Endo, *Polym. Bull.* **1993**, *30*, 179.
[21] J. R. Nitschke, S. Zurcher S, T. D. Tilley, *J. Am. Chem. Soc.*, **2000,** *122*, 10345.
[22] H. Yamazaki, N. Hagihara, *Bull. Chem. Soc. Jpn.* **1971**, *44*, 2260.

[23] Y. Wakatsuki, H.Yamazaki, *J. Chem. Soc., Chem. Commun.* **1973**, 280.
[24] Y. Wakatsuki, T. Kuramitsu, H. Yamazaki, *Tetrahedron Lett.* **1974**, 4549.
[25] H. Bönnenman, *Angew. Chem., Int. Ed. Engl.* **1985**, *24*, 248.
[26] K. M. Nicholas, M. O. Nestle, D. Seyferth, in "Transition Metal Organometallics in Organic Synthess", H. Alper Ed., Academic Press, New York, 1978, Vol. 2.
[27] H. Nishihara, T. Shimura, A. Ohkubo, N. Matsuda, K. Aramaki, *Adv. Mater.* **1993**, *5*, 752.
[28] T. Shimura, A. Ohkubo, K. Aramaki, H. Uekusa, T. Fujita, S. Ohba, H. Nishihara, *Inorg. Chim. Acta* **1995**, *230*, 215.
[29] T. Fujita, H. Uekusa, A. Ohkubo, T. Shimura, K. Aramaki, H. Nishihara, S. Ohba, *Acta Cryst.* **1995**, *C51*, 2265.
[30] Y. Wakatsuki, O. Nomura, K. Kitaura, K. Morokuma, H. Yamazaki, *J. Am. Chem. Soc.* **1983**, *105*, 1907.
[31] N. Matsuda, T. Shimura, K. Aramaki, H. Nishihara, *Synth. Metals* **1995**, *69*, 559.
[32] T. Shimura, A. Ohkubo, N. Matsuda, I. Matsuoka, K. Aramaki, H. Nishihara, *Chem. Mater.* **1996**, *8*, 1307.
[33] I. Matsuoka, K. Aramaki, H. Nishihara, *Mol. Cryst. Liq. Cryst.*, **1996**, *285*, 199.
[34] I. Matsuoka, K. Aramaki, H. Nishihara, *J. Chem. Soc., Dalton Trans.*, **1998**, 147.
[35] I. Matsuoka, H. Yoshikawa, M. Kurihara, H. Nishihara, *Synth. Metals* 1999, **102**, 1519.
[36] I. Kudmar, T. Seidel, *J. Appl. Phys.* **1962**, *33*, 771.
[37] R. S. Kelly, W. E. Geiger, *Organometallics* **1987**, *6*, 1432.
[38] B. T. Donovan, W. E. Geiger, *J. Am. Chem. Soc.* **1988**, *110*, 2335.
[39] B. T. Donovan, W. E. Geiger, *Organometallics* **1990**, *9*, 865.
[40] A. Ohkubo, T. Fujita, S. Ohba, K. Aramaki, H. Nishihara, *J. Chem. Soc., Chem. Commun.* **1992**, 1553.
[41] K. Aoki, J. Chen, *J. Electroanal. Chem.*, 1995, **380**, 35.
[42] T. Hirao, M. Kurashina, K. Aramaki, H. Nishihara, *J. Chem. Soc., Dalton Trans.*, **1996**, 2929.
[43] H. Nishihara, T. Hirao, K. Aramaki, K. Aoki, *Synth. Metals*, **1997**, *84*, 935.
[44] H. Nishihara, A. Ohkubo, K. Aramaki, *Synth. Metals*, **1993**, *55*, 821.
[45] K. Yoshino, S. Morita, T. Kawai, H. Araki, X. H. Yin, A. A. Zakhidov, *Synth. Met.* **1993**, *56*, 2991.
[46] N. S. Sariciftci, L. Smilowitz, D. Braun, G. Srdaniov, V. Srdanov, F. Wudl, A. J. Heeger, *Synth. Met.* **1993**, *56*, 3125.
[47] M. Murata, T. Hoshi, I. Matsuoka, T. Nankawa, M. Kurihara, H. Nishihara, *J. Inorg. Organomet. Polym.* **2000**, *10*, 209.
[48] M. O. Albers, D. J. A. deWaal, D. C. Lies, D. J. Robinson, E. Singleton, M. B. Wiege, *J. Chem. Soc., Chem. Commun.* **1986**, 1681.
[49] C. Ernst, O. Walter, E. Dinjus, S. Arzberger, H. Görls, *J. Pract. Chem.* **1999**, *8*, 341.
[50] Y. Yamada, J. Mizutani, M. Kurihara, H. Nishihara, *J. Organomet. Chem.* **2001**, *80-83*, 637.
[51] M. Kurashina, MS thesis, The Univesisty of Tokyo, 2002.

Interactions of Ferrocenoyl-Peptides in Solution and on Surfaces

Heinz-Bernhard Kraatz

Department of Chemistry, University of Saskatchewan, 110 Science Place, Saskatoon, Saskatchewan, Canada S7N 5C9
E-mail: kraatz@skyway.usask.ca

Summary: The Ferrocenoyl-peptide-cystamines, such as $[Fc\text{-}Gly\text{-}CSA]_2$ (Gly = glycine, CSA = cystamine), $[Fc\text{-}Ala\text{-}CSA]_2$ (Ala = alanine) and Fc-conjugates involving collagen models, such as $[Fc\text{-}(Pro_2Gly)_n\text{-}CSA]_2$ (Pro = proline, n = 1-6) are readily prepared by solution methods. In solution and the solid state, these systems exhibit intermolecular hydrogen bonding between adjacent peptide chains. For $[Fc\text{-}Gly\text{-}CSA]_2$ this results in the formation of a supramolecular helicate with two different H-bonding patterns. Ferrocenoyl-collagen-cystamines form assemblies in solution, which melt at elevated tempertures. All systems for monolayers on gold surfaces, which show a well-behaved ferrocene-based electrochemistry, which allows the determination of the spatial requirements of the peptides on the surface.

Keywords: ferrocene; hydrogen bonding; peptide; self-assembled monolayer; supramolecular assembly

Introduction

In solution and in the crystalline state, amino acids and peptides often assemble into extended supramolecular three-dimensional structures. These often form as a consequence of hydrogen bonding between individual molecules.[1] Interestingly, the properties of these peptide supramolecular assemblies are related to the molecular arragement of the subunits. Considerable effort has focussed on the design of secondary structural elements,[2] and on the design of new peptidic materials, such as nanotubes[3] and hydrogels,[4] with potential applications in drug delivery and biomedical engineering. In many cases, scaffolds are used to assist the design and guide formation of a particular peptide structural mimic. Recent effort have been directed at equipping non-covalent supramolecular peptide assemblies with redox-active groups, such as ferrocenes,[5] and give them specific electric properties that may be exploited for biosensing or may have potential for the design of bioelectronic circuitry. Our efforts have been guided by our

© 2003 WILEY-VCH Verlag GmbH & KGaA, Weinheim CCC 1022-1360/00/$ 17,50+.50/0

desire to investigate the electron transfer properties in peptides.[6]

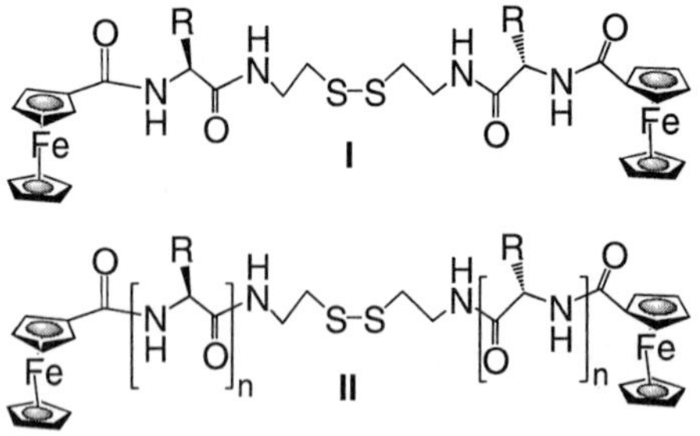

For this purpose, we have developed a synthetic pathway allowing the synthesis of ferrocenoyl-labeled amino acid (**I**) and peptide cystamines (**II**), which in turn can be used to prepare ordered two-dimensional arrays on a gold surface.

Hydrogen-Bonding

$[Fc\text{-}Gly\text{-}CSA]_2$ readily crystallizes from chloroform in the chiral space group $P4_3$. A view of the helical structure is shown in Figure 1.[7]

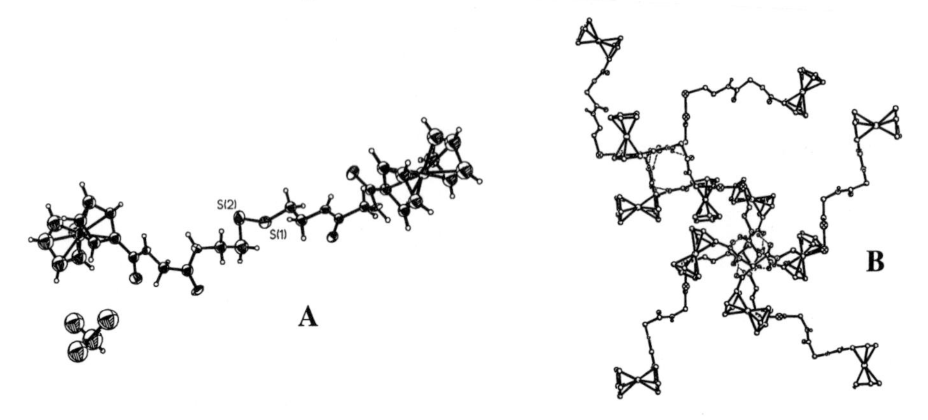

Figure 1. Molecular view $[Fc\text{-}Gly\text{-}CSA]_2$ (A) of the double helicity in the crystalline state (B) giving a square helix and a twisted helix. The compound crystallizes as a $CHCl_3$ solvate. The solvate does not exhibit any significant interactions with the Fc-peptide.

The structure shows two sets of H-bionding interactions – one which is commonly found in parallel peptide β-sheets and another one which involves a pair of cis-amides. The net result of this complex intermolecular H-bonding interaction is that invividual molecules are forced to turn with respect to each other, resulting in a helical arrangement. Importantly, the two side of [Fc-Gly-CSA]$_2$ are different and each one is involved in a different supramolecular helical arrangement. The result is an arrangement of two propeller-shaped helices with H-bonded cores linked to each other through a disulphide bridge. The redox active ferrocenoyl moieties are on the outside of a central H-bonded peptide core. Both helices (Figure 1 B) have a pitch height of ca. 14 Å. Although, peptide disulfides often exhibit unusal structural features, the presence of two different chiral helical arrangements in a peptide conjugate is unique and not been described before. Importantly, Fc-amino acid and peptide cystamines exhibit strong H-bonding even in solution. Figure 2 shows the Amide A region of the IR spectrum of [Fc-Ala-CSA]$_2$ in chloroform solution. It clearly shows the presence of two N-H stretching vibrations, typical of H-bonded and non-hydrogen bonded amides. This is readily rationalized by the equilibirium between H-bonded and non-associated molecules.

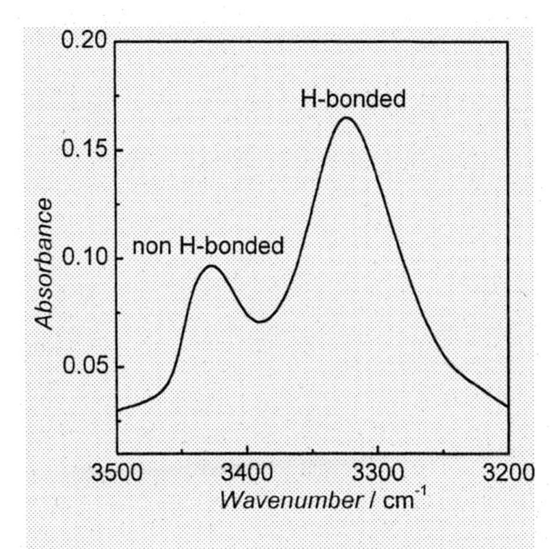

Figure 2. FT-IR absorption bands in the Amide A (NH) region from 3500-3200 cm^{-1} for a 50mM solution of compound [Fc-Ala-CSA]$_2$ in $CHCl_3$ showing hydrogen bonded and non-bonded amide.

Ferrocenoyl-collagen cystamines of the general formula [Fc-(Pro$_2$Gly)$_n$-CSA]$_2$ (n = 1-3) exhibit solution association, typical for collagens. The melting curve for [Fc-(Pro$_2$Gly)$_3$-CSA]$_2$ is shown in Figure 3, having a T_m of 339(2) K. The corresponding „dicollagen“ [Fc-(Pro$_2$Gly)$_2$-CSA]$_2$ has a melting point that is only slightly lower (336(1) K)).

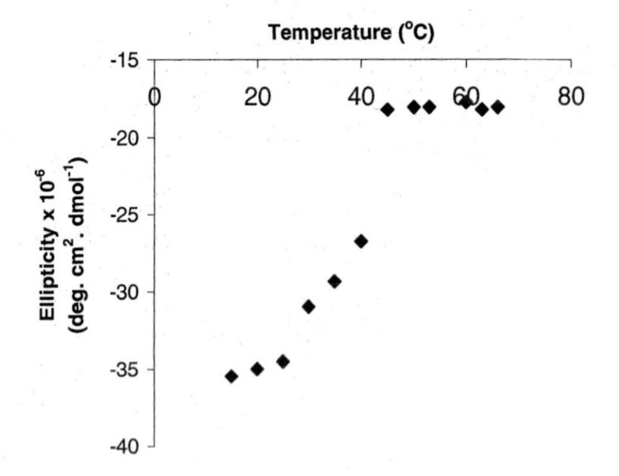

Figure 3. Melting curve for $[Fc\text{-}(Pro_2Gly)_3\text{-}CSA]_2$ determined by CD spectroscopy.

Monolayers

Using Fc-amino acid and peptide cystamines we were able to prepare stable and well-behaved monolayers on gold surfaces.[6] The synthetic strategy employed for the synthesis of these systems is shown in Figure 4.

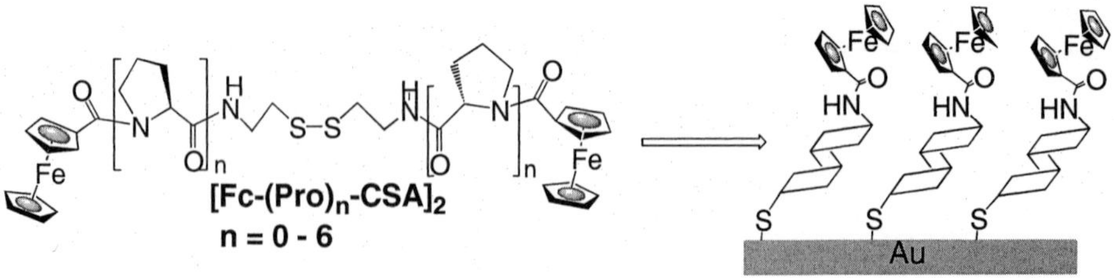

Figure 4. Preparation of monolayers of Fc-peptide cysteamines. Shown is the example of a series of Fc-oligoproline cystamines resulting in stable monolayers.

The redox properties are summaried in Table 1. All monolayers exhibit a single reversible one-electron oxidation. The area occupied by individual Fc-amino acid and peptide molecules on the surface, onbtained from the integration of the oxidative peak currents in the cyclic voltammogram, is given in Table 1. Our results compare well with other helical peptides.[8]

Shorter Fc-peptides which are not able to adopt a helical conformation require less space.

Table 1. Specific Area (in Å^2) cccupied by individual Fc-amino acids and peptide cystamines cbtained from the integrated oxidative peak currents.

Entry	Compound	Specific Area
1	$[Fc\text{-}CSA]_2$	85(26)
2	$[Fc\text{-}Gly\text{-}CSA]_2$	70(3)
3	$[Fc\text{-}Ala\text{-}CSA]_2$	81(9)
4	$[Fc\text{-}Pro\text{-}CSA]_2$	123(38)
5	$[Fc\text{-}Ala_2\text{-}CSA]_2$	103(20)
6	$[Fc\text{-}Pro_2\text{-}CSA]_2$	157(20)
7	$[Fc\text{-}Pro_3\text{-}CSA]_2$	180(21)
8	$[Fc\text{-}Pro_4\text{-}CSA]_2$	190(27)
9	$[Fc\text{-}Pro_5\text{-}CSA]_2$	220(40)
10	$[Fc\text{-}Pro_6\text{-}CSA]_2$	240(13)

In general, the oligoproline monolayers are well blocked whereas the other monolayers appear disordered and do not efficiently block the direct electron transfer between the electrode surface and electro active molecules in solution. Thus, there is a fundamental difference between oligoproline monolayers and the other systems, which is most likely related to a better packing on the surface, preventing defects, such as pinholes. Our studies show that H-bonding of the surface supported species is crucial for the formation of a tight monlayer.

Conclusion

Fc-amino acid and peptide cystamines having available amide NH groups assemble to give larger H-bonded assemblies in the solid state and in solution. Expectedly, at higher temperature, these assemblies melt. Monolayers are readily formed. Shorter Fc-amino acid systems tend to produce „leaky“ monolayers which allow access of molecules from the solution to the surface. It is suspected that an increase in the intermolecular H-bonding may result in better blocked surfaces.

Detailed investigations (ellipsometry, RAIRS, AFM) are currently under way to obtain more information about the structural arrangement on the surface.

[1] a) A. Aggeli, I.A. Nyrkova, M. Bell, R. Harding, L. Carrick, T.C.B. McLeish, A.N. Semenov and N. Boden, *Proc. Natl. Acad. Sci. USA*, **98**, 11857 (2001); W.A. Petka, J.L. Harden, K.P. McGrath, D. Wirtz and D.A. Tirrell, *Science*, **281**, 389 (1998).
[2] a) J.P. Schneider and J.W. Kelly, *Chem. Rev.*, **95**, 2169 (1995); b) J.S. Nowick, *Acc. Chem. Res.*, **32**, 287 (1999).
[3] for example: M.R. Ghadiri, J. R. Granja, R.A. Milligan, D.E. McRee and Khazanovich, *Nature*, **366**, 324 (1993).
[4] see for example: R.P. Lyon and W.M. Atkins, *J. Am. Chem. Soc.*, **123**, 4408 (2001).
[5] see for example: T. Moriuchi, A. Nomoto, K. Yoshida, A. Ogawa and T. Hirao, *J. Am. Chem. Soc.*, 2001, **123**, 68.
[6] M.M. Galka and H.-B. Kraatz, *ChemPhysChem*, 2002, **4**, 356.
[7] I. Bediako-Amoa, R. Silerova, H.-B Kraatz, Chem. Commun. 2002, 2430.
[8] N. Higashi, T. Koga, M. Nina, Langmuir, 2000, 16, 3482.

Poly(ferrocenylsilanes) as Etch Barriers in Nano and Microlithographic Applications

*Mark A. Hempenius, Rob G. H. Lammertink, Mária Péter, G. Julius Vancso**

MESA+ Research Institute, University of Twente, P.O. Box 217, 7500 AE Enschede, The Netherlands
E-mail: g.j.vancso@ct.utwente.nl

Summary: Thin films of organic-organometallic block copolymers are shown to be efficient self-assembled templates for nanolithography. Block copolymers composed of organic blocks such as polyisoprene or polystyrene and a poly(ferrocenylsilane) block microphase separate to form a monolayer of densely packed organometallic spheres in an organic matrix. The high resistance of the organometallic phase to reactive ion etching enables the nanoscale patterns to be transferred into silicon substrates, forming nanostructured surfaces. Electrostatic self-assembly of poly(ferrocenylsilane) polyanions and polycations is discussed as a means to form laterally structured organometallic multilayer thin films by area-selective adsorption onto chemically patterned substrates.

Keywords: block copolymers; nanolithography; organometallic polymers; polyelectrolytes; reactive ion etching

Introduction

Macromolecules containing inorganic elements or organometallic units in the main chain combine potentially useful chemical, electrochemical, optical and other interesting characteristics with the properties and processability of polymers.[1] Poly(ferrocenylsilanes), composed of alternating ferrocenyl and alkylsilyl units in the main chain, belong to this class of organometallic polymers. With the discovery of the anionic ring-opening polymerization (ROP) of silicon bridged ferrocenophanes,[1] well-defined poly(ferrocenylsilanes) and block copolymers featuring corresponding organometallic blocks have become accessible.[2]

As we discovered that poly(ferrocenyldimethylsilane) was an effective resist in reactive ion etching processes,[3] it became of interest to employ this polymer in surface patterning of silicon substrates, which has relevance in the fabrication of *e.g.* electrooptical and magnetic storage devices, and sensors. Patterns on the (sub)micrometer scale can be introduced by soft lithography,

© 2003 WILEY-VCH Verlag GmbH & KGaA, Weinheim CCC 1022-1360/00/$ 17,50+.50/0

using poly(ferrocenylsilanes) as ink. Block copolymers featuring poly(ferrocenylsilane) blocks form nanoperiodic microdomain structures upon phase separation.[4] Thin films of such block copolymers, *e.g.* poly(ferrocenyldimethylsilane-*block*-isoprene), can serve as self-assembling templates, enabling nanometer-sized patterns to be transferred directly into silicon or silicon nitride substrates by reactive ion etching.[5] Ferrocenylsilane-styrene block copolymer thin films were successfully used as templates in the fabrication of arrays of nanometer-sized cobalt magnetic dots.[6]

Water soluble poly(ferrocenylsilane) polyions, belonging to the rare class of main chain organometallic polyelectrolytes, have recently been reported by us and others.[7-10] Polyelectrolytes can be employed in electrostatic self-assembly processes to form ultrathin multilayer films.[11,12] In this process, a substrate is immersed alternatingly in polyanion and polycation solutions, leading to the assembly of a multilayer thin film with controlled thickness and composition. We recently reported the synthesis of the first poly(ferrocenylsilane) polyanion.[13] This development enabled the fabrication of all-organometallic multilayers,[14,15] which are of interest e.g. due to their redox activity. In addition to forming continuous organometallic multilayer thin films, we explored the layer-by-layer deposition of poly(ferrocenylsilane) polyions onto hydrophilically/hydrophobically modified substrates with the aim of building two-dimensionally patterned organometallic multilayers.[14] Such films may be of interest as ultrathin resists in reactive ion etch processes.

Nanostructured Surfaces by a Combination of Block Copolymer Self-Assembly and Reactive Ion Etching

Block copolymer self-assembly is an attractive method for introducing nanoperiodic patterns on substrates, due to the simplicity and low cost of the process. The domain size and domain spacing obtained by phase segregation can be controlled by tuning the chemistry, composition and molar mass of the diblock copolymer. As domains obtained by block copolymer phase separation typically have sub-100 nm dimensions, applications in electronic, optoelectronic and magnetic devices can be envisaged. The applicability of block copolymers in forming nanostructured surfaces increases significantly if the self-assembled pattern can act as a lithographic template, enabling patterns to be transferred into substrates. Block copolymers featuring poly(ferrocenylsilane) (PFS) blocks have great potential as etch masks: the high resistance of the organometallic phase to reactive ion etching compared to the organic phase enables one to form nanopatterned surfaces in a one-step etching process.

Various approaches have been employed to enhance the etch selectivity in thin block copolymer films. A common feature of these approaches is to selectively load one of the phases with suitable inorganic components. Park *et al.*[16] used an OsO_4-stained microphase-separated thin film of poly(styrene-*block*-butadiene), PS-*b*-PB, which produced holes upon RIE in silicon nitride substrates, resulting in an etch selectivity of 2:1. Möller *et al.* in a series of papers discussed the use of poly(styrene-*block*-2-vinylpyridine), PS-*b*-P2VP, to prepare masks for nanolithography either by loading the P2VP domains with gold particles[17] or by selective growth of Ti on top of PS domains.[18] These approaches achieve etching contrast by selective introduction of inorganic components in one of the phases.

The advantage of our approach is that the inorganic components are inherently present in the block copolymer, thus eliminating the need for a loading step by a metal and allowing for simple one-step lithography.[19,20]

A thin film (approximately 30 nm) of isoprene-*block*-ferrocenylsilane copolymer (IF 36/12, 36 kg/mol PI -*block*- 12 kg/mol PFS) forms lateral patterns when cast on silicon substrates.[5] At suitable volume fractions and film thicknesses it is possible to obtain a regular morphology, which consists of hexagonally packed organometallic domains (see Figure 1). After spin-casting, the entire silicon wafer is covered by this regular morphology, without the need for subsequent annealing.

Atomic concentrations of carbon (1s), oxygen (1s), silicon (2p), and iron (2p), obtained by XPS for a PI-*b*-PFS diblock and PFS homopolymer before and after oxygen reactive ion etching (O_2-RIE), provide information on the etching process. For the unetched PFS homopolymer we find atomic concentration ratios for C:Si:Fe of 12:0.86:0.97 which is close to the theoretical relative amounts of 12:1:1 ($C_{12}H_{14}SiFe$ for a repeat unit). After O_2-RIE the oxygen concentration increased significantly, accompanied by a decrease in the carbon concentration. A large amount of carbon is removed by RIE compared to the relative amounts of silicon and iron. Upon etching the PFS homopolymer, the Si/C ratio increased from 0.07 to 0.24. At the same time the Fe/C ratio increased from 0.08 to 0.51. It is interesting to see that following O_2-RIE, the relative amount of Fe with respect to Si increased. This suggests that Fe is more stable towards O_2-RIE treatments than Si. XPS further shows that both the Si2p and Fe2p binding energies increase following the oxygen plasma treatment, indicating the conversion of Si and Fe into a complex oxide.[3] These results demonstrate that poly(ferrocenylsilane) forms an etch barrier when exposed to an oxygen plasma due to the formation of an oxide layer at the surface of the polymer. This leads to a very low etch rate of the organometallic material compared to organic polymers such as polyisoprene,

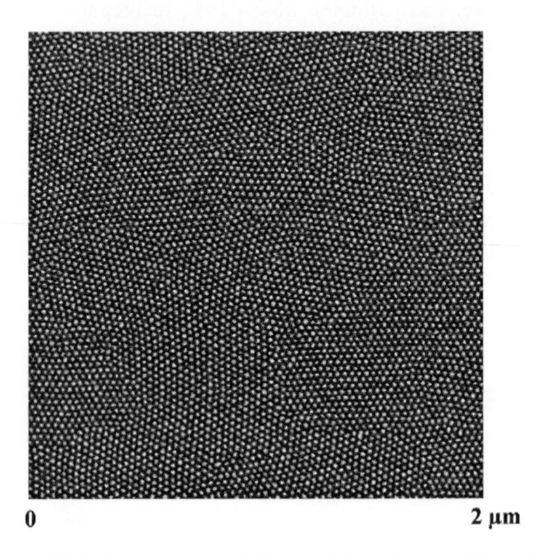

Figure 1. Tapping mode AFM phase image of 30 nm thin spin cast film of IF 36/12. The film consists of laterally separated poly(ferrocenylsilane) domains in an organic matrix.

since the oxide protects the underlying polymer from the plasma. Under the conditions employed here, we observed an etch selectivity for PI:PFS of approximately 40:1, as determined from the corresponding homopolymer etch rates (ca. 7.3 nm/s and 0.18 nm/s for PI and PFS respectively). Figure 2 shows a representative AFM height image of a thin film (~30 nm) of IF 36/12 (Figure 1) after O_2-RIE for 10 s. The organometallic domains, which were converted into an oxide by O_2-RIE, are still arranged in their original hexagonal structure. The height contrast in the

AFM image increased significantly upon etching, due to the removal of the organic matrix. The line scan in Figure 2 (top right) indicates a domain height of approximately 7 nm. According to the Fourier transform of the AFM image (bottom right in Figure), no change in domain spacing occurred upon etching. This simple one-step etching process results in a nanostructured surface with an area density of approximately $1.2 \cdot 10^{11}$ dots/cm^2. After etching, the nanostructured surface is stable when stored at room temperature or rinsed with common organic solvents, in contrast to the thin diblock copolymer film.

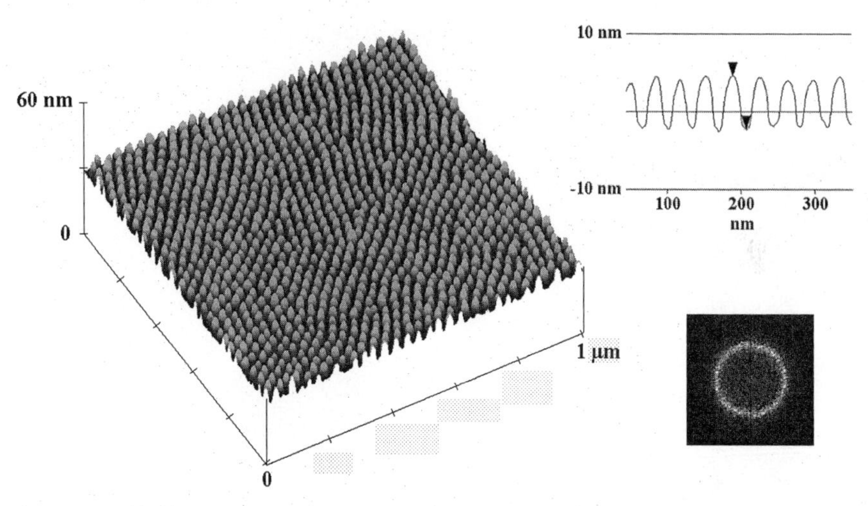

Figure 2. Tapping mode AFM height image (left) of a 30 nm thin film of IF 36/12 after O_2-RIE. A single line scan (top right) and the 2D Fourier transform (bottom right) demonstrate the regularity of the structure. Reproduced with permission from ref. 5, © 2000 ACS.

The aspect ratio of the domains, however, is limited. Due to the removal of material from the organometallic phase, the features shrink which leads to somewhat flatter domains. To obtain a good visualization of the domains, a cross-sectional transmission electron micrograph is shown in Figure 3. Since the cross-section contains more than one row of domains, the sample had to be tilted so that the corresponding rows were aligned. The dots are regularly spaced which confirms the AFM observations. Furthermore, selected area diffraction indicated that the inorganic nanodomains were essentially amorphous.

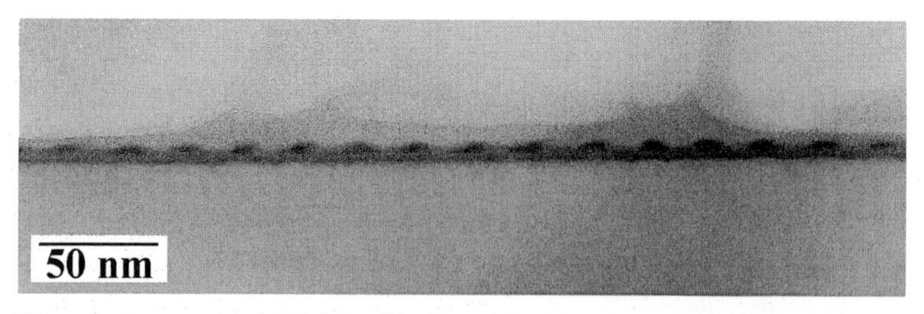

Figure 3. Cross sectional TEM of a thin film of IF 36/12 after O_2-RIE treatment. The Fe-Si-oxide domains appear dark in the TEM, as does the silicon oxide layer at the surface of the silicon substrate.

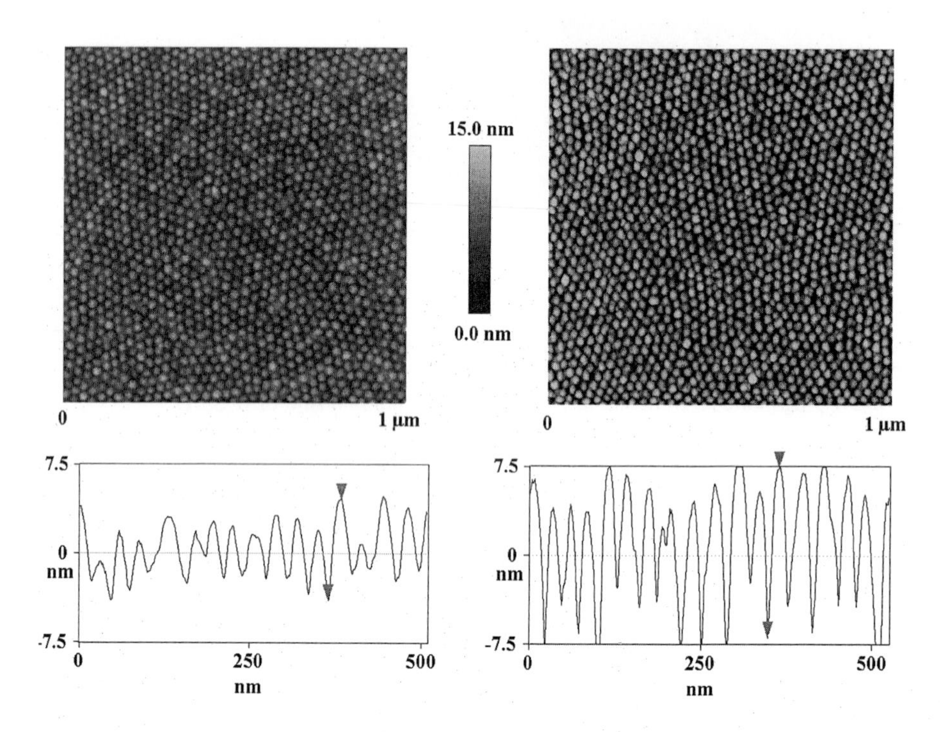

Figure 4. AFM height images of a thin IF 36/12 film on a silicon nitride substrate. The left image was taken after O_2-RIE of the film. The film displayed on the right was subsequently treated for 30 seconds with a CF_4/O_2 plasma. The line scans below each image clearly demonstrate the increase in aspect ratio of the domains.

Due to the presence of iron in the inorganic domains, the pattern can be transferred into the underlying substrate by the action of CF_4/O_2 reactive ion etching. Figure 4 shows the AFM images of a thin IF 36/12 film after O_2-RIE treatment (left) followed by a CF_4/O_2-RIE treatment (right). As can be concluded from the image scale and the line scans, the height of the features has increased upon the etching treatment. This is a result of the removal of the substrate in between the domains combined with the high etch resistance of the inorganic domains.

Recently, it was demonstrated that the organic-organometallic block copolymer domain patterns described here can be ordered over large areas by using sidewall constraints.[21] This development is of particular importance with regard to applications where long-range positional order of the domains is required.

Water-Soluble Poly(ferrocenylsilanes), Synthesis and Directed Self-Assembly

Any route to high molar mass poly(ferrocenylsilanes) incorporates a ring-opening polymerization step of a strained silicon-bridged ferrocenophane.[1] Such ferrocenophanes are obtained by treating 1,1'-dilithioferrocene with a dichlorosilane of choice. Many functionalities, however, do not tolerate the highly basic dilithioferrocene or the reactive chlorosilane moieties, are incompatible with the reactive strained monomer itself, or hinder monomer purification. Therefore, functionalization after polymerization by side-group modification reactions will be required if polar or even ionic moieties are to be introduced.

We introduced and employed a poly(ferrocenylsilane), featuring chloropropylmethylsilane repeat units, as an organometallic main chain which already has reactive pendant groups in place for further functionalization by nucleophilic substitution, see Scheme 1.[10]

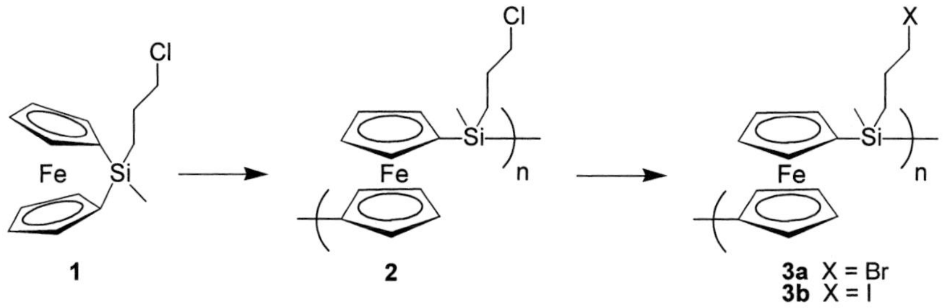

Scheme 1. Poly(ferrocenylsilanes) featuring haloalkyl side groups.

Poly(ferrocenyl(3-chloropropyl)methylsilane) **2** was readily accessible by transition-metal catalyzed ring-opening polymerization[22] of the corresponding (3-chloropropyl)methylsilyl[1] ferrocenophane **1**.[10] The chloropropyl groups are linked to silicon by Si-C bonds, which are stable to hydrolytic cleavage, as opposed to groups linked by e.g. Si-O bonds. By means of halogen exchange,[23] **2** can be converted quantitatively into its bromopropyl (**3a**) or iodopropyl analogues (**3b**),[13] which are particularly suitable for functionalization by nucleophilic substitution. Thus, a wider range of nucleophiles can be employed to prepare functional poly(ferrocenylsilanes).

As an example, reaction of **2** with potassium 1,1,3,3-tetramethyldisilazide[24] and dicyclohexano-18-crown-6 in THF afforded a N,N-bis(dimethylsilyl)-protected poly(ferrocenyl(3-aminopropyl)methylsilane) **4**, which was hydrolyzed to the desired polycation **5** in aqueous acid,[10] see Scheme 2.

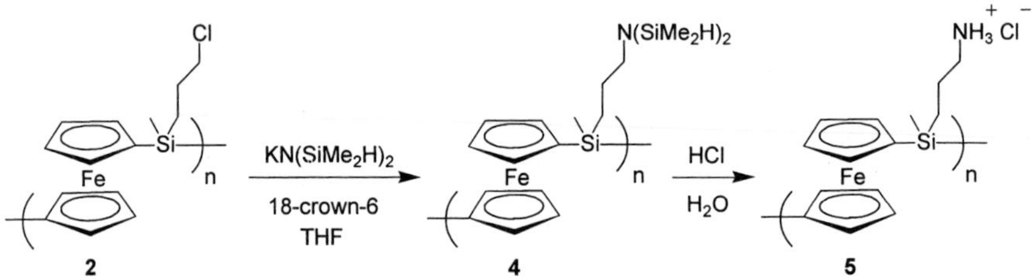

Cl
Si
Fe
n
2
$KN(SiMe_2H)_2$
18-crown-6
THF
$N(SiMe_2H)_2$
Si
Fe
n
4
HCl
H_2O
$\overset{+}{N}H_3\ \overset{-}{Cl}$
Si
Fe
n
5

Scheme 2. Synthesis of a poly(ferrocenylsilane) polycation.

The more reactive poly(ferrocenyl(3-bromopropyl)methylsilane) **3a** can be converted into polycation **5** using lithium 1,1,3,3-tetramethyldisilazide, without the aid of a crown ether.

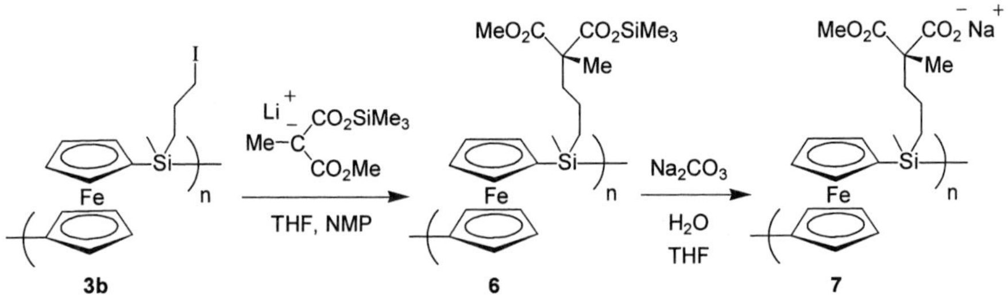

I
Si
Fe
n
3b
Li^+ $Me-\overset{-}{C}(CO_2SiMe_3)(CO_2Me)$
THF, NMP
MeO_2C CO_2SiMe_3
Me
Si
Fe
n
6
Na_2CO_3
H_2O
THF
MeO_2C $CO_2^-\ Na^+$
Me
Si
Fe
n
7

Scheme 3. Synthesis of a poly(ferrocenylsilane) polyanion.

In case of functionalization by carbon nucleophiles, poly(ferrocenyl(3-iodopropyl)methylsilane) **3b** is a suitable starting material. Malonic ester enolates such as dimethyl methylmalonate anion or dibenzyl methylmalonate anion smoothly react to produce the corresponding polyesters with quantitative conversions. By using hydrolytically labile ester enolates, such as methyl trimethylsilyl methylmalonate anion,[25] one can easily convert the corresponding pendant ester groups into carboxylate salts, thus forming a polyanion[13] (Scheme 3). This polyelectrolyte is, to our knowledge, the first reported organometallic polyanion. The material is highly water soluble: it could be dissolved to concentrations exceeding 100 mg/mL.

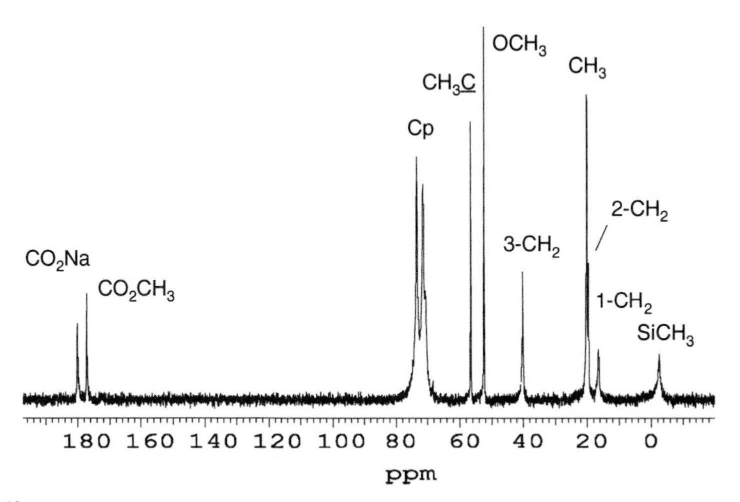

Figure 5. ^{13}C NMR spectrum of the poly(ferrocenylsilane) polyanion **7** in D_2O. Reproduced with permission from ref. 13, © 2002 ACS.

^{1}H and ^{13}C NMR spectroscopy support the high efficiency of the side-group modifications. As an example, the ^{13}C NMR spectrum of **7** is shown in Figure 5. No signals associated with residual iodopropyl moieties could be identified in the spectrum. In addition, elemental analysis of all polymers had excellent agreement between the expected and measured compositions. Polymers **2**, **3a** and **3b** were characterized using Gel Permeation Chromatography (GPC) in THF, using polystyrene standards, to ensure that no molar mass decline had occurred during the halogen exchange reaction. Based on GPC, poly(ferrocenylsilanes) **2** and **3a,b** have a degree of polymerization DP_n = 80 and a polydispersity of Mw/Mn = 1.8-1.9. GPC measurements on polycation **5** and polyanion **7** had to be carried out in water, precluding a direct comparison with their precursors, but the corresponding GPC traces showed a single maximum, indicating that the organometallic main chain had remained intact.

Patterned Multilayer Thin Films

The availability of poly(ferrocenylsilane) polyanions and polycations enabled us to fabricate fully organometallic multilayer thin films on substrates such as quartz, silicon wafers and gold.[14] In addition, we were interested in forming patterned organometallic multilayer structures, i.e. to confine the deposition of the poly(ferrocenylsilane) polyelectrolytes to selected areas on substrates, as this would broaden the applicability of such multilayers. The selective deposition of polyelectrolytes on hydrophilically/hydrophobically patterned gold substrates has been described.[26,27] In this case, patterned self-assembled monolayers consisting of e.g. methyl-terminated and oligo(ethyleneglycol)-terminated alkanethiols were introduced on gold substrates, using microcontact printing.[28] Areas covered with oligo(ethyleneglycol)-terminated alkanethiols were found to prevent adsorption of polyelectrolytes. Here, as a demonstration, a gold substrate was patterned with 5 μm wide methyl-terminated alkanethiol lines, separated by 3 μm, by microcontact printing of 1-octadecanethiol. The uncovered areas were subsequently filled in with 11-mercapto-1-undecanol, resulting in a hydrophilically/hydrophobically patterned substrate. AFM height and friction force images of these patterned self-assembled monolayers (Figure 6, top images) show minimal height contrast but a large contrast in friction force, with the hydroxyl-terminated lines corresponding to the high friction areas.

The patterned substrate was then coated with 12 bilayers of organometallic polyions (**5**/**7**) and again examined by contact mode AFM. Clearly, after deposition, the height contrast increased and the contrast in friction force was reversed, which shows that the multilayers grow selectively on the broad, methyl-terminated stripes (Figure 6, lower images).[14]

The resistivity of the hydroxyl-terminated areas to polyion deposition was demonstrated by forming a monolayer of 11-mercapto-1-undecanol on a gold substrate, which was then processed similarly as the patterned substrates, and subsequently analyzed by XPS. Fe 2p signals, indicating

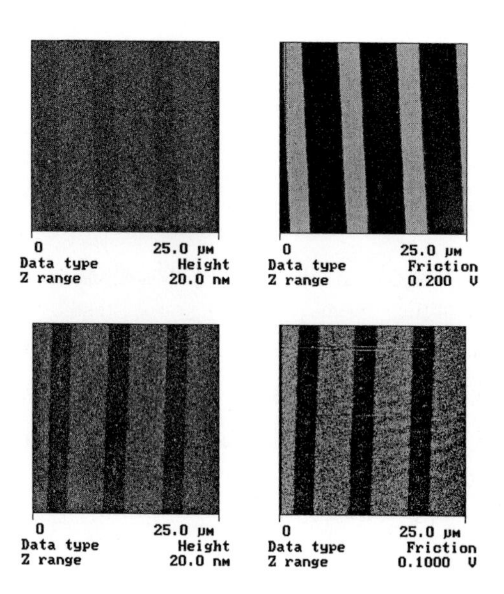

Figure 6. Multilayer deposition on a hydrophilically/hydrophobically patterned gold substrate. Upper AFM images: height (left) and friction force (right) images of patterned methyl- and hydroxyl- alkanethiol self-assembled monolayers. Adsorption of poly(ferrocenylsilane) polyions (**5**/**7**, 12 bilayers) occurs selectively on the broad methyl-terminated stripes (lower AFM images). Reproduced with permission from ref. 14, © 2002 ACS.

adsorbed polyions, were absent in the survey scan. The selective adsorption of the polyions on the methyl-terminated regions of the surface is most likely driven by favorable hydrophobic interactions between these areas and the hydrophobic poly(ferrocenylsilane) backbone. Such favorable secondary interactions[29] with the hydrophilic regions are excluded. Furthermore, the charged segments of the polyions cannot compete with water in forming hydrogen bonds with the hydroxyl-terminated regions, which are hydrated under the processing conditions.[30] Thus, deposition occurs selectively on the methyl-terminated domains.

Further support for the role of hydrophobic interactions in the area-selective adsorption was obtained by reversing the dipping sequence. A single bilayer was adsorbed on a methyl- and hydroxyl-terminated stripe pattern (10 and 5 µm, respectively), in one case starting with the polycation, in the other case with first adsorbing the polyanion. In both cases the bilayer had been deposited selectively on the methyl-terminated areas.

Acknowledgement

The University of Twente and the MESA$^+$ Research Institute are acknowledged for financial support.

[1] For a recent review on poly(ferrocenylsilanes) see K. Kulbaba, I. Manners, *Macromol. Rapid Commun.* **2001**, *22*, 711.
[2] Y. Ni, R. Rulkens, I. Manners, *J. Am. Chem. Soc.* **1996**, *118*, 4102.
[3] R. G. H. Lammertink, M. A. Hempenius, V. Z.-H. Chan, E. L. Thomas, G. J. Vancso, *Chem. Mater.* **2001**, *13*, 429.
[4] R. G. H. Lammertink, M. A. Hempenius, E. L. Thomas, G. J. Vancso, *J. Pol. Sci. Part B: Pol. Phys.* **1999**, *37*, 1009.
[5] R. G. H. Lammertink, M. A. Hempenius, J. E. Van den Enk, V. Z.-H. Chan, E. L. Thomas, G. J. Vancso, *Adv. Mater.* **2000**, *12*, 98.
[6] J. Y. Cheng, C. A. Ross, V. Z.-H. Chan, E. L. Thomas, R. G. H. Lammertink, G. J. Vancso, *Adv. Mater.* **2001**, *13*, 1174.
[7] E. W. Neuse, F. B. D. Khan, *Macromolecules* **1986**, *19*, 269.
[8] S. Kelch, M. Rehahn, *Macromolecules* **1999**, *32*, 5818.
[9] Z. Wang, A. Lough, I. Manners, *Macromolecules* **2002**, *35*, 7669.
[10] (a) M. A. Hempenius, N. S. Robins, R. G. H. Lammertink, G. J. Vancso, *Macromol. Rapid Commun.* **2001**, *22*, 30. (b) M. A. Hempenius, N. S. Robins, R. G. H. Lammertink, G. J. Vancso, *IUPAC MACRO 2000 Proceedings*, Vol. 2, 900, Warsaw, July 2000, ISBN 83-904741-7-4.
[11] G. Decher, *Science* **1997**, *277*, 1232.
[12] P. Bertrand, A. Jonas, A. Laschewsky, R. Legras, *Macromol. Rapid Commun.* **2000**, *21*, 319.
[13] M. A. Hempenius, G. J. Vancso, *Macromolecules* **2002**, *35*, 2445.
[14] M. A. Hempenius, N. S. Robins, M. Péter, E. S. Kooij, G. J. Vancso, *Langmuir* **2002**, *18*, 7629.
[15] J. Halfyard, J. Galloro, M. Ginzburg, Z. Wang, N. Coombs, I. Manners, G. A. Ozin, *Chem. Commun.* **2002**, 1746.
[16] M. Park, C. Harrison, P. M. Chaikin, R. A. Register, D. H. Adamson, *Science* **1997**, *276*, 1401.
[17] J. P. Spatz, T. Herzog, S. Mössmer, P. Ziemann, M. Möller, *Adv. Mater.* **1999**, *11*, 149.
[18] J. P. Spatz, P. Eibeck, S. Mössmer, M. Möller, T, Herzog, P. Ziemann, *Adv. Mater.* **1998**, *10*, 849.
[19] A. H. Gabor, E. A. Lehner, G. Mao, L. A. Schnegenburger, C. K. Ober, *Chem. Mater.* **1994**, *6*, 927.
[20] (a) A. Avgeropoulos, V. Z.-H. Chan, V. Y. Lee, D. Ngo, R. D. Miller, N. Hadjichristidis, E. L. Thomas, *Chem. Mater.* **1998**, *10*, 2109. (b) V. Z.-H. Chan, J. Hoffman, V. Y. Lee, H. Iatrou, A. Avgeropoulos, N. Hadjichristidis, R. D. Miller, E. L. Thomas, *Science* **1999**, *286*, 1716.
[21] J. Y. Cheng, C. A. Ross, E. L. Thomas, H. I. Smith, G. J. Vancso, *Appl. Phys. Lett.* **2002**, *81*, 3657.
[22] P. Gómez-Elipe, P. M. Macdonald, I. Manners, *Angew. Chem. Int. Ed. Engl.* **1997**, *36*, 762.
[23] W. E. Willy, D. R. McKean, B. A. Garcia, *Bull. Chem. Soc. Japan* **1976**, *49*, 1989.
[24] S. Itsuno, T. Koizumi, S. Okumura, K. Ito, *Synthesis* **1995**, 150.
[25] J. W. F. K. Barnick, J. L. van der Baan, F. Bickelhaupt, *Synthesis* **1979**, 787.
[26] P. T. Hammond, G. M. Whitesides, *Macromolecules* **1995**, *28*, 7569.
[27] S. L. Clark, P. T. Hammond, *Adv. Mater.* **1998**, *10*, 1515.
[28] Y. Xia, G. M. Whitesides, *Angew. Chem. Int. Ed.* **1998**, *37*, 550.
[29] S. L. Clark, P. T. Hammond, *Langmuir* **2000**, *16*, 10206.
[30] M. Sprik, E. Delamarche, B. Michel, U. Röthlisberger, M. L. Klein, H. Wolf, H. Ringsdorf, *Langmuir* **1994**, *10*, 4116.

Macromol. Symp. ***196***, *57–62 (2003)*

Polymer Science with Main Group Elements and Transition Metals

Ian Manners

Department of Chemistry, University of Toronto, 80 St. George Street, Toronto, Ontario, M5S 3H6, Canada
E-mail: imanners@chem.utoronto.ca

Summary: The efficient formation of soluble, processable polymers with main chains containing inorganic elements provides a synthetic challenge but represents potentially important approach to new macromolecular and supramolecular materials with interesting properties. This talk will survey some of the recent research performed in our group concerning the development of controlled routes to a variety of different inorganic polymer systems and the use of the resulting materials in self-assembly processes.

Keywords: nanolithography; nanostructures; photonic materials; polyferrocene; ring-opening polymerization

Introduction

Because of their processing advantages over ceramics and metals, organic polymeric materials are ubiquitous. The incorporation of inorganic elements such as transition metals into a polymer main chain offers unique potential for the preparation of processable materials with properties which differ significantly from those of conventional carbon-based polymers [1]. For example, the diverse range of coordination numbers and geometries which exist for transition elements offer the possibility of accessing polymers with unusual conformational, mechanical, and morphological characteristics. Metallocenters may also generate interesting electronic, optical, and magnetic properties. Transition metal-based polymers might also be expected to function as convenient and processible thermal or photochemical precursors to metal-containing ceramic films, fibers, and coatings with high stability and desirable and useful physical properties [1-3]. However, the development of the area of metal-containing polymers has been held up by considerable synthetic difficulties. Much of the early work in the field in the 1950's and 1960's targeted polymetallocenes. However, most of the attempted polymer syntheses utilized polycondensation reactions and the resulting products were generally of low molecular weight or insoluble and poorly-characterized [2,4].

© 2003 WILEY-VCH Verlag GmbH & KGaA, Weinheim CCC 1022-1360/00/$ 17,50+.50/0

In the early 1990s we reported the discovery of a ring-opening polymerization (ROP) route to high molecular weight polyferrocenylsilanes from strained [1]silaferrocenophanes [5]. The latter species were first prepared in the mid 1970's via the reaction of the dilithioferrocene•tetramethylethylenediamine complex with organodichlorosilanes. ROP processes generally occur via a chain growth route which involve a highly reactive intermediate that reacts rapidly with monomer molecules. This allows the facile formation of long polymer chains making high molecular weight polymers easily accessible. This is vital as high molecular weights (M_n > ca.10,000) are needed in order to realize the chain entanglement needed to access the advantageous properties of macromolecules such as ease of fabrication into films.

Polyferrocenylsilanes (PFSs)

A wide range of silicon-bridged [1]ferrocenophanes **1** have been prepared and similarly polymerized to high molecular weight ($M_w = 10^5 - 10^6$, $M_n > 10^5$) polyferrocenylsilanes (PFSs) **2**. Substituents such as hydrogen, chlorine, alkyl, aryl, trifluoropropyl, norbornenyl alkoxy, aryloxy, amino, and ferrocenyl groups have also been introduced [6].

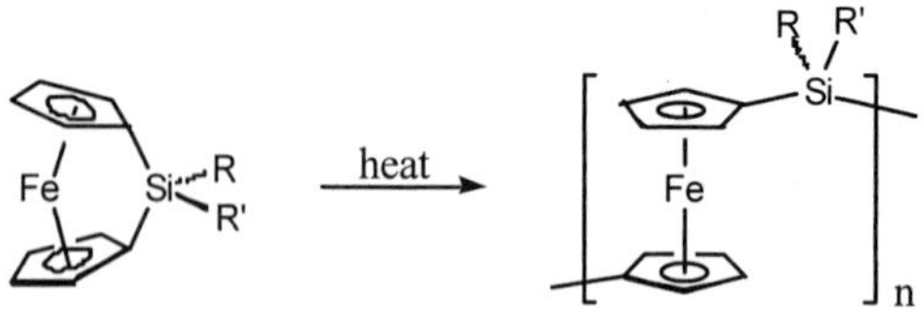

Considerable effort has also been directed towards investigating and understanding the properties of the resulting PFS materials, the vast majority of which are soluble in organic solvents or water despite their very high molecular weights [6]. Cyclic voltammetric studies of the high polymers generally show the presence of two reversible oxidation waves in a 1: 1 ratio providing clear evidence for the existance of interactions between the iron atoms [7]. The presence of metal-metal interactions has led to studies of the charge transport properties of these materials. Oxidative doping of amorphous samples of poly(ferrocenyldimethylsilane) with I_2 have been shown to yield semiconducting materials. Recent collaborative work with the group of E. W. Sargent at Toronto has identified holes as the charge carriers [8]. Charge transport properties such as hole mobility, which are important for xerography and other device applications, have also been measured and are appreciable. Polyferrocenes also function as charge dissipation coatings which provide protection with respect to ionizing radiation such as electrons. The films of PFSs are believed to function in this regard via both radiation scattering and conduction mechanisms [9]. The magnetic properties of oxidized PFSs have also attracted some attention [10]. In addition, cationic and anionic water-soluble PFSs are useful in the preparation of organic-organometallic and all organometallic electrostatic superlattices by layer-by-layer assembly approaches [11].

PFS materials also exhibit interesting morphology and several of the symmetrically substituted derivatives will crystallize [6]. Alkoxy-substituted PFS materials such as **2** (R = R' = n-hexyloxy) possess T_g values down to - 51°C, a remarkable testament to the flexibility of polyferrocene chains which probably arises from the freely-rotating "molecular ball bearing" nature of the ferrocene unit. The first liquid crystalline PFS materials have also been prepared via hydrosilylation [12]. These show a nematic mesophase and the influence of redox state on the liquid crystalline properties is of significant interest.

Thermal ROP of metallocenophanes at elevated temperatures leads to virtually no control over molecular weight and the molecular weight distribution is broad. We have shown that silicon-bridged [1]ferrocenophanes undergo living anionic ROP [13]. This has permitted the synthesis of PFSs with controlled molecular weights and narrow polydispersities and has also allowed functionalization of polymer chain ends and the preparation of the first block copolymers containing skeletal transition metal atoms. Transition metal-catalyzed ROP of silicon-bridged [1]ferrocenophanes, which occurs in solution at room temperature, has also been reported using Pd, Pt, and Rh catalysts. Molecular weight control can also be achieved and materials with novel architectures such as star and graft polymers are accessible [14].

PFS Applications

PFSs function as preceramic polymers and have been shown to yield interesting magnetic Fe-containing ceramic composites at 500 - 1000°C [15-18]. The use of such involatile but processable polymeric precursors to ceramics is potentially attractive way of circumventing the difficulty of processing ceramic materials into desired shapes. In collaboration with the group of G.A. Ozin at

Toronto we have recently prepared PFS inverse opals and magnetic ceramic replicas using the preceramic polymer approach. The resulting materials possess periodic variations in refractive index at the micron scale [19]. The relatively high refractive indices possible with PFS materials ($n > 1.6$) make the structures of interest as redox-tunable photonic crystals. In addition, magnetic ceramic photonic crystals may allow photonic properties to modified with a magnetic field.

PFS block copolymers allow access to nanostructured materials via self-assembly processes in the solid state or solution [20]. In collaboration with the group of M. Winnik at Toronto we have shown that interesting architectures can be generated via the solution aggregation of PFS-based block copolymers [20]. Thus, block copolymers such as poly(ferrocenyldimethylsilane) - b - poly(dimethylsiloxane) ($n : m = 1 : 6$) with long polysiloxane blocks have been found to afford cylindrical, wormlike micelles in hexanes. These structures can be readily visualized by Atomic Force Microscopy or by TEM. These cylinders consist of a wire-like core of semiconducting PFS surrounded by a sheath or corona of insulating poly(dimethylsiloxane). Pyrolysis of such structures, particularly if crosslinked, offers the possibility of generating magnetic wire-like structures. The structures can also be used as nanoscopic etching resists to yield patterns of width 10 – 20 nm [19]. The analogous use of cylindrical micelles with PFS cores but an easily etchable organic corona (e.g. derived from polyisoprene) allows one to obtain structures of smaller width ca. 8 nm [21].

In summary, ring-opening polymerization has allowed facile access to a variety of high molecular weight PFSs. Much has been accomplished in terms of understanding the properties of these novel and interesting materials. Future work will be directed towards extending the polymerization approach to develop new polymers [22].

New Polymers Based on Main Group Elements: Polythionylphosphazenes and Polyphosphinoboranes

Our group has also been interested in the possibility of creating new macromolecular chains based on main group elements [23]. We have found, for example, that cyclic thionylphosphazene $[NSOCl(NPCl_2)_2]$ thermally polymerizes at 165°C to yield poly(pentachlorothionylphosphazene) $[NSOCl(NPCl_2)_2]_n$ which undergoes halide substitution in a manner similar to poly(dichlorophosphazene) (e.g. with amines) to form hydrolytically stable sulfur-nitrogen-phosphorus polymers [24]. Mechanistic studies of the ROP process suggested a cationic, chain growth ROP mechanism. Significantly, by using halide accepting, Lewis acids as catalysts we have found that the ROP can be performed at room temperature [25]. Interestingly, the polymerization shows a dramatic dependence on monomer concentration and no polymer is formed a below a concentration of 0.15 M. This is characteristic of a polymerization with a very small ΔH value which indicates that, unlike the case of [1]ferrocenophanes, monomer $[NSOCl(NPCl_2)_2]$ is virtually unstrained. The properties of polythionylphosphazenes are currently being studied. The backbone is stable to hydrolysis and indeed water soluble derivatives have been made. The high gas permeability of some of these materials has led to interest in their use as matrices for luminescent, oxygen sensors [26].

The preparation of polymers with backbones of alternating phosphorus and boron atoms attracted significant attention in the 1950's and early 1960's as a consequence of the envisaged high thermal stability and possible flame retardancy of these materials. The main synthetic route explored at that time involved thermally-induced dehydrocoupling of phosphine-borane adducts. However, these reactions only proceeded at 180-200°C and the main products were six-membered rings (which, promisingly, were resistant to hydrolysis by HCl at temperatures as high as 300°C in some cases) [27]. Only negligable yields of low molecular weight, partially characterized polymers were claimed, mainly in patents. Recently we have shown that the dehydrocoupling process is effectively catalyzed by late transition metal complexes (especially Rh species) which has permitted the formation of high molecular weight polyphosphinoboranes $[RPH\text{-}BH_2]_n$ ($M_w > 30{,}000$) [28].

The properties of these interesting, recently prepared polymers are now beginning to be explored. Polyphenylphosphinoborane, for example, is a white, air-stable powder and is formally an analog of polystyrene with a phosphorus-boron backbone [28]. Improved dehydropolymerization catalysts for P-B bond formation are highly desirable in order to further raise molecular weights and to hopefully allow more controlled polymerizations.

[1] I. Manners *Science* **2001**, *294,* 164.
[2] P. Nguyen, P. Gómez-Elipe, I. Manners, *Chem. Revs* **1999**, 99, 1515.
[3] Archer R.D, *Inorganic and Organometallic Polymers*, Wiley-VCH (2001).
[4] E.W. Neuse, H. Rosenburg, *J. Macromol. Sci.* **1970**, *C4*, 110.
[5] D.A. Foucher, B.Z. Tang, I. Manners, *J. Am. Chem. Soc.*, **1992**, *114*, 6246.
[6] K. Kulbaba, I. Manners *Macromol. Rapid. Commun.* **2001**, *22*, 711.

[7] Foucher, D.A., Nelson, J.M., Honeyman, C., Tang, B.Z., and Manners, I. *Angew. Chem. Int. Ed. Engl.* **1993**, *32,* 1709.

[8] L. Bakueva, E.H. Sargent, R. Resendes A, Bartole I. Manners *J. Mat. Sci. Mater. Elec.* **2001**, *12*, 21.

[9] R. Resendes, A. Berenbaum, G. Stojevic, F. Jakle, A. Bartole, F. Zamanian, G. Dubois, C. Hersom, K. Balmain, I. Manners, *Adv. Mater.* **2000**, *12*, 327.

[10] M. Hmyene, A. Yasser, M. Escorne, A. Percheron-Guegan, F. Garnier *Adv. Mater.* **1994**, *6*, 564.

[11] M. Ginzburg, J. Galloro, F. Jäkle, K.N. Power-Billard, S. Yang, I. Sokolov, C.N.C Lam, A.W. Neumann, I. Manners, G.A. Ozin *Langmuir* **2000**, *16*, 9609.

[12] X-H Liu, D.W. Bruce, D. W., I. Manners *Chem. Commun.*, **1997**, 2890.

[13] R. Rulkens, Y. Ni, I. Manners, *J. Am. Chem. Soc.* **1996,** *118*, 4102.

[14] P. Gómez-Elipe, P.M. Macdonald, I. Manners, *Angew. Chem. Int. Ed. Engl.* **1997**, *36*, 762.

[15] B-Z Tang, R. Petersen, D.A. Foucher, A.J. Lough, N. Coombs, R. Sodhi, I. Manners, *J. Chem. Soc. Chem. Comm.* **1993**, *6*, 523.

[16] R. Petersen, D.A, Foucher, B-Z Tang, A.J. Lough, N. P. Raju, J.E. Greedan, I. Manners *Chem. Mater.* **1995**, *7*, 2045.

[17] M.J. MacLachlan, M. Ginzburg, N. Coombs, T.W. Coyle, N.P. Raju, J.E. Greedan, G.A. Ozin, I. Manners, *Science* **2000**, *287*, 1460.

[18] Kulbaba, K.; Resendes, R.; Cheng, A.; Bartole, A.; Safa-Sefat, A.; Coombs, N.; Stover, H. D. H.; Greedan, J. E.; Ozin, G. A.; Manners, I. *Adv. Mater.* **2001**, *13*, 732.

[19] J.Galloro, M. Ginzburg, H. Míguez, S-M. Yang, N. Coombs, A. Safa-Sefat, J. E. Greedan, I. Manners, G. A. Ozin *Adv. Funct. Mater.* **2002**, *12*, 382.

[20] a) J. Massey, K.N. Power,. M.A.Winnik, I. Manners, *Adv. Mater* **1998**, *10,* 1559 b) K. Temple, K. Kulbaba, K.N. Power-Billard, I. Manners, K.A. Leach, T. Xu, T.P. Russell, C.J. Hawker *Adv. Mater.* **2003**, *15*, 297.

[21] J. Massey, M.A. Winnik, I. Manners, V.Z-H. Chan, J.M. Ostermann, R. Enchelmaier, J.P. Spatz, M. Möller, *J. Am. Chem. Soc.* **2001**, *123*, 3147.

[22] a) R. Rulkens, D.P. Gates, J.K. Pudelski, D. Balaishis, D.F. McIntosh, A.J. Lough, I. Manners, *J. Am. Chem. Soc.* **1997**, *119*, 10976 b) Braunschweig, H., Dirk, R., Müller, M. Nguyen, P., Resendes, R., Gates, D.P., and Manners, I. *Angew. Chem. Int. Ed. Engl.* **1997**, *36*, 2338.

[23] I. Manners *Angew. Chem. Int. Ed. Engl.* **1996**, *35,* 1602.

[24] a) M. Liang, I. Manners, *J. Am. Chem. Soc.,* **1991**, *113*, 4044. b) Ni, Y.Z., Park, P., Liang, M, Massey, J., Waddling, C., Manners, I. *Macromolecules* **1996**, *29*, 3401.

[25] a) Gates, D.P., Edwards, M., Liable-Sands, L.M., Rheingold, A.L., Manners, I. *J. Am. Chem. Soc.* **1998,** *120,* 3249 b) A.R. McWilliams, D.P. Gates, M. Edwards, L.M. Liable-Sands, I. Guzei, A.L. Rheingold, I. Manners *J. Am. Chem. Soc.* **2000,** *122*, 8848.

[26] a) Pang, Z.; Gu, X.; Yekta, A.; Masoumi, Z.; Coll, J. B.; Winnik, M. A.; Manners, I. *Adv. Mater.* **1996**, *8*, 768. b) Wang, Z.; McWilliams, A.R.; Evans, C.E.B.; Lu, X.; Chung, S.; Winnik, M.A.; Manners, I. *Adv. Func. Mater.* **2002**, *12*, 415.

[27] Burg, A. B. *J. Inorg. Nucl. Chem.* **1959**, 11, 258.

[28] a) Dorn, H.; Singh, R.A.; Massey, J.A. Lough, A.J.; Manners, I. *Angew. Chem. Int. Ed. Engl.* **1999**, *38,* 3321. b) H. Dorn, R.A. Singh, J.A. Massey, J.M. Nelson, C. Jaska, A.J. Lough, I. Manners, *J. Am. Chem. Soc.* **2000**, *122*, 6669. c) H. Dorn, J.M. Rodezno, B. Brunnhofer, E. Rivard, J.A. Massey, I. Manners, *Macromolecules* **2003**, *36*, 291.

Towards Photonic Ink (P-Ink): A Polychrome, Fast Response Metallopolymer Gel Photonic Crystal Device

*André C. Arsenault, Hernán Míguez, Vladimir Kitaev, Geoffrey A. Ozin,**
*Ian Manners**

Department of Chemistry, University of Toronto, 80 St George Street, Toronto, Ontario, Canada, M5S 3H6
E-mail: aarsenau@chem.utoronto.ca

Summary: We demonstrate here a new kind of planarized and oriented colloidal photonic crystal device whose optical stop-band position, width and intensity can be reversibly redox and solvent tuned over a broad wavelength range by an anisotropic expansion of the photonic lattice. The material is composed of silica microspheres in a matrix of crosslinked polyferrocenylsilane[1], a metallopolymer network with a continuously variable state of oxidation[2]. Optical data was fitted using a scalar wave approximation, with a congruence to experimental data, allowing facile extraction of information concerning polymer swelling behaviour. The chemomechanical polychrome optical response of the material was exceptionally fast, attaining its fully swollen state from the dry shrunken state on a sub-second time-scale.

Keywords: colloids; gels; photonic crystals; polyferrocenylsilanes; swelling

Introduction

Photonic crystals are a class of materials which interact with electromagnetic radiation through a periodic spatial modulation in their refractive indices, when this periodicity coincides with the scale of the radiation wavelength[3-4]. These materials can easily be assembled *via* the self-assembly of a collection of monodisperse spherical colloids into a long-range ordered lattice of well-defined geometry. Spherical colloids of both inorganic and polymeric composition have been studied extensively[5], can be routinely obtained as monodisperse suspensions, and are relatively inexpensive to produce. Recent success in assembling these colloids into face-centered cubic (fcc) crystals with very low defect concentration have thrust them to the forefront of photonic crystal research [6,7].

Since many materials have been incorporated into these architectures through templating approaches[8], research has begun to focus on systems capable of being tuned in a variety of ways by external stimuli[9]. One of the ways in which colloidal photonic crystal properties can

© 2003 WILEY-VCH Verlag GmbH & KGaA, Weinheim CCC 1022-1360/00/$ 17,50+.50/0

be tuned is *via* a mechanical response, that is changing external conditions such that the crystal changes in shape or dimensions. Previous studies have experimentally demonstrated this concept by fixing in a hydrogel matrix an array of highly charged latex spheres assembled through mutually repulsive electrostatic interactions in a rigorously deionized medium[10]. These so-called polymerized crystalline colloidal arrays (PCCA's) could take advantage of the well-known swelling properties of acrylamide-based gels[11], as well as perform sensing functions by incorporating receptors for specific analytes[12]. However, factors such as poor mechanical stability due to high solvent content, slow response to stimuli, and the polycrystalline nature of the samples, which prevents one from accurately controlling features of the Bragg peak such as frequency, width and intensity, may be stumbling blocks for their implementation into optical devices.

Silica-Polyferrocenylsilane Gel Composite Photonic Crystals

PFS is a polymer whose main chain is composed of alternating substituted silicon atoms and ferrocene groups connected at the 1- and 1'- positions of the cyclopentadienyl (Cp) rings. Thermal ring-opening polymerization (ROP) of strained, ring-tilted [1]-silaferrocenophanes affords high molecular weight polymer[13]. The monomer used here was (ethylmethyl)sila-[1]-ferrocenophane[14] unsymmetrically substituted at silicon, avoiding the possibility of micro-crystallization within the polymer network. The crosslinker was sila(cyclobutyl)-[1]-ferrocenophane, constituted of two strained rings, both of which can be ring-opened thermally[15].

The procedure and materials used to fabricate the samples have been recently described[16]. In essence, colloidal crystal films of silica spheres [17] on glass substrates were synthesized by an evaporative deposition method[18]. Following surface treatment, the films were infiltrated with a mixture of monomer and crosslinker and thermally polymerized to yield the desired silica-PFS composite colloidal crystal film. In the dry state these samples are extremely stable, and can be handled without any special precautions. A cross-sectional SEM image of one of the samples is shown in Figure 1.

Optical properties of both dry and solvent swollen samples in transmission mode were measured in a standard IR cell mount, which allowed for injection and removal of solvent. The peaks investigated represent selective reflection of a narrow band of wavelengths due to Bragg diffraction from the (111) crystal planes of the colloidal photonic crystal. In order to estimate the effect of the swelling on the structure of the colloidal photonic crystal, the results

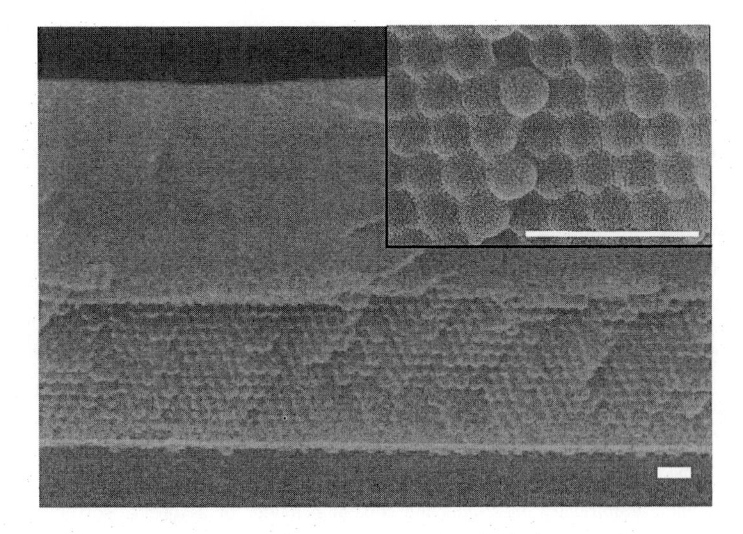

Figure 1. Scanning Electron Micrograph (SEM) image of a cross-section of the PFS-silica composite colloidal crystal film. Both scale bars represent 1 μm.

of the optical characterization were analyzed employing a theoretical model based on a scalar wave approximation[19]. This simple model has been proven to be valid to describe the transmittance and reflectance of finite size photonic crystals at the spectral region corresponding to that of the lower energy photonic bands such as the ones we investigate here[20,21,22,23]. In our case, knowing the refractive index of both the silica microspheres and the infiltrated cross-linked polymer, n=1.425 and n=1.65 respectively[24,25], as well as that of the solvent we are able to obtain the increase of the (111) interplanar distance $d_{(111)}$ resulting from the swelling process. We take into account the variation of the refractive index of the colloidal photonic crystal interstitial sites due to incorporation of the solvent to the polymer network when the swelling occurs, as well as include the presence of a substrate (glass slide) and a superstrate (polymer excess) of different refractive indices[26]. By using this approach, simulated transmission spectra were fitted to the experimental ones considering $d_{(111)}$ as the only adjustable parameter. In doing so, we reproduce the particular optical features observed for each one of the solvents, such as reflectance peak position, width, and intensity. Both the model and the experiment show explicitly how it is possible, by choosing the appropriate pair of solvents, to modulate the width of the photonic stop band keeping the spectral position of the maximum constant. For instance, both chlorobenzene and carbon disulfide swollen samples display a stop band with a spectral maximum of 843nm, but the width is varied from 13 to 15% (respectively). This is a consequence of the dual character of the tuning process. On one hand, the distance between microspheres increases during the swelling, which enlarges the lattice constant and simultaneously decreases the filling fraction of the silica microspheres in the structure. On the other hand, the refractive index contrast between microspheres and background is modified as a result of the incorporation of the solvent into the polymer network.

To demonstrate how this system allows the determination of swelling behaviour of the matrix material, the values of $d_{(111)}$ attained from this simulation were plotted *versus* the solubility parameter (δ_s) of various solvents. The greatest obtained degree of swelling results in a 21% increase in the lattice constant of the composite colloidal crystal. The solubility parameter is a numerical value empirically describing the solvent behavior of a specific solvent[27]. The obtained plot shows a maximum value of $d_{(111)}$ at δ_s=19.5, and these results are in good agreement with those obtained using a weight-difference technique on a similar polymer[12]. The straightforward analysis by the scalar wave approximation of these types of planarized photonic crystals could conceivably by used to quickly and accurately measure this important physical property for a wide range of polymers.

A kinetics experiment was conducted to evaluate the chemo-mechanical response time of the polymer-silica colloidal photonic crystal nanocomposite upon exposure to solvent. The results of the experiment are shown in Figure 2. In order to determine when the sample had reached its fully swollen state, the maximum in intensity of the first Bragg peak of the swollen sample was first noted. This wavelength was then monitored with respect to time, and carbon disulfide was injected into the cell. Before injecting the solvent, the absorbance value was low since a background had been taken of the dry sample, and this value was constant from 0 to 7 seconds. We see then a brief increase, which is due to scattering by the air-liquid interface created by the solvent rising in the sample cell and crossing the incident beam. Next, we see a sharp drop in absorbance due to a decrease in refractive index contrast between the fluid content of the cell (solvent instead of air) and the quartz plates (n_{quartz}/n_{air}=1.54 ; $n_{carbon\ disulfide}/n_{quartz}$=1.06), as well as with the glass substrate (Δn_{old}=1.45; Δn_{new}=1.12) and PFS superstrate (Δn_{old}=1.65; Δn_{new}=1.01). This decreases the light reflected at these interfaces and results in more light reaching the detector. After this increase in transmitted light intensity we see a rapid decrease representing the swelling event, after which the sample attains a constant steady state transmittance value. The whole process, from initial injection of solvent onto the dry polymer-silica colloidal photonic crystal sample to the completely swollen state, occurs in 0.4 seconds. We believe this value is a lower limit for the swelling process, and preliminary results using a Fourier-transform CCD array spectrometer indicate an optical response within 50 milliseconds of exposure to solvent with crude non-optimized solvent injection. Although we were impressed by the response speed of our materials, a fast response could have been expected given the proportionality of swelling time for polymer gels to the inverse square of the smallest dimension of the gel[28], which in our case is microscale. We were able to easily and reproducibly synthesize these materials in a very thin form which confers to them speed due to their size. In addition, we believe that the washing out of unpolymerized oligomers from the polymer network creates a certain amount of microporosity in the gel (not visible by SEM) which would reduce its effective dimensions and lead to an improvement of swelling kinetics versus a bulk dense gel.

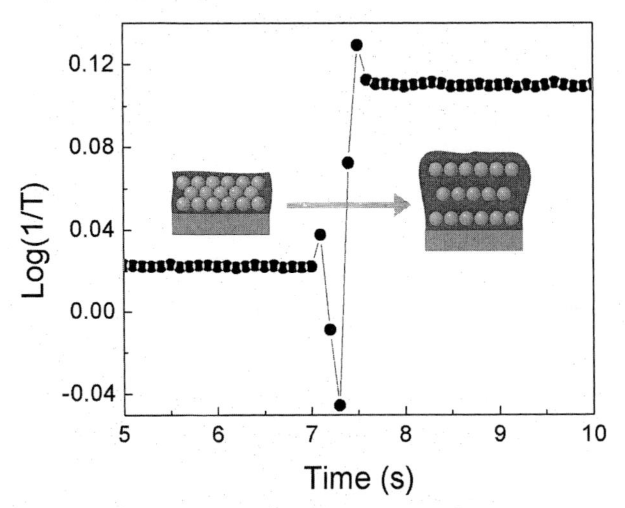

Figure 2. Kinetics plot showing the response of the silica-PFS gel composite opal upon exposure to carbon disulfide. As can be seen, the sample reaches its equilibrium swelling value within 0.2 to 0.4 seconds.

To ensure that the swelling and shrinking of these samples was indeed reversible, we conducted deswelling experiments in which the solvent was rapidly removed by application of vacuum to the flow-cell while monitoring optical properties with *in situ* optical spectroscopy.

The time required for the sample to fully deswell to its initial dimensions was comparable to the time required for the swelling process, confirming the rapid and reversible nature of the process.

Each repeat unit of PFS contains a redox-active ferrocene group, and consequently the polymer itself is redox-active. Charges on the individual iron atoms in each repeat unit can be controlled between a 0 and +1 state[29]. Therefore, PFS can have continuously variable interactions with any solvent it contacts. As an experimental demonstration of this phenomenon, the PFS-silica composite photonic crystals were subjected to multiple partial oxidations using a solution in dichloromethane (DCM) of *tris*(4-bromophenyl)aminium hexafluorophosphate (5mg/mL), a one-electron oxidant which is known to cleanly oxidize ferrocene derivatives[30]. After thorough washing and drying, the oxidized sample was swollen in carbon disulfide and an absorbance spectrum taken. From this spectrum, the degree of polymer oxidation could be estimated based on the intensity of the characteristic ferrocenium LMCT. After drying the sample, it was again oxidized, and these steps repeated until the LMCT transition no longer increased in intensity. The spectra obtained show clearly that the diffracted wavelength in the swollen state monotonically decreases with an increase in the degree of oxidation, since a charged polymer cannot be effectively solvated by a non-polar solvent. Not only can the diffracted wavelength in this system be varied between two extremes, but all intermediate states can be individually and controllably accessed.

When a sample was oxidized in the above fashion then reduced with a solution of decamethylferrocene in DCM, in was possible to obtain a spectrum virtually identical to that in the same solvent before oxidation. This not only demonstrates reversibility but also illustrates the potential use of these materials as redox sensors for species being oxidizing or reducing with respect to the redox potential of the PFS gel (tuned by the degree of oxidation) such as biologically relevant redox-proteins.

Preliminary electrochemical results indicate that when supported on a conductive substrate, this material can be repeatedly cycled between its oxidized and reduced forms within a certain degree of oxidation and in appropriate media. An electrochemical cell has been designed for this purpose, and will implemented into prototype devices such as visible light pixilated displays and near-infrared optical telecommunication switches and attenuators by using, respectively, smaller and larger spheres than the ones used in this work. A detailed spectroelectrochemical study is in progress.

Conclusion

This report demonstrates a conceptually new approach to the fabrication of rapid-response chemo-mechanical colloidal photonic crystal device with redox and solvent tunability based on the anisotropic expansion of the crosslinked PFS matrix surrounding a close-packed array of unconnected silica spheres[16]. The samples are planarized and oriented, and present a very low defect concentration, portending the use of this system in future optical communication systems and electro-photonic displays.

[1] K. Kulbaba, M.J. MacLachlan, C.E.B. Evans, I. Manners, *Macromol. Chem. Phys.* **2001**, *202*, 1768.
[2] I. Manners, *Science* **2001**, *294*, 1664.
[3] E. Yablonovitch, *Phys. Rev. Lett.* **1987**, *58*, 2059.
[4] S. John, *Phys. Rev. Lett.* **1987,** *58*, 2486.
[5] P.C. Hiemenz, R. Rajagopalan "*Principles of Colloid and Surface Chemistry*", Marcel Dekker Inc., New York 1997.
[6] Y. Xia, B. Gates, Z-Y. Li, *Adv. Mater.* **2001**, *13*, 409.
[7] V.L. Colvin, *MRS Bull.* **2001**, *26*, 637.
[8] B.T. Holland, C. Blanford, A. Stein, *Science* **1998**, 281, 538.
[9] K. Busch, S. John, *NATO Sci. Ser., Ser. C: Math. Phys. Sci.* **2001**, 563, 41.
[10] J.M. Weissman, H.B. Sunkara, A.S. Tse, S.A. Asher, *Science* **1996**, 274, 959.
[11] T. Tanaka, *Phys. Rev. Lett.* **1978**, *40*, 820.
[12] J.H. Holtz, S.A. Asher, *Nature* **1997**, *389*, 829.
[13] D.A. Foucher, B.Z. Tang, I. Manners, *J. Am. Chem. Soc.* **1992**, *114*, 6246.
[14] K. Temple, J.A. Massey, Z. Chen, N. Vaidya, A. Berenbaum, M.D. Foster, I. Manners, *J. Inorg. Organomet. Poly.* **1999**, *9*, 189.
[15] M.J. MacLachlan, A.J. Lough, I. Manners, *Macromolecules* **1996**, *29*, 8562.
[16] A.C. Arsenault, H. Míguez, V. Kitaev, G.A.Ozin, I. Manners, *Adv. Mater* **2003**, *15*, 503.
[17] W. Stöber, A. Fink, E.J. Bohn, *J. Colloid Interface Sci.* **1968**, *26*, 62.
[18] P.Jiang, J.F. Bertone, K.S. Hwang, V.L. Colvin, *Chem. Mater.* **1999**, *11*, 2132.
[19] K.W.-K. Shung, Y.C. Tsai, *Phys. Rev. B* **1993**, *48*, 11265.
[20] I.I. Tarhan, G.H. Watson, *Phys. Rev. B* **1996**, *54*, 7593.
[21] J. F. Bertone, P. Jiang, K. S. Hwang, D. M. Mittleman, V. L. Colvin, *Phys. Rev. Lett.* **1999**, *83*, 300.
[22] H. Miguez, S.-M. Yang, G.A. Ozin, *Appl. Phys. Lett.* **2002**, *81*, 2493.
[23] Y.A.Vlasov, M. Deutsch, D.J. Norris, *Appl. Phys. Lett.* **2000**, *76*, 1627.
[24] F. Garcia-Santamaria, H. Miguez, M. Ibisate, F. Meseguer, C. Lopez, *Langmuir* **2002**, *18*, 1942.
[25] J. Galloro, M. Ginzburn, H. Miguez, S.M. Yang, N. Coombs, A. Safa-Sefat, J.E. Greedan, I. Manners, G.A. Ozin, *Adv. Funct. Mater.* **2002**, *12*, 382.
[26] D.M. Mittleman, J.F. Bertone, P. Jiang, K.S. Hwang, V.L. Colvin, *J. Chem. Phys.* **1999**, *111*, 345.
[27] J.H. Hildebrand, R.L. Scott, "The Solubility of Non-Electrolytes", 3rd ed. Rienhold Publishing Corporation New York (1950); Dover Publications Inc. New York (1964).
[28] J.M.G. Cowie, "*Polymers: Chemistry & Physics of Modern Materials*", 2nd ed. Chapman & Hall, London 1991.
[29] R. Rulkens, A.J. Lough, I. Manners, S.R. Lovelace, C. Grant, W.E. Geiger, *J. Am. Chem. Soc.* **1996**, *118*, 12683.
[30] Steckhan, E. *Angew. Chem., Int. Ed. Engl.* **1986**, *25*, 683.

Lithographic Applications of Highly Metallized Polyferrocenylsilanes

*Scott B. Clendenning, Ian Manners**

Department of Chemistry, University of Toronto, 80 St. George Street, Toronto, Ontario, M5S 3H6, Canada
E-mail: sclenden@chem.utoronto.ca

Summary: Organometallic polymers are excellent candidates for the introduction of metals into nanostructures using lithographic techniques due to their inherently high and uniform metal loadings and processibility. Soluble, high molecular weight polyferrocenylsilanes possess unique physical properties and function as excellent ceramic precusors. Recent advances in the use of lithographic techniques with a highly metallized PFS resist to form new nanopatterned materials will be presented.

Keywords: lithography; nanopatterned materials; organometallic polymers; polyferrocenylsilanes; resists

Introduction

The patterning of surfaces on the nanometer scale with metals offers the possibility of fabricating materials with novel catalytic, optical, sensing, electrical and magnetic properties. Patterning methods can vary from soft lithography, scanning probe lithography, electron beam lithography and photolithography. Recent reports include the use of microcontact printing (μ-CP) to order monodisperse nanoparticles of iron oxide[1] and nanotransfer printing (nTP) to transfer a gold pattern with 75 nm feature sizes from a gold coated GaAs stamp to an appropriately primed PDMS substrate.[2] Scanning probe techniques also offer precise control of patterning. An AFM tip has been used to pen 35 nm wide lines of MoO_3 through local oxidation of a Mo film[3] while Pt lines 30 nm wide have been drawn *via* the reduction of H_2PtCl_6 at an AFM tip using electrochemical AFM dip-pen lithography.[4] It is also possible to employ a film of an organometallic polymer as a resist for electron-beam lithography. Johnson and co-workers have used thin films of the organometallic cluster polymer $[Ru_6C(CO)_{15}Ph_2PC_2PPh_2]_n$ as negative electron-beam resists to direct write conducting wires of metal nanoparticles.[5] This example illustrates the convenience and utility of combining conventional lithographic techniques with easily processible organometallic polymer resists. Moreover, post-development treatments of the

© 2003 WILEY-VCH Verlag GmbH & KGaA, Weinheim CCC 1022-1360/00/$ 17,50+.50/0

patterned organometallic resist such as pyrolysis or reactive ion etching (RIE) offer additional control over the chemical and physical properties of the surface. The ring-opening polymerization of sila[1]ferrocenophanes (**1**) by thermal,[6] anionic[7] and transition metal-catalyzed[8] routes yields high molecular weight, soluble polyferrocenylsilanes (**2**, PFS) containing covalently bonded iron atoms in the main chain (Scheme 1). The incorporation of PFS into patterned surfaces has already yielded materials with tunable magnetic properties that may find applications as protective coatings, magnetic recording media and anti-static shielding.[9] Furthermore, the low plasma etch rates of polymers containing organometallic moities in comparison to their purely organic counterparts[10] suggest their use as etch masks which can also deposit interesting materials. The introduction of additional metals into the PFS chain can increase metal loadings and allow access to binary or higher metallic species.

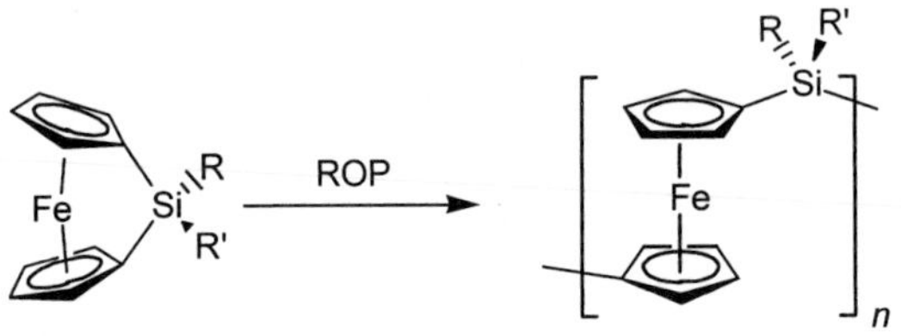

Fe Si R R' ROP Fe Si R R' n

1 **2**

Scheme 1. Ring-opening polymerization of a sila[1]ferrocenophane (**1**) to afford a polyferrocenylsilane (**2**).

Polyferrocenylsilanes as Ceramic Precursors

The synthesis of shaped ceramics from processible polymer precursors is currently receiving much attention. For example, Sneddon and co-workers have prepared BN[11] and SiBNC[12] ceramic fibers from polymer precursors. The primary obstacle to this process is the scarcity of high ceramic yield polymers with interesting chemical compositions. One promising avenue of research is the incorporation of transition metals into the polymer precusor. In collaboration with Ozin, we have found that a highly cross-linked PFS (**2a**) network can be formed through the thermal ROP of the spirocyclic sila[1]ferrocenophane **1a** (Scheme 2).[9,13] Furthermore, pyrolysis of the cross-linked polymer network under a nitrogen atmosphere affords

$Fe/SiC/C/Si_3N_4$ ceramics in greater than 90% ceramic yield with good shape retention and approximately 10% reduction in dimensions. Most interestingly, these ceramics contain α-Fe nanoparticles ranging in size from 1.5 to 70 nm in diameter with smaller superparamagnetic particles formed at temperatures below 900 °C and larger ferromagnetic particles resulting from pyrolysis at 1000 °C. On the road to nano-patterned magnetic materials, the soft lithographic technique of micromolding has been used to create micron scale patterned polymer channels housed inside silicon wafers and their derived magnetic ceramics.[13] It has also proved possible to control the size of the α-Fe nanoclusters through the thermal ROP and subsequent pyrolysis of **1a** within the hexagonal *ca.* 3-4 nm channels of mesoporous silica (MCM-41).[14] In this case only superparamagnetic clusters calculated to have diameters between 5.0-6.4 nm coated with a 0.4-0.6 nm Fe_3O_4 layer were present. Similarly, a templating process has been employed to fabricate inverse opal magnetic ceramics.[15] Crosslinked PFS microspheres formed by precipitation polymerization may be oxidized and electrostatically self-assembled with silica spheres to form core-shell particles. Their pyrolysis affords ceramic microspheres which can be magnetically arranged.[16] Finally, we have reported that pyrolysis of a thin film of self-assembled PS-*b*-PFS block copolymer affords an array of ceramic dots *ca.* 20-25 nm in diameter on a silicon substrate.[17] Such an array may find applications in magnetic data storage or as an etch resist for the formation of high aspect ratio features. This work illustrates the potential of organometallic polymers to function as metal sources for the growth of metallic nanoclusters within a larger patterned structure.

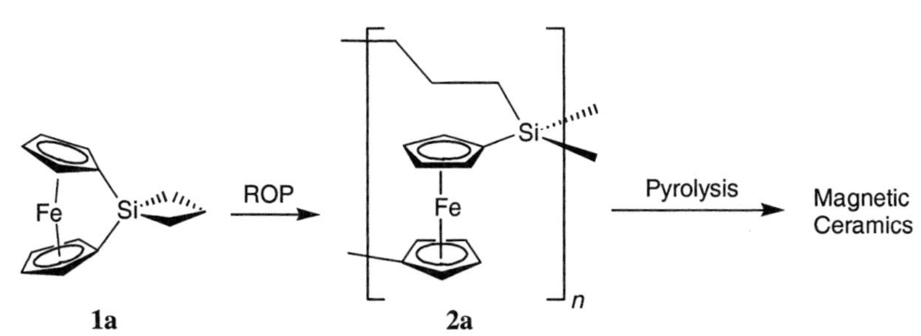

Scheme 2. Thermal ROP of a spirocyclic sila[1]ferrocenophane (**1a**) to yield a cross-linked preceramic polyferrocenylsilane network (**2a**).

Polyferrocenylsilanes as Plasma Etch Resists

The ability of block copolymers containing immiscible blocks to self-assemble into regular morphologies on the nanometer scale in both solution and the solid state is of great interest for block copolymer lithography when a difference in etch rates between the two blocks exists. It is know that the plasma etch rate of polymers can be greatly reduced through the incorporation of metals which tend to form an involatile refractory layer.[10] In the past this has been achieved by selective reaction of a metal reagent with one block. For example, Möller and co-workers selectively vaporized Ti onto the self-assembled PS clusters protruding from a thin films of polystyrene-*b*-poly(2-vinylpyridine) to form a high contrast mask for ion etching.[18] The direct incorporation of metal into the polymer chain ensures a high and uniform loading and avoids post-resist fabrication metallization steps. We were able to exploit both the solution self-assembly and plasma chemistry of inorganic polymers to form nanowires on a semiconducting substrate.[19] Living anionic polymerization was employed to prepare a poly(ferrocenyldimethylsilane-*b*-dimethylsiloxane) block copolymer with suitable block ratios for self-assembly into cylinderical micelles with an iron-rich PFS core in *n*-hexane. The micelles are *ca.* 20 nm in diameter and their length can be controlled from between 70 nm to 10 μm. It proved possible to align the micelles using capillary forces on an electron-beam patterned silicon substrate. Subsequent treatment with a hydrogen plasma stripped away the hydrocarbon component leaving behind iron-silicon-carbon-oxide ceramic lines approximately 4 nm high. Thus self-assembled block copolymers containing PFS have great potential as etch resists for nanopatterning metal-containing structures through plasma RIE.

Polyferrocenylsilanes would be expected to have low plasma etch rates due to the formation of a protective layer of involatile iron and silicon compounds upon reaction with the plasma.[20] X-ray photoelectron spectroscopy (XPS) and Auger electron spectroscopic (AES) depth profiling have shown that polyferrocenyldimethylsilane (**2b**) reacts with an oxygen plasma to form a *ca.* 10 nm thick carbon-depleted layer rich in iron and silicon oxides.[21] In addition, by varying etching conditions it was possible to obtain a remarkably high etch ratio of almost 50:1 between PI and PFS (**2b**).[22] The same study demonstrated that PFS (**2b**) is an effective etch barrier in CF_4/O_2-RIE used to etch silicon wafers. Thus a self-assembled organic-PFS block copolymer film is an ideal candidate for the patterning of large aspect ratio nanometer-sized features in an underlying substrate using plasma RIE.[20] This process was illustrated using a poly(isoprene-*b*-ferrocenyldimethylsilane) block copolymer which phase segregated into a monolayer of PFS spheres in a PI matrix with a domain spacing of 30 nm.[21] Under the oxygen plasma conditions employed the etch rate of the PFS domains was 40 times less than that of the PI matrix. Following O_2-RIE, an AFM height image revealed 7 nm high organometallic regions with short-range hexagonal order and the same domain spaces as found in the initial resist. The ordering of the organometallic domains can be greatly improved using graphoepitaxy in which film growth is ordered by artificial topographic patterning of the substrate.[23] Near perfect alignment of the self-assembled 20 nm spherical PFS domains in a PS matrix in an annealed thin film of PS-*b*-PFS was achieved in 240 nm wide grooves made in an oxidized silicon wafer using interference lithography.[24] Reactive ion etching of the film with a CHF_3 plasma led to the formation of ordered silica posts 20 nm wide with aspect ratios of three or greater. This process has the potential to be generalized to the patterning of many substrate materials.

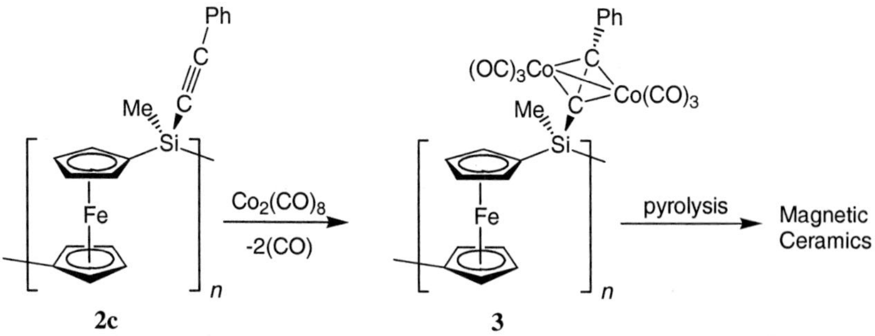

Scheme 3. Cobalt clusterization of the acetylenic substituent of a polyferrocenylsilane (**2c**) to yield the highly metallized Co-PFS (**3**), a precursor to magnetic ceramics.

The incorporation of additional metals into the PFS chain is particularly attractive for the metal patterning of surfaces using standard lithographic processes. We have found that polyferrocenylsilanes with acetylenic substituents at silicon (**2c**)[25] can be clusterized by treatment with dicobalt octacarbonyl to yield the highly metallized, soluble, air stable cobalt-clusterized polyferrocenylsilane (Co-PFS) (**3**) (Scheme 3).[26] We have also shown that the pyrolysis of Co-PFS affords ceramics containing magnetic Co/Fe alloy nanoparticles in high yield. Thin films of Co-PFS are readily accessible *via* spin coating. Results on the lithographic patterning of this novel highly metallized organometallic resist will be presented.

[1] Q. Guo, X. Teng, S. Rahman, H. Yang, *J. Am. Chem. Soc.* **2003**, *125*, 630.
[2] Y. Loo, R.L. Willett, K.W. Baldwin, J.A. Rogers, *J. Am. Chem. Soc.* **2002**, *124*, 7654.
[3] M. Rolandi, C.F. Quate, H. Dai, *Adv. Mater.* **2002**, *14*, 191.
[4] Y. Li, B.W. Maynor, J. Liu, *J. Am. Chem. Soc.* **2001**, *123*, 2105.
[5] B.F.G. Johnson, K.M. Sanderson, D.S. Shephard, D. Ozkaya, W. Zhou, H. Ahmed, M.D.R. Thomas, L. Gladden, M. Mantle, *Chem. Commun.* **2000**, 1317.
[6] D.A. Foucher, B.Z. Tang, I. Manners, *J. Am. Chem. Soc.* **1992**, *114*, 6246.
[7] Y. Ni, R. Rulkens, I. Manners, *J. Am. Chem. Soc.* **1996**, *118*, 4102.
[8] (a)Y. Ni, R. Rulkens, J.K. Pudelski, I. Manners, *Macromol. Rapid Commun.* **1995**, *16*, 637. (b) N.P. Reddy, H. Yamashita, M. Tanaka, *Chem. Commun.* **1995**, 2263.
[9] M.J. MacLachlan, M. Ginzburg, N. Coombs, T.W. Coyle, N.P. Raju, J.E. Greedan, G.A. Ozin, I. Manners, *Science* **2000**, *287*, 1460.
[10] (a) G.N. Taylor, T.M. Wolf, L.E. Stillwagon, *Solid State Technol.* **1984**, *27*, 145. (b) I. Brodie, J.J. Muray, *The Physics of Micro/Nano-Fabrication*, Plenum Press, New York, **1992**.
[11] T. Wideman, E.E. Remsen, E. Cortez, V.L. Chlanda, L.G. Sneddon, *Chem. Mater.* **1998**, *10*, 412.
[12] T. Wideman, E. Cortez, E.E. Remsen, G.A. Zank, P.J. Carroll, L.G. Sneddon, *Chem. Mater.* **1997**, *9*, 2218.
[13] M. Ginzburg, M.J. MacLachlan, S.M. Yang, N. Coombs, T.W. Coyle, N.P. Raju, J.E. Greedan, R.H. Herber, G.A. Ozin, I. Manners, *J. Am. Chem. Soc.* **2002**, *124*, 2625.
[14] M.J. MacLachlan, M. Ginzburg, N. Coombs, N.P. Raju, J.E. Greedan, G.A. Ozin, I. Manners, *J. Am. Chem. Soc.* **2000**, *122*, 3878.
[15] J.Galloro, M. Ginzburg, H. Miguez, S.M. Yang, N. Coombs, A. Safa-Sefat, J.E. Greedan, I. Manners, G.A. Ozin, *Adv. Funct. Mater.* **2002**, *12*, 382.
[16] K. Kulbaba, A. Cheng, A. Bartole, S. Greenberg, R. Resendes, N. Coombs, A. Safa-Sefat, J.E. Greedan, H.D.H. Stöver, G.A. Ozin, I. Manners, *J. Am. Chem. Soc.* **2002**, *124*, 12522.
[17] K. Temple, K. Kulbaba, K. N. Power-Billard, I. Manners, K.A. Leach, T. Xu, T.P. Russell, C.J. Hawker, *Adv.Mater.* **2003**, *15*, 297.
[18] J.P. Spatz, P. Eibeck, S. Mössmer, M. Möller, T. Herzog, P. Ziemann, *Adv. Mater.* **1998**, *10*, 849.
[19] (a) J.A. Massey, M.A. Winnik, I. Manners, V.Z.-H. Chan, J.M. Ostermann, R. Enchelmaier, J.P. Spatz, M. Möller, *J. Am. Chem. Soc.* **2001**, *123*, 3147. (b) L. Cao, J.A. Massey, M.A. Winnik, I. Manners, S. Riethmüller, F. Banhart, M. Möller, *Adv. Funct. Mater.* **2003**, *13*, 271.
[20] I. Manners, *Pure Appl. Chem.* **1999**, *71*, 1471.
[21] R.G.H. Lammertink, M.A. Hempenius, J.E.van den Enk, V.Z.-H. Chan, E.L. Thomas, G.J. Vancso, *Adv. Mater.* **2000**, *12*, 98.
[22] R.G.H. Lammertink, M.A. Hempenius, V.Z.-H. Chan, E.L. Thomas, G.J. Vancso, *Chem. Mater.* **2001**, *13*, 429.
[23] H.I. Smith, M.W. Geis, C.V. Thompson, H.A. Atwater, *J. Cryst. Growth* **1983**, *63*, 527.
[24] J.Y. Cheng, C.A. Ross, E.L. Thomas, H.I. Smith, G.J. Vancso, *Appl. Phys. Lett.* **2002**, *81*, 3657.
[25] A. Berenbaum, A.J. Lough, I. Manners, *Organometallics* **2002**, *21*, 4415.
[26] A. Berenbaum, M. Ginzburg-Margau, N. Coombs, A.J. Lough, A. Safa-Sefat, J.E. Greedan, G.A. Ozin, I. Manners, *Adv. Mater.* **2003**, *15*, 51.

Polyethers and Thioethers Incorporating Neutral and Cationic Organoiron Complexes

Alaa S. Abd-El-Aziz, Erin K. Todd, Rawda M. Okasha*

Department of Chemistry, The University of Winnipeg, Winnipeg, Manitoba, Canada R3B 2E9
E-mail: a.abdelaziz@uwinnipeg.ca

Summary: The synthesis of linear and star-shaped oligomers containing cationic and neutral organoiron groups in their structures was achieved by reaction of cationic arene complexes of cyclopentadienyliron containing terminal hydroxyl groups with 1,1'-ferrocenedicarbonyl chloride or ferrocene carboxylic acid. The use of chloroarene complexes allowed for the formation of triiron complexes that were subsequently polymerized via nucleophilic aromatic substitution with various oxygen- and sulfur-based dinucleophiles. The corresponding polyethers and thioethers were isolated in good yields and these materials exhibited excellent solubilities in polar organic solvents. Cyclic voltammetric investigations revealed that the cationic iron centers pendent to the polymer backbones underwent reversible reduction steps, while the neutral iron centers within the polymer backbones underwent reversible oxidation steps. Photolysis of these polymers resulted in the removal of the cationic cyclopentadienyliron moieties pendent to the polymer backbones. Thermogravimetric analysis (TGA) revealed that the cationic iron complexes were cleaved from the polymers at approximately 210 °C. Differentials scanning calorimetry (DSC) revealed that the glass transition temperatures of the cationic polymers occurred at higher temperatures than their neutral analogs.

Keywords: arene cyclopentadienyliron complexes; ferrocene; organoiron polymers; soluble polymers; star polymers

Introduction

The synthesis of organoiron polymers has been the focus of numerous investigations in recent years in light of the interesting properties that this class of material possesses.[1-4] Two of the most prevalent organoiron groups that have been incorporated into polymers are the ferrocene and arene cyclopentadienyliron complexes. The ferrocene group is a very stable organometallic functionality and undergoes reversible oxidation processes.[5] On the other hand, cationic arene

© 2003 WILEY-VCH Verlag GmbH & KGaA, Weinheim CCC 1022-1360/00/$ 17,50+.50/0

cyclopentadienyliron complexes undergo reversible reduction steps.[5, 6] Astruc has reported the synthesis of star and dendritic complexes containing arene complexes of cyclopentadienyliron and ferrocene moieties in their structures.[7, 8] The synthesis of a bimetallic amide complex incorporating one ferrocene and one cobaltocenium complex in the backbone has been described by Cuadrado and coworkers.[9] The ferrocene complex was oxidized at 0.56 V, while the cobaltocenium unit was reduced at –0.75 V vs. SCE. The synthesis of a dendrimer containing these isoelectronic metallocene groups at their periphery was also reported.[10] It was found that the cationic cobalt-based metallocenes functionalized with carboxylic acid groups reacted with amines more rapidly than their neutral iron-based analogs.[9, 10]

Our research focuses on the use of arene complexes in monomer and polymer synthesis. Mono- and di-chloroarene complexes undergo nucleophilic aromatic substitution reactions with a large variety of oxygen, sulfur and nitrogen nucleophiles to produce novel materials.[11-15] Polymerization of dichloroarene complexes of cyclopentadienyliron with dinucleophiles readily affords soluble cationic polymers.[12] Photolysis of these polymers allows for the removal of the organoiron moieties and isolation of the corresponding organic polymers. Monomers containing unsaturated centers have been subjected to radical and ring-opening metathesis polymerization reaction, yielding high molecular weight cationic organoiron polymers.[16-18]

Our success in the design of polymeric materials containing arene cyclopentadienyliron complexes combined with the attractive properties that ferrocene-based polymers possess, prompted our investigation into the development of polymers containing these two organoiron groups. To that end, we recently reported the synthesis of oligomers and polymers containing cationic cyclopentadienyliron moieties and neutral ferrocenyl moieties in their structures.[15] While the neutral ferrocene groups were incorporated as integral parts of these materials, the cationic cyclopentadienyliron moieties were introduced as pendent groups. This article will highlight our results in this area and describe some of our recent findings.

Results and Discussion

Oligomers

Our initial investigations into the synthesis of complexes containing arene complexes of cyclopentadienyliron and ferrocene led to the synthesis of diiron and triiron complexes. Scheme

1 shows the synthesis of some of these complexes by reaction of arene complexes containing terminal -OH groups with carboxylic acid and diacid chloride derivatives of ferrocene. While reaction of complexes **1a** and **1b** with ferrocene carboxylic acid (**2**) in the presence of dicyclohexylcarbodiimide (DCC) and *N,N*-dimethylamino pyridine (DMAP) resulted in the production of bimetallic complexes **3a** and **3b**, reaction of **1a** and **1b** with ferrocene dicarbonyl chloride (**4**) in the presence of pyridine produced the trimetallic complexes **5a** and **5b**, respectively.

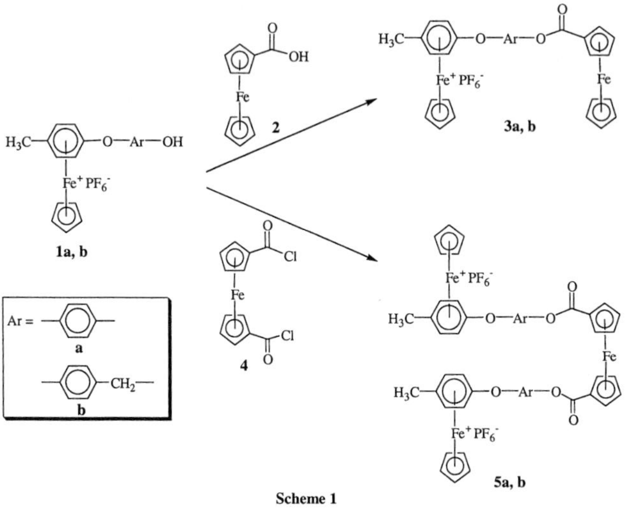

Scheme 1

The electrochemical properties of these complexes were examined using cyclic voltammetry. It was possible to distinguish the two types of iron complexes by their distinct redox behaviors. Figure 1 shows the cyclic voltammogram of the trimetallic complex **5a**. The neutral iron centers in this complex underwent reversible oxidation at $E_{1/2}$ = 1.00 V and the cationic iron centers pendent to the aromatic rings underwent reversible reduction at $E_{1/2}$ = -1.34 V. It is also important to note that the redox wave corresponding to the two cationic iron centers is twice as large as the redox wave corresponding to the central ferrocenyl moiety. In contrast, the cyclic voltammograms for complexes **3a** and **3b** exhibited reduction and oxidation processes of equal current intensities. For penta- and hepta-metallic complexes containing one central ferrocene group each, the ratios of the current intensities for the oxidation to reduction peaks were 1 to 4 and 1 to 6, respectively.

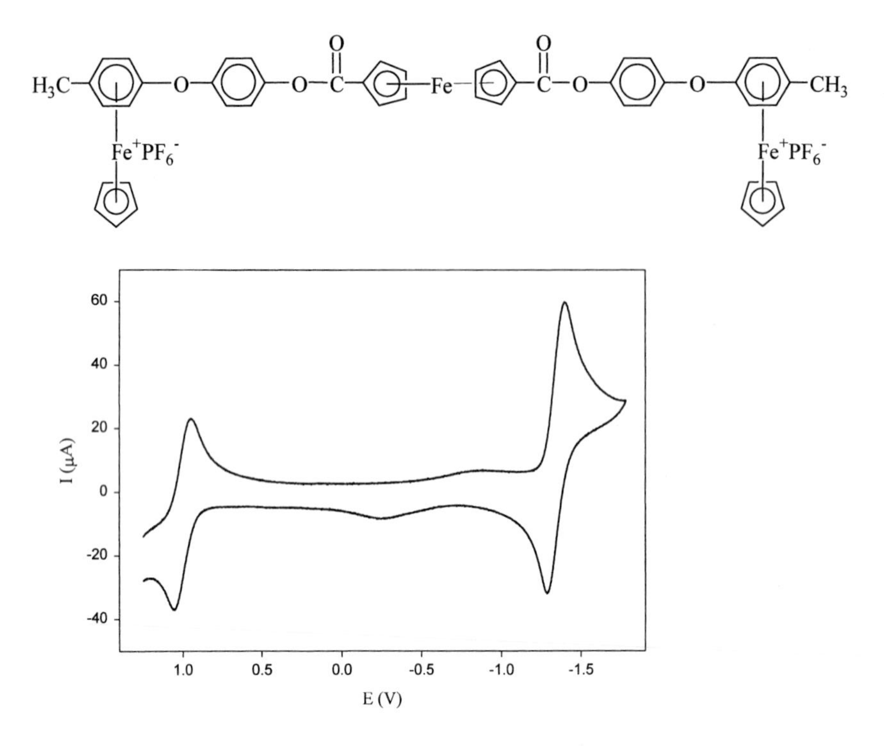

Figure 1. Cyclic voltammogram at glassy carbon of 0.002 M 5a in 0.1 M TBAP in propylene carbonate, ν = 0.1 V/s at 0 °C.

Star-Shaped Complexes

It was also of interest to prepare star-shaped complexes containing these two different redox active organoiron groups. The synthesis of a hexametallic complex containing neutral and cationic cyclopentadienyliron moieties in its structure is shown in Scheme 2. We have recently reported the synthesis of complex **6**, which served as a core molecule in the synthesis of star-shaped polyaromatic ethers.[13] Reaction of trimetallic star complex **6** with ferrocene carboxylic acid (**2**) allowed for the isolation of the hexametallic star complex **7**. A slight excess of complex **2** was utilized in the reaction in order to ensure complete coupling of the carboxylic acid groups to the phenolic groups.

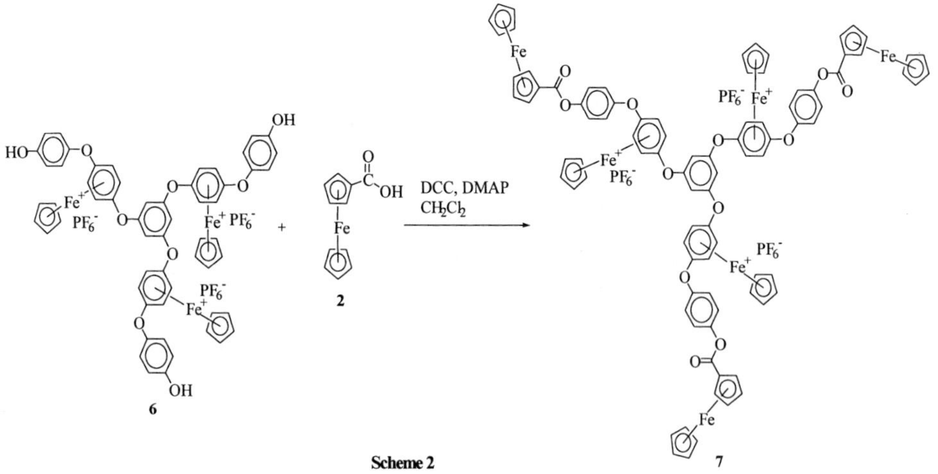

Scheme 2

The ^{1}H NMR spectrum of complex **7** is shown in Figure 2a. The cyclopentadienyl protons of the unsubstituted ferrocenyl groups appear as a singlet at 4.33 ppm, while the protons of the substituted cyclopentadienyl ring appear as two sets of triplets at 4.40 and 4.94 ppm. The protons of the cyclopentadienyl ring coordinated to the cationic iron center appear as a singlet further downfield at 5.35 ppm. The complexed aromatic protons appear as two sets of doublets at 6.41 and 6.63 ppm, which is consistent with the proposed structure. The protons of the central aromatic ring appear as a singlet at 7.35 ppm, while the protons of the three other aromatic rings appear as a singlet at 7.38 ppm. The ^{13}C NMR (APT) spectrum displayed in Figure 2b shows the cyclopentadienyl carbons (CH) of the ferrocenyl unit at 70.68, 71.23 and 72.95 ppm, while the *ipso* carbons appear at 70.46 ppm. The cyclopentadienyl resonance of the rings coordinated to the cationic iron centers appear at 78.99 ppm. The complexed aromatic carbons (CH) appear at 75.84 and 76.87 ppm, while the *ipso* complexed aromatic carbons resonate at 130.70 and 131.94 ppm. The carbons *alpha* to the ether linkages in the central arene and the branches appear at 110.71 ppm, and 122.17 and 124.84 ppm, respectively. The three *ipso* aromatic carbons appear at 149.57, 151.70 and 157.62 ppm. The peak at 170.35 ppm corresponds to the carbonyl carbons of the ester functionalities.

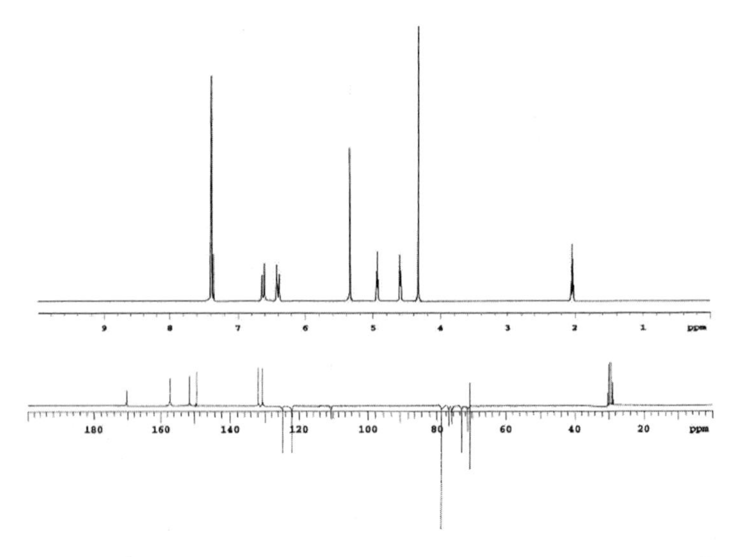

Figure 2. (a) top, ^{1}H NMR spectrum of complex **7** in acetone-d_6 (a) bottom, ^{13}C NMR spectrum of complex **7** in acetone-d_6.

The electrochemical properties of complex **7** were examined in order to determine the redox potentials of the two cationic and neutral iron centers within its structure. These cyclic voltammograms were obtained in propylene carbonate from –30 to +10 °C and were found to be reversible. The cyclic voltammogram in Figure 3 shows the oxidation of the neutral iron centers at $E_{1/2} = 0.694$ V and reduction of the cationic iron centers at $E_{1/2} = -1.29$ V.

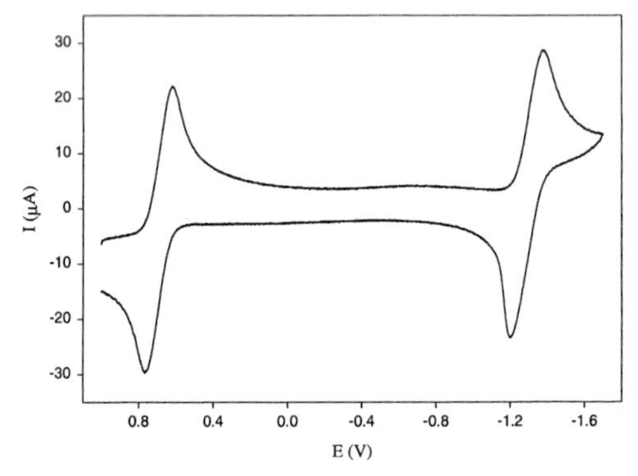

Figure 3. Cyclic voltammogram at glassy carbon of 0.001 M of the star-shaped hexametallic complex **7** in 0.1 M TBAP in propylene carbonate, ν = 0.2 V/s at –30 °C.

Polymers

The synthesis of polymers containing both neutral and cationic cyclopentadienyliron complexes in their structures has recently been communicated.[15] Reaction of trimetallic complexes containing terminal chloroarene complexes of cyclopentadienyliron were reacted with various dinucleophiles in the presence of potassium carbonate. Scheme 3 shows the reaction of complex **8** with bisphenol A (**9a**) and 4,4'-thiobisbenzenethiol (**9b**), producing the mixed-charge organoiron polymers **10a** and **10b**, respectively. NMR analysis of these polymers was consistent with successful polymerization. In order to measure the molecular weights of these materials, the cationic cyclopentadienyliron moieties were first cleaved from the polymer backbones due to interactions between these complexes and GPC columns. Irradiation of **10a** and **10b** in acetonitrile/dichloromethane solutions allowed for the isolation of the novel ferrocene-containing polymers **11a** and **11b**.

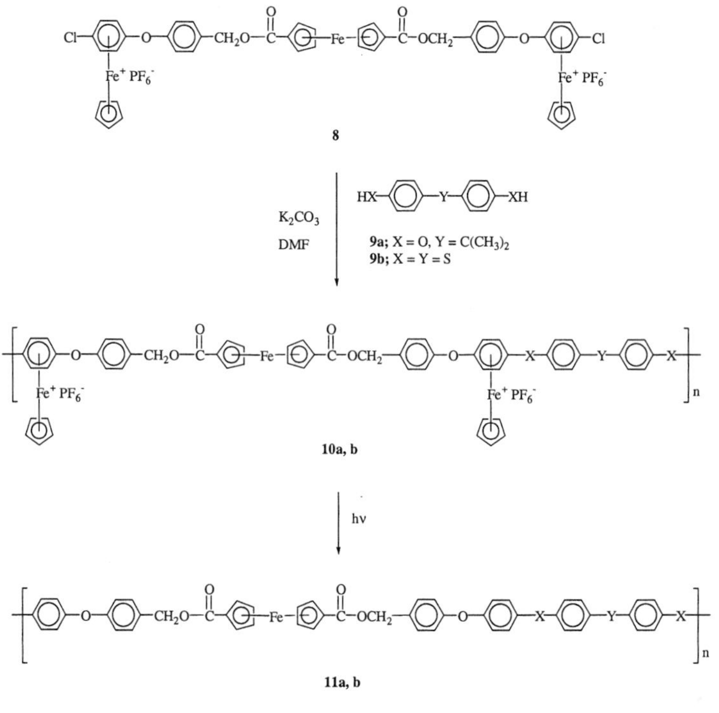

Cl–O–CH_2O–C(=O)–Fe–C(=O)–OCH_2–O–Cl
Fe^+ PF_6^- Fe^+ PF_6^-

8

HX–Y–XH

K_2CO_3
DMF

9a; X = O, Y = $C(CH_3)_2$
9b; X = Y = S

–O–CH_2O–C(=O)–Fe–C(=O)–OCH_2–O–X–Y–X– n
Fe^+ PF_6^- Fe^+ PF_6^-

10a, b

hν

–O–CH_2O–C(=O)–Fe–C(=O)–OCH_2–O–X–Y–X– n

11a, b

Scheme 3

^1H and ^{13}C NMR analysis of these polymers clearly showed that the cyclopentadienyl resonances corresponding to the cyclopentadienyliron cations were no longer present in the spectra of **11a** and **11b**, however the cyclopentadienyl resonances corresponding to the neutral organoiron groups were still observed. Figure 4 shows the ^1H NMR spectrum of polymer **11a**. It is clear that the ferrocenyl cyclopentadienyl resonances are present at 4.24 and 4.75 ppm, while the cyclopentadienyl resonance of the cationic complex is not visible. The peak at 5.18 ppm corresponds to the methylene protons and the singlet at 1.63 ppm corresponds to the methyl protons of the isopropylidene groups. The aromatic protons appear as three sets of doublets and two singlets between 6.86 and 7.34 ppm.

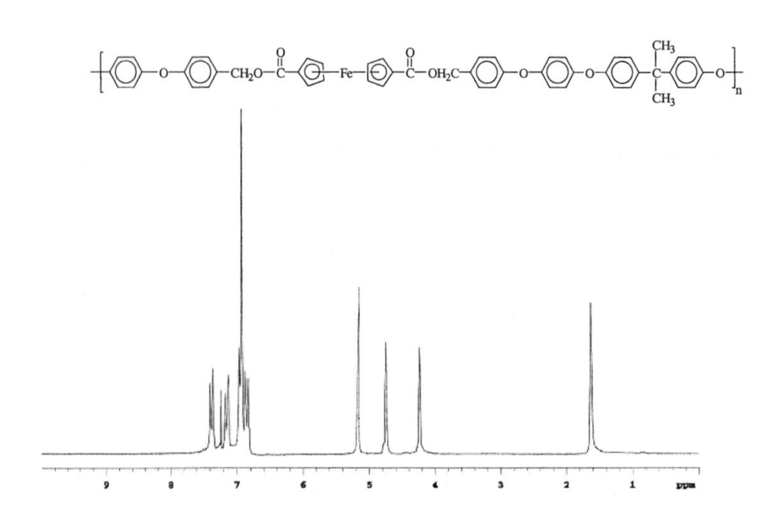

Figure 4. ^1H NMR spectrum of the ferrocene-based polymer **11a** in $CDCl_3$.

The cyclic voltammogram of polymer **10b** is shown in Figure 5. It can be seen that the neutral iron centers in the polymer backbone were oxidized at $E_{1/2}$ = 1.05 V, while the cationic iron centers pendent to the polymer backbone were reduced at $E_{1/2}$ = -0.995 V. While the redox patterns of the polymers were similar to the smaller oligomeric complexes, it can be seen that the oxidation and reduction waves are broader in the polymeric materials.

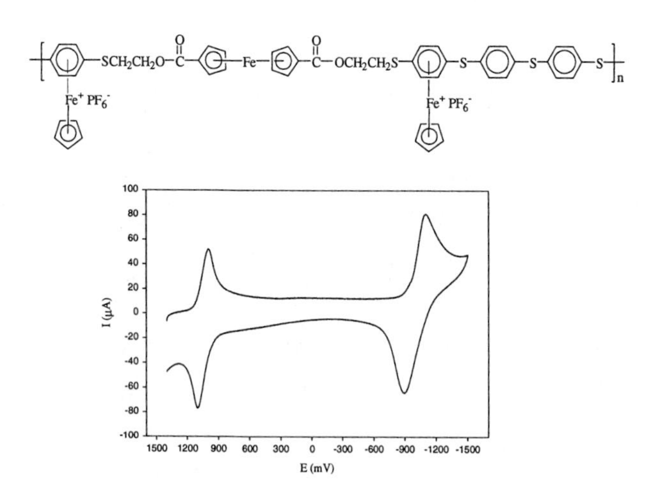

Figure 5. Cyclic voltammogram at glassy carbon of 0.002 M of the organoiron polymer **10b** in 0.1 M TBAP in propylene carbonate, ν = 0.5 V/s at –20 °C.

Electrochemical analysis of the photolyzed polymers was also performed. Since polymers **11a** and **11b** still contained redox active groups, cyclic voltammetry was a useful method to compare the properties of the cationic and neutral organoiron polymers. Figure 6 shows the cyclic voltammogram of polymer **11b**, which was obtained in dichloromethane. It can be seen that oxidation of the iron centers in this polymer occurs at $E_{1/2}$ = 0.609 V. In contrast, the CV of polymer **10b** showed that the neutral iron centers were oxidized at a much more positive potential.

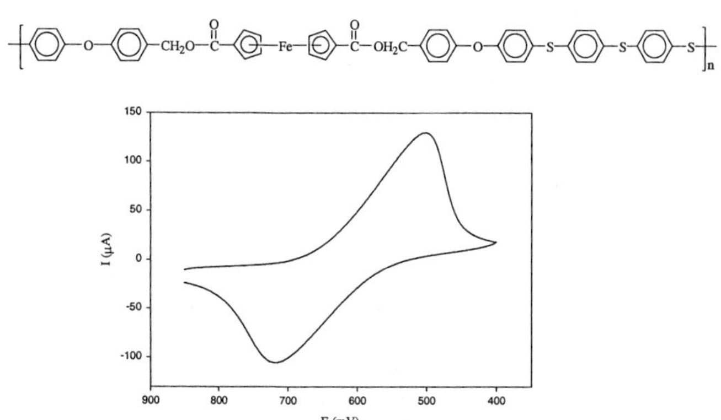

Figure 6. Cyclic voltammogram at glassy carbon of 0.001 M of the neutral organoiron polymer **4.18** in 0.1 M TBAP in dichloromethane, ν = 0.1 V/s at –30 °C.

The molecular weights of polymers **11a** and **11b** were measured using gel permeation chromatography. It was not possible to measure the molecular weights of polymers **10a** and **10b** directly due to interactions of the cationic iron centers with GPC columns. Therefore, the molecular weights of the organometallic polymer were calculated based on the average weights determined for their corresponding neutral analogs. The number average molecular weights of polymers **10a** and **10b** were determined to be 9,700 and 9,100 with polydispersities of 1.13 and 2.19.

Thermogravimetric analysis of the organoiron polymers showed that the cyclopentadienyliron cations were cleaved from the polymer backbones at about 210 °C. These values are consistent with our other thermal tests of cyclopentadienyliron-coordinated polyethers and thioethers.[12] Degradation of the polymer backbones began at about 440 °C. Following photolytic cleavage of the cyclopentadienyliron moieties from the backbones of the polymers, the weight losses at 210 °C were not present. However, two weight losses were present in the thermograms of **11a** and **11b**. Small weight loss steps were observed at about 290 °C and more significant decomposition began at about 400 °C.

Differential scanning calorimetry of polymers **10a,b** and **11a,b** was performed in order to examine the glass transition temperatures of these organoiron polymers. The glass transition temperatures of polymers **10a** and **10b** were 141 and 161 °C, respectively. Upon removal of the cationic cyclopentadienyliron moieties pendent to the polymer backbones, the T_gs of these polymers decreased to 92 to 84 °C, for **11a** and **11b**, respectively. These results indicate that the presence of bulky cationic cyclopentadienyliron moieties pendent to the backbones of polymers causes significant increases in their glass transition temperatures.

Conclusion

New classes of oligomers and polymers containing neutral ferrocene units in their backbones and cationic cyclopentadienyliron moieties pendent to their backbones were synthesized via nucleophilic aromatic substitution reactions. Photolysis of these polymers resulted in cleavage of the pendent cationic iron groups, allowing for the isolation of novel ferrocene-containing polymers. Thermal analysis indicated that these polymers possessed higher thermal stability but lower glass transition temperatures than their cationic analogs. Cyclic voltammetry of the complexes and polymers incorporating neutral and cationic organoiron complexes in their structures showed that the iron centers underwent reversible oxidation and reduction processes, respectively.

Acknowledgement

Financial support provided by the Natural Sciences and Research Council of Canada (NSERC) and Manitoba Hydro are gratefully acknowledged. R.M.O. would also like to thank the Department of Chemistry, University of Manitoba.

[1] A. S. Abd-El-Aziz, *Macromol. Rapid Commun.* **2002**, *23*, 995.
[2] A. S. Abd-El-Aziz, *Coord. Chem. Rev.* **2002**, *233-234*, 177.
[3] R. D. Archer, in "*Inorganic and Organometallic Polymers*", Wiley-VCH, New York, 2001.
[4] P. Nguyen, P. Gomez-Elipe, I. Manners, *Chem. Rev.* **1999**, *99*, 1515.
[5] D. Astruc, "*Electron Transfer and Radical Processes in Transition-Metal Chemistry*", VCH Publishers Inc., New York, 1995.
[6] A. S. Abd-El-Aziz, C. R. de Denus, K. M. Epp, S. Smith, R. J. Jaeger, D. T. Pierce, *Can. J. Chem.* **1996**, *74*, 650.
[7] J.-L. Fillaut, J. Linares, D. Astruc, *Angew. Chem. Int. Ed. Engl.* **1994**, *33*, 2460.
[8] J.-L. Fillaut, D. Astruc, *New J. Chem.* **1996**, *20*, 945.
[9] B. Gonzalez, I. Cuadrado, C. M. Casado, B. Alonso, C. J. Pastor, *Organometallics* **2000**, *19*, 5518.
[10] C. M. Casado, B. Gonzalez, I. Cuadrado, B. Alonso, M. Moran, J. Losada. *Angew. Chem. Int. Ed.* **2000**, *39*, 2135.
[11] A. S. Abd-El-Aziz, S. Bernardin, *Coord. Chem. Rev.* **2000**, *203*, 219.
[12] A. S. Abd-El-Aziz, E. K. Todd, G. Z. Ma, *J. Polym. Sci., Part A: Polym. Chem.* **2001**, *39*, 1216.
[13] A. S. Abd-El-Aziz, E. K. Todd, T. Afifi, *Macromol. Rapid Commun.* **2002**, *23*, 113.
[14] A. S. Abd-El-Aziz, T. H Afifi, W. R. Budakowski, K. J. Friesen, E. K. Todd, *Macromolecules* **2002**, *35*, 8929.
[15] A. S. Abd-El-Aziz, E. K. Todd, R. M. Okasha, T. E. Wood, *Macromol. Rapid Commun.* **2002**, *23*, 743.
[16] A. S. Abd-El-Aziz, E. K. Todd, G. Z. Ma, J. DiMartino, *J. Inorg. Organomet. Polym.* **2000**, *10*, 265.
[17] A. S. Abd-El-Aziz, L. J. May, J. A. Hurd, R. M. Okasha, *J. Polym. Sci., Part A: Polym. Chem.* **2001**, *39*, 2716.
[18] A. S. Abd-El-Aziz, R. M. Okasha, T. H. Afifi, E. K. Todd, *Macromol. Chem. Phys.* **2003**, 204, 555.

Macromol. Symp. ***196***, *89–99 (2003)*

Organoiron Polymers Containing Azo Dyes

Alaa S. Abd-El-Aziz, Erin K. Todd, Rawda M. Okasha, Tarek H. Afifi*

Department of Chemistry, The University of Winnipeg, Winnipeg, Manitoba, Canada R3B 2E9
E-mail: a.abdelaziz@uwinnipeg.ca

Summary: The synthesis of cationic organoiron polymers with azobenzene moieties in their side chains has been accomplished via nucleophilic aromatic substitution and ring-opening metathesis polymerization (ROMP) reactions. Polyaromatic ethers and thioethers with azobenzene moieties in their side chains were functionalized with different chromophores to yield yellow-, orange- and red-colored polymers. Polynorbornenes with azobenzene-containing side chains were isolated following ROMP of their monomeric analogs. All of the organoiron polymers were soluble in polar organic solvents and underwent reversible electrochemical reduction processes. Photobleaching of the azobenzene-containing polymers was achieved in the presence of hydrogen peroxide. The metallated polymers had glass transition temperatures approximately 50 to 80 °C higher than their organic analogs.

Keywords: azobenzene; cyclopentadienyliron complexes; organoiron polymers; poly(aromatic ethers); polynorbornene

Introduction

Organoiron polymers are one of the most important classes of metal-containing polymers.[1-4] This class of polymer has been the topic of numerous studies in light of their interesting properties and applications. The focus of our research is on the use of cyclopentadienyliron-coordinated chloroarenes in the design of iron-based monomers and polymers.[2, 5] The mild reaction conditions associated with the nucleophilic aromatic substitution reactions of complexed chloroarenes as well as the variety of dinucleophiles used in these reactions has allowed for an efficient route to the synthesis of functionalized monomers and polymers. Photolytic demetallation of organoiron monomers resulted in the decoordination of the cyclopentadienyliron moieties and the isolation of the free organic monomers. Our research has utilized both the cationic organoiron, and their corresponding organic monomers to produce polymeric materials. For example, oligomeric ether complexes containing terminal naphthyl groups were demetallated

© 2003 WILEY-VCH Verlag GmbH & KGaA, Weinheim CCC 1022-1360/00/$ 17,50+.50/0

and subsequently polymerized using ferric chloride to yield polyaromatic ethers.[6] Substituted norbornene monomers were also synthesized using cationic arene cyclopentadienyliron complexes.[7-10] These monomers were either demetallated and then subjected to ring-opening metathesis polymerization, or the organoiron monomers were polymerized directly. The organoiron-coordinated polynorbornenes were also demetallated using photolytic techniques in order to conduct molecular weight measurements on these materials. Radical polymerization of methacrylate monomers functionalized with cyclopentadienyliron complexes has also produced high molecular weight cationic organoiron polymethacrylates.[11, 12] The direct polymerization of dichloroarene complexes with bisphenols and bisthiols produced polyaromatic ethers and thioethers with cyclopentadienyliron cations pendent to the polymer backbone.[13, 14] It was found that the organoiron groups increased the solubility of these polymers relative to their corresponding organic analogs. The electrochemical properties of oligomeric and polymeric materials containing arene complexes of cyclopentadienyliron have been examined using cyclic voltammetry and it was found that the cationic eighteen-electron iron complexes underwent reversible reduction steps.[15] We have also found that cationic organoiron polymers exhibit polyelectrolyte effects in polar organic solvents.[16]

Currently, there is considerable interest directed towards the synthesis of polymers containing azobenzene moieties in their backbones and side chains.[17-27] While a number of different classes of organic polymers have been studied, there has been very little research detailing azobenzene-modified organometallic polymers. However, Manners has reported the synthesis and liquid crystalline properties of polyferrocenylsilanes incorporating azobenzene groups in their side chains.[28] Recently, we have been investigating the incorporation of azobenzene chromophores into the side chains of cationic organoiron polymers.[29, 30] This article will describe the synthesis and properties of azobenzene-containing organoiron polymers via metal-mediated nucleophilic aromatic substitution reactions and ring-opening metathesis polymerization.

Results and Discussion

Aromatic Polymers with Azobenzene Side Chains

We have recently reported the synthesis of a bimetallic complex containing a pendent carboxylic acid group.[9] These organoiron complexes were reacted through the carboxylic acid groups with a number of azo-containing compounds functionalized with hydroxyl groups to produce azobenzene-functionalized complexes.[29] Scheme 1 shows the reaction of complex **1** with the azo chromophores **2a-c** in the presence of dicyclohexylcarbodiimide (DCC) and *N,N*-dimethylamino pyridine (DMAP). Complexes **3a-c** were isolated in good yields as orange to red solids. As the R group on the azobenzene moieties became more electron-withdrawing, there was a concurrent increase in the red shifts that these complexes experienced. For example, the λ_{max} values for the dyes increased from 421 to 452 to 491 nm in DMF, when the R groups changed from H, to $COCH_3$ to NO_2.

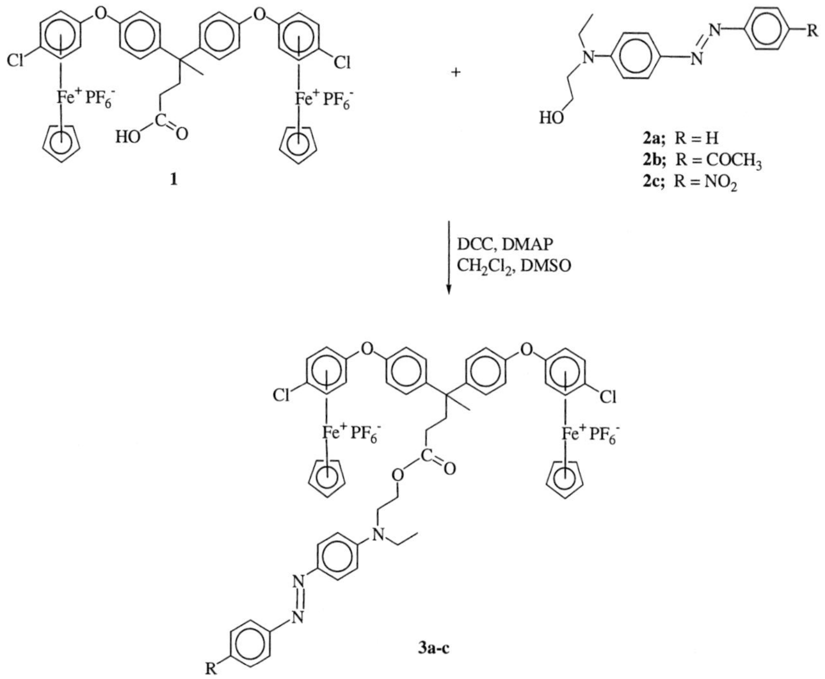

2a; R = H
2b; R = $COCH_3$
2c; R = NO_2

DCC, DMAP
CH_2Cl_2, DMSO

Scheme 1

Polymerization of complexes **3a-c** was achieved in the presence of potassium carbonate in DMF using a number of different dinucleophiles. Scheme 2 shows the reaction of monomers **3a-c** with bisphenol A (**4**). Polymerization reactions were carried out at 60 °C for a period of 18 hours. The resulting aromatic polyethers (**5a-c**) were isolated in good yields and were soluble in common polar organic solvents. In order to determine the molecular weights of polymers **5a-c**, it was first necessary to remove the cyclopentadienyliron cations pendent to their backbones due to interactions between these polar groups and GPC columns. Photolytic demetallation of polymers **5a-c** produced polymers **6a-c**.

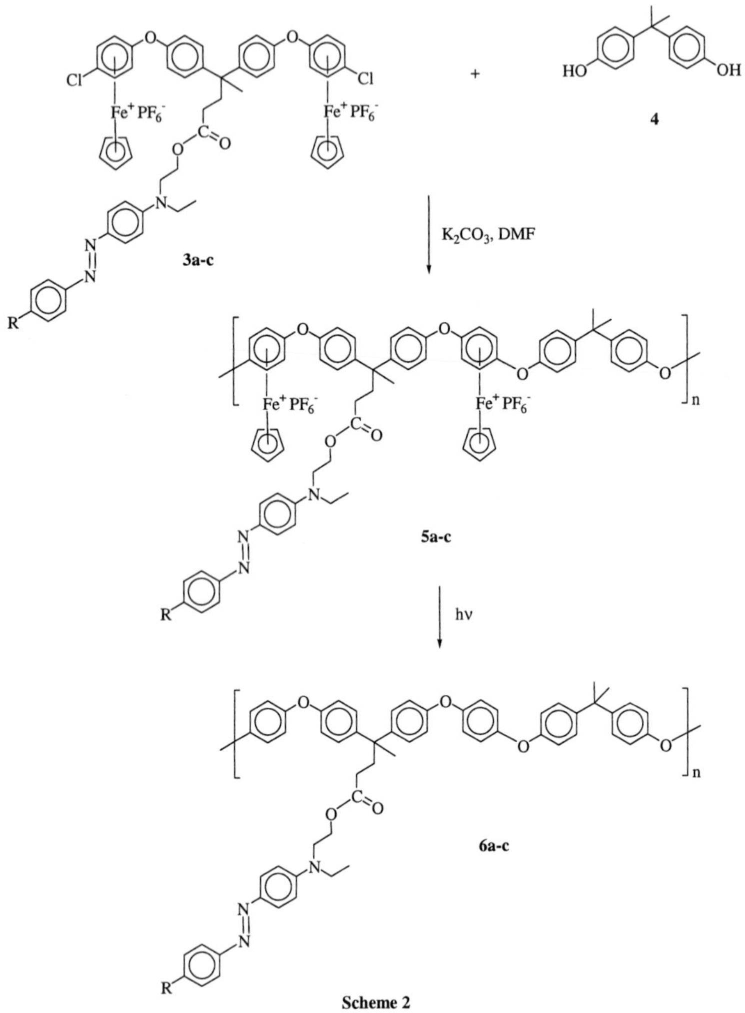

Cl Cl $Fe^+ PF_6^-$ $Fe^+ PF_6^-$ + HO OH **4** **3a-c** R K_2CO_3, DMF **5a-c** hν **6a-c**

Scheme 2

The weight average molecular weights of polymers **6a-c** were estimated to range from 8,600 to 11,500 with polydispersities ranging from 1.2 to 2.0. From these values, the molecular weights of the corresponding organoiron polymers were estimated to range from 13,400 to 18,200. Differential scanning calorimetry showed that the glass transition temperatures (T_g) of polymers **5a-c** ranged from 167 to 173 °C, while the T_g values of polymers **6a-c** ranged from 111 to 123 °C. Following the photolytic demetallation reactions, NMR, IR and visible spectroscopic investigations of the organic polyethers **6a-c** verified that the azobenzene chromophores were still present in the polymer side chains. Figure 1 shows the visible spectra of polymers **6a**, **6b** and **6c** with R = H, $COCH_3$ and NO_2. The λ_{max} values for these polymers were measured to be 417, 452 and 489 nm, respectively, and it was found that there were not significant shifts in the λ_{max} values of these polymers in comparison to their organoiron analogs or their corresponding monomers.

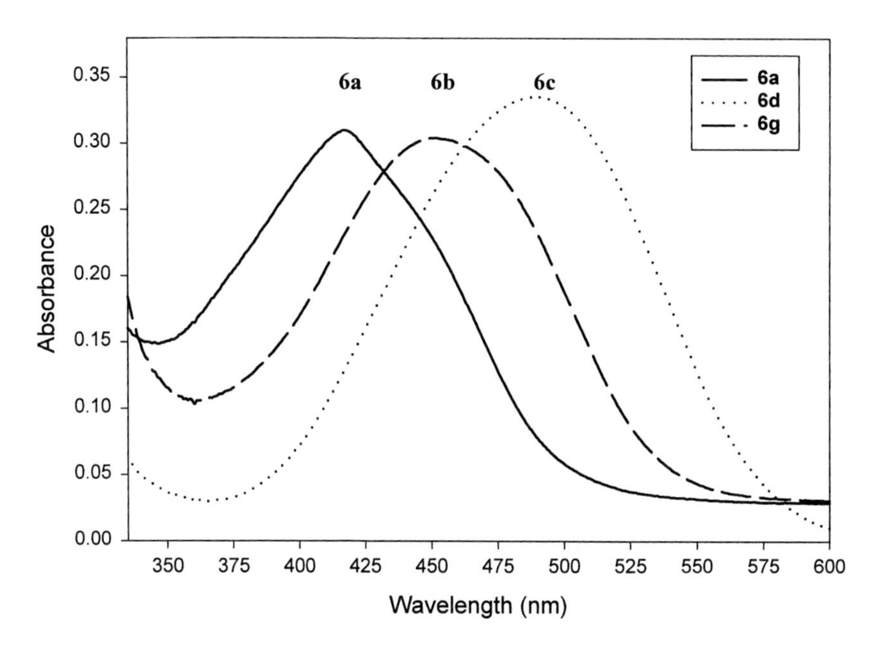

Figure 1. Visible spectra of organic polymers **6a-c** in ethanol (Reproduced with permission from *Macromolecules* **2002**, *35,* 8929-8932. Copyright 2002 Am. Chem. Soc.).

Polymer **7** was also synthesized using our previously described procedure.[29] The visible spectrum of polymer **7** showed a λ_{max} value at 422 nm in DMF. Upon acidification of the solution with HCl, the spectrum showed a bathochromic shift to 534 nm.

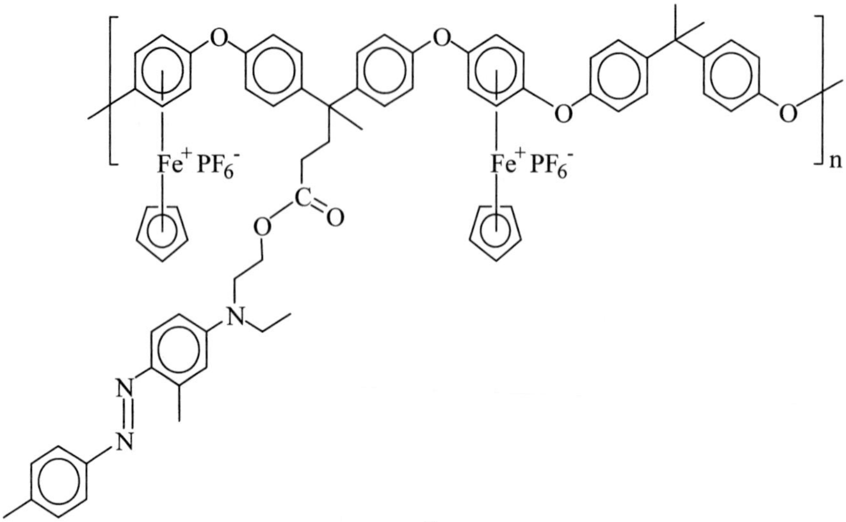

7

While it was found that irradiation of the polymers in acetonitrile solutions removed the cyclopentadienyliron moieties pendent to the polymer backbones, irradiation of the organic polymers in the presence of hydrogen peroxide resulted in their discoloration.[29] The photobleaching reactions were followed using UV-Vis spectroscopy. Figure 2 shows the absorption spectra of polymer **6c** in the presence of excess hydrogen peroxide following different irradiation times. It can be seen that the absorption of the polymer decreased as the length of time increased, and after 90 minutes of photolysis, the polymer solution was colorless.

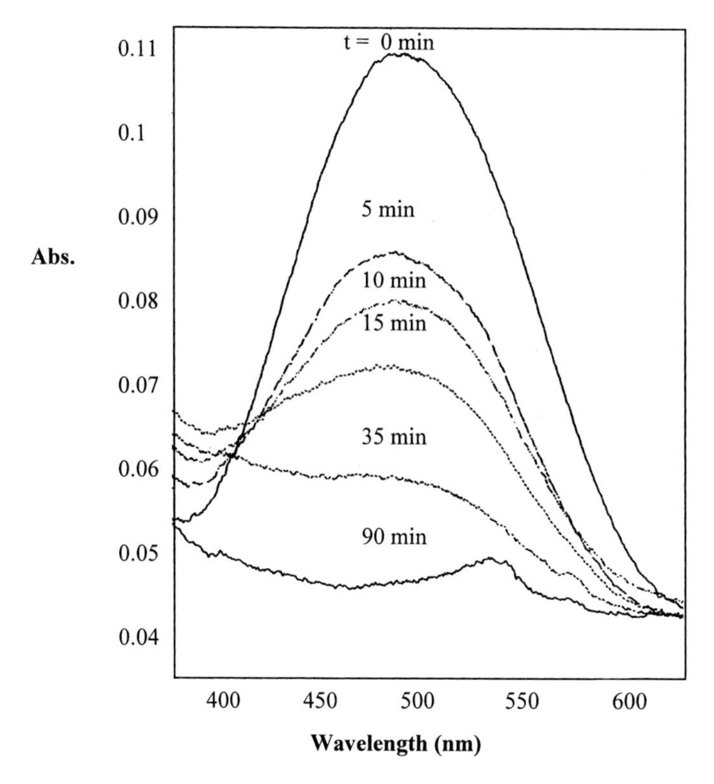

Figure 2. Absorption spectra following irradiation of a H_2O_2 solution of **6c** with a 300 nm light source.

Polynorbornenes with Azobenzene Side Chains

We have also been interested in synthesizing azobenzene-substituted polymers by ring-opening metathesis polymerization of substituted norbornene monomers.[30] The synthesis of these monomers was accomplished via the reaction of azobenzene-substituted arene complexes containing terminal hydroxyl groups with 5-norbornene-2-carboxylic acid. These organoiron monomers were subjected to ROMP using the Grubbs catalyst in dichloromethane. Scheme 3 shows the synthesis of some of the ether and thioether complexes substituted with azobenzene chromophores that we have examined.

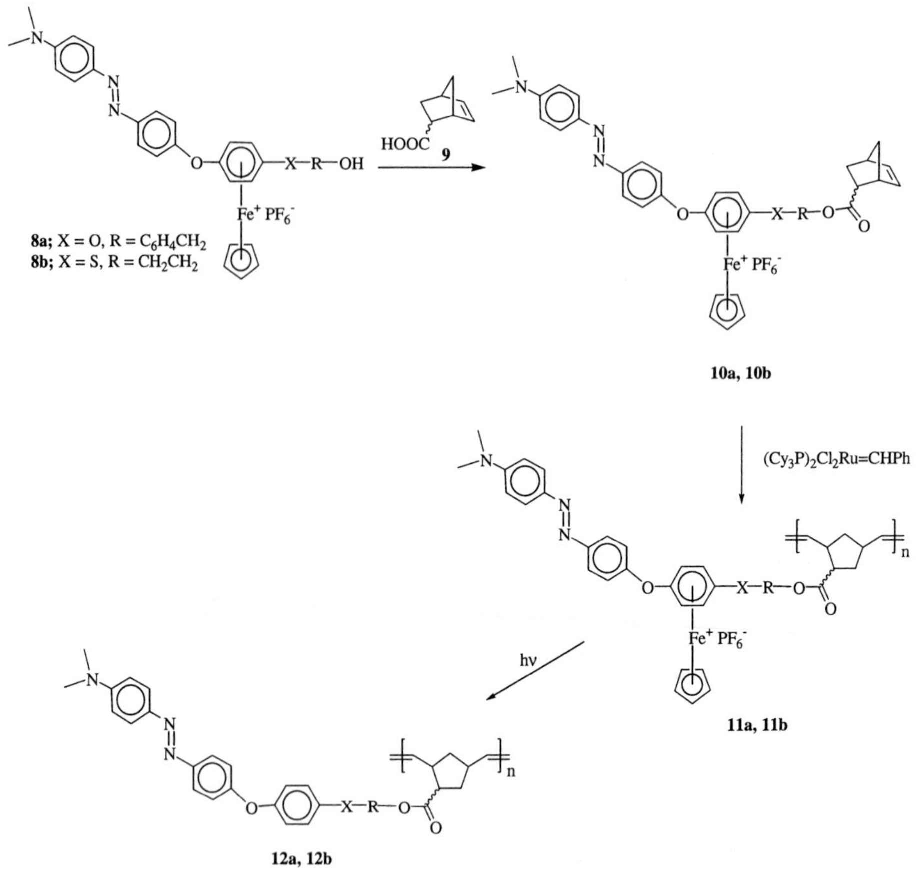

HOOC 9

$(Cy_3P)_2Cl_2Ru{=}CHPh$

hν

8a; X = O, R = $C_6H_4CH_2$
8b; X = S, R = CH_2CH_2

10a, 10b

11a, 11b

12a, 12b

Scheme 3

All of the polynorbornenes were orange solids that had λ_{max} values between 420 and 430 nm in DMF. Figure 3 shows the visible spectrum of the organoiron polynorbornene **13** in DMF and DMF/HCl solutions. Protonation of the azo groups by the addition of HCl to these solutions caused a red shift of about 90 nm.[30]

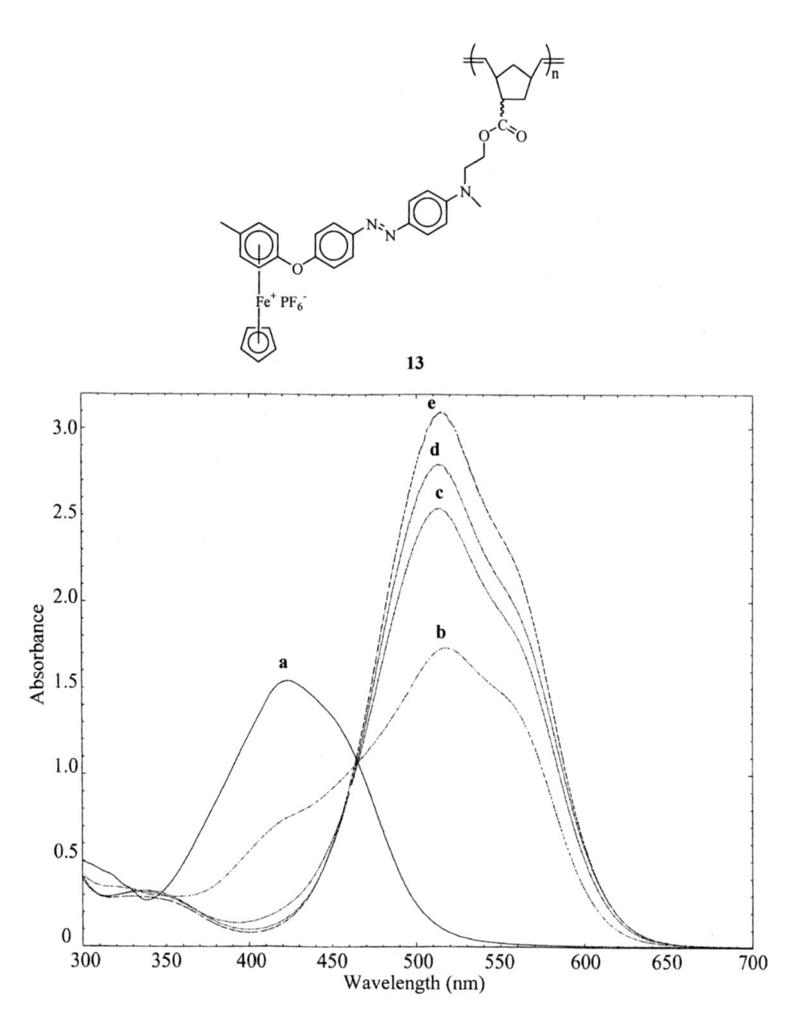

Figure 3. UV-Vis spectra of **13** in) DMF, b) DMF/10% HCl, c) DMF/30% HCl, d) DMF/50% HCl and e) DMF/100% HCl. (Reproduced from, “A New Class of Cationic Organoiron Polynorbornenes Containing Azo Dyes”, A.S. Abd-El-Aziz, R.M. Okasha, T.H. Afifi, E.K. Todd, *Macromol. Chem. Phys.*, **204**, 555 (2003) Copyright, © [2003] Wiley Periodicals, Inc.).

It was found that the cationic iron centers pendent to the side chains of these polynorbornenes underwent reversible reduction processes between –1.2 and –1.4 V versus Ag/AgCl. The cyclic voltammogram of polymer **11a** obtained in propylene carbonate containing tetrabutylammonium perchlorate (0.1 M) as the supporting electrolyte is shown in Figure 4. The cathodic and anodic peak potentials were measured at –1.4 and –1.1 V, respectively, and the half wave potential was calculated to be –1.25 V.

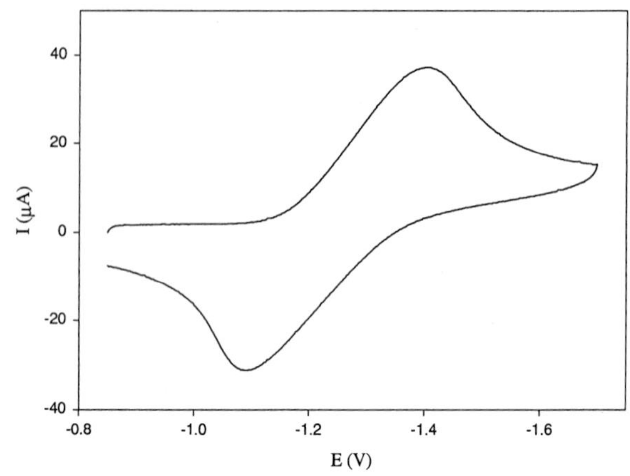

Figure 4. Cyclic voltammogram of 0.002 M polymer **11a** obtained in propylene carbonate at -40 °C using a scan rate of 0.5 V/s.

The thermal properties of the polynorbornenes were analyzed using thermogravimetric analysis and differentials scanning calorimetry. The metallated polymers underwent weight losses starting between 225 and 231 °C, which were assigned to degradation of the cyclopentadienyliron moieties and to partial decomposition of the polymer side chains. Second weight losses that began between 400 and 450 °C were caused by decomposition of the polymer backbones. Polynorbornenes with cationic cyclopentadienyliron moieties pendent to their side chains had T_g values that were much higher than their organic counterparts. For example, the organoiron polymer **11a** had a T_g at 178 °C, while its organic analog **12a** had a T_g at 99 °C. This large difference in glass transition temperatures was attributed to the bulkiness and polarity of the cyclopentadienyliron cations.

Conclusions

A number of different classes of cyclopentadienyliron-coordinated polymers functionalized with azobenzene moieties have been synthesized. These polymers were prepared either via metal-mediated nucleophilic aromatic substitution reactions of chloroarene complexes or via ring-opening metathesis polymerization of organoiron norbornenes. The colors of these organoiron polymers could be tuned by changing the functional groups on the azobenzene chromophores. Photolytic demetallation reactions allowed for the isolation of the corresponding organic polymers. These polymers could be bleached through irradiation in the presence of hydrogen peroxide.

Acknowledgements

Financial support provided by the Natural Sciences and Research Council of Canada (NSERC) and Manitoba Hydro are gratefully acknowledged. R.M.O. would also like to thank the Department of Chemistry, University of Manitoba.

[1] A. S. Abd-El-Aziz, *Macromol. Rapid Commun.* **2002**, *23*, 995.
[2] A. S. Abd-El-Aziz, *Coord. Chem. Rev.* **2002**, *233-234*, 177.
[3] R. D. A. Hudson, *J. Organomet. Chem.* **2001**, *637-639*, 47.
[4] P. Nguyen, P. Gomez-Elipe, I. Manners, *Chem. Rev.* **1999**, *99*, 1515.
[5] A. S. Abd-El-Aziz, S. Bernardin, *Coord. Chem. Rev.* **2000**, *203*, 219.
[6] A.S. Abd-El-Aziz, C.R. de Denus, E.K. Todd, S.A. Bernardin, *Macromolecules* **2000**, *13*, 5000.
[7] A. S. Abd-El-Aziz, L. May, A. L. Edel, *Macromol. Rapid Commun.* **2000**, *21,* 598.
[8] A. S. Abd-El-Aziz, A. L. Edel, L. May, K. M. Epp, H. M. Hutton, *Can. J. Chem.* **1999**, *77*, 1797.
[9] A. S. Abd-El-Aziz, L. J. May, J. A. Hurd, R. M. Okasha, *J. Polym. Sci., Part A: Polym. Chem.* **2001**, *39*, 2716.
[10] A. S. Abd-El-Aziz, R. M. Okasha, J. Hurd, E. K. Todd, *Polym. Mater. Sci. Eng.* **2002**, *86(1)*, 91.
[11] A. S. Abd-El-Aziz, E. K. Todd, G. Z. Ma, J. DiMartino, *J. Inorg. Organomet. Polym.* **2000**, *10*, 265.
[12] A. S. Abd-El-Aziz, E. K. Todd, *Polym. Mater. Sci. Eng.* **2002**, *86(1)*, 103.
[13] A. S. Abd-El-Aziz, E. K. Todd, G. Z. Ma, *J. Polym. Sci., Part A: Polym. Chem.* **2001**, *39*, 1216.
[14] A. S. Abd-El-Aziz, E. K. Todd, R. M. Okasha, T. E. Wood, *Macromol. Rapid Commun.* **2002**, *23*, 743.
[15] A. S. Abd-El-Aziz, C. R. de Denus, K. M. Epp, S. Smith, R. J. Jaeger, D. T. Pierce, *Can. J. Chem.* **1996**, *74*, 650.
[16] A. S. Abd-El-Aziz, T. C. Corkery, E. K. Todd, T. H. Afifi, G. Z. Ma *J. Inorg. Organomet. Polym.* In press.
[17] A. Natansohn, P. Rochon, *Chem. Rev.* **2002**, *102*, 4139.
[18] M. Hasegawa, T. Ikawa, M. Tsuchimori, O. Watanabe, *J. Appl. Polym. Sci.* **2002**, *86*, 17.
[19] L. A. Howe, G. D. Jaycox, *J. Polym. Sci., Part A: Polym. Chem.* **1998**, *36*, 2827.
[20] K. Huang, H. Qiu, M. Wan, *Macromolecules* **2002**, *35*, 8653.
[21] S. H. Kang, H.-D. Shin, C. H. Oh, D. H. Choi, K. H. Park, *Bull. Korean Chem. Soc.* **2002**, *23*, 957.
[22] P. Uznanski, J. Pecherz, *J. Appl. Polym. Sci.* **2002**, *86*, 1456.
[23] C. Samyn, T. Verbiest, A. Persoons, *Macromol. Rapid Commun.* **2000**, *21*, 1.
[24] G. Iftime, F. L. Labarthet, A. Natansohn, P. Rochon, K. Murti, *Chem. Mater.* **2002**, *14*, 168.
[25] G. H. Kim, C. D. Keunm S. J. Kim, L. S. Park, *J. Polym. Sci., Part A: Polym. Chem.* **1999**, *37*, 3715.
[26] K. Tawa, K. Kamada, K. Kiyohara, K. Ohta, D. Yasumatsu, Z. Sekkat, S. Kawata, *Macromolecules* **2001**, *34*, 8232.
[27] L. Wu, X. Tuo, H. Cheng, Z. Chen, X. Wang, *Macromolecules* **2001**, *34*, 8005.
[28] X.-H. Liu, D. W. Bruce, I. Manners, *Chem. Commun.* **1997**, 289.
[29] A. S. Abd-El-Aziz, T. H Afifi, W. R. Budakowski, K. J. Friesen, E. K. Todd, *Macromolecules* **2002**, *35*, 8929.
[30] A. S. Abd-El-Aziz, R. M. Okasha, T. H. Afifi, E. K. Todd, *Macromol. Chem. Phys.* **2003**, 204, 555.

Self-Assembly of Metal-Organometallic Coordination Networks

*Moonhyun Oh, Gene B. Carpenter, Dwight A. Sweigart**

Department of Chemistry, Brown University, Providence, Rhode Island 02912, USA

Summary: A range of neutral metal-organometallic coordination networks (MOMNs) containing both backbone and pendant metal sites have been synthesized and characterized. These materials consist of metal ions or metal ion clusters as nodes that are linked by the bifunctional "organometalloligand" (η^4-benzoquinone)$Mn(CO)_3^-$, which binds through the quinone oxygen atoms. The resulting MOMNs can be rationally designed to a significant extent based upon a knowledge of the electronic and geometrical requirements of the metal ion nodes, the solvent, organometalloligand substituents, and the presence or absence of organic spacers. An impressive range of architectures have been accessed in this manner, suggesting that the use of π-organometallics in coordination directed self-assembly holds much promise.

Keywords: coordination networks; organometallics; quinone complexes; self-assembly; supramolecular structures

Introduction

Supramolecular chemistry involves the study of finite and infinite (polymeric) nanostructures consisting of self-assembled modular units.[1] Implicit in the formation of well defined nanostructures is the ability of the modular components to participate in highly efficient molecular recognition events. These involve a variety of possible intermodular interactions, ranging in energy from very weak to fairly strong, with the most common ones being hydrogen-bonding and metal-ligand bond formation. The driving force behind such work is the desire to fabricate new functional solids that have applications in areas such as magnetics, optics, molecular recognition, sensing, catalysis, separations, guest-host chemistry, etc.

Metal-coordination-directed formation of metal-organic or "coordination" supramolecular networks (MONs) have received much recent attention. Both discrete (finite)[2] and infinite[3-5] MONs have been investigated. The nodes in the majority of infinite MONs reported are metal ions, which are bonded to relatively simple multifunctional organic ligands that serve as spacers

© 2003 WILEY-VCH Verlag GmbH & KGaA, Weinheim CCC 1022-1360/00/$ 17,50+.50/0

within the network. In more complex systems, in particular ones that most frequently afford porous structures, the nodes consist of metal carboxylate clusters as secondary building units (SBUs).[6] MON coordination polymers hold much promise, albeit this is tempered somewhat by a generally poor ability to predict structure (and hence function). Difficulty in predicting structure is often due to the existence of supramolecular isomers of similar energies. Even when a particular structural architecture can be pre-designed with some confidence, the product often consists of several interpenetrating networks. Overall, there is a need to develop a better understanding of the factors controlling structure, so that resources can be invested in making new materials with desired properties.

Herein we are concerned with MONs of a special type, namely, metal-*organometallic* coordination networks (MOMNs). The essential difference between MONs and MOMNs is illustrated in Figure 1. In the former, the spacer molecules connecting the metal nodes consist of simple organic ligands, whereas in the latter the spacers consist of *organometallic* π*-complexes* that function as "organometalloligands".

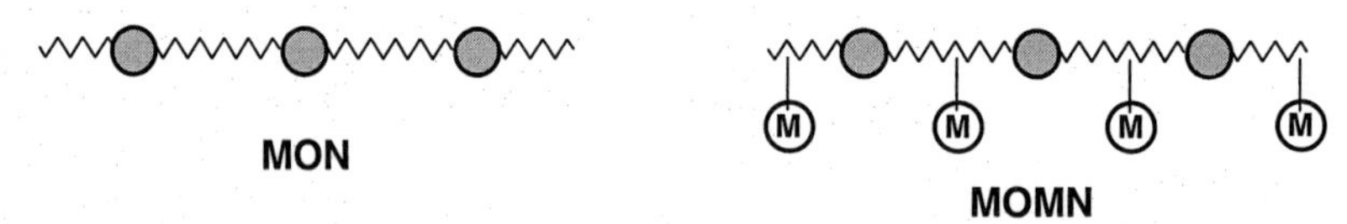

Figure 1. One-dimensional coordination networks with organic and organometallic spacers.

It can be seen that MOMNs may exhibit metal-metal interactions not available in corresponding MONs. Of equal significance is the fact that organometalloligands offer the opportunity to incorporate truly novel types of spacers, especially ones with pendant metal atoms, into coordination networks. We anticipate that MOMNs would most generally be synthesized directly from intact "organometalloligand" spacers, and not by "coordinating" metal fragments to the organic spacers in intact MONs. Overall, it seems that the possibility of constructing novel metal-organometallic coordination networks that have interesting magnetic, electronic and catalytic properties is good.

Substituted hydroquinones are of fundamental importance in mediating electron and proton transfers in biological reactions. Our work with organometallic coordination networks is related to the reversible reactions of hydroquinone to afford semiquinone and quinone oxidation products, as indicated in Scheme 1. The members of this series have a propensity[7] to σ-bond to metals through the oxygen atoms rather than to π-bond through the carbocyclic ring. There are, however, several examples[8,9] of stable complexes containing a hydroquinone *π-bonded* to a transition metal, suggesting that in general such species can be sufficiently stable for isolation and utulization. One would may anticipate that the attachment of a metal fragment to the π-system would facilitate proton and, perhaps, electron transfer.

Scheme 1

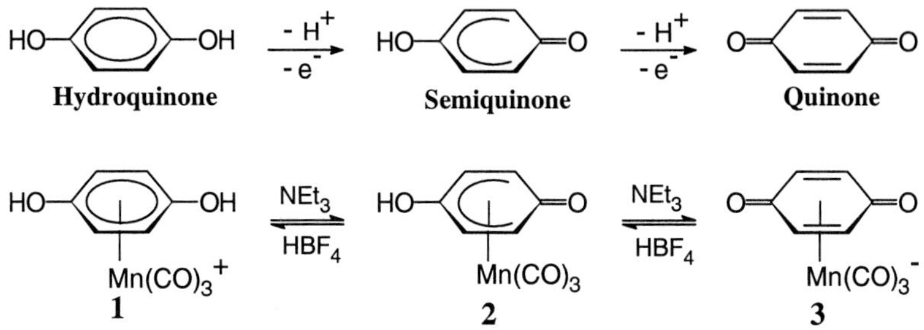

Hydroquinone — $-H^+$, $-e^-$ → **Semiquinone** — $-H^+$, $-e^-$ → **Quinone**

$Mn(CO)_3^+$ **1** ⇌ (NEt_3 / HBF_4) $Mn(CO)_3$ **2** ⇌ (NEt_3 / HBF_4) $Mn(CO)_3^-$ **3**

By using (η^6-naphthalene)$Mn(CO)_3^+$ as a manganese tricarbonyl transfer reagent (MTT), we prepared in high yield the thermally stable π-complex (η^6-hydroquinone)$Mn(CO)_3^+$ (**1**), as well as the catechol and resorcinol analogues.[9] The key to the utility of **1** in the construction of MOMNs is the great ease with which the -OH protons are removed to give η^5-semiquinoine and η^4-quinone complexes according to Scheme 1. Although the sequence **1** ↔ **2** ↔ **3** in Scheme 1 can be viewed as simple deprotonations, the analogy to the proton *and electron* transfers occurring with free quinones is evident if it is considered that each proton loss is accompanied by electron transfer to the metal, which acts as an internal oxidizing agent or electron sink. The neutral η^5-semiquinone complex **2** was structurally characterized and found[10] to exist in linear polymeric arrays with the structure dictated by strong intermolecular hydrogen bonding.[11] By comparison, and as might be expected, the catechol analogue exists as discrete hydrogen bonded dimers.

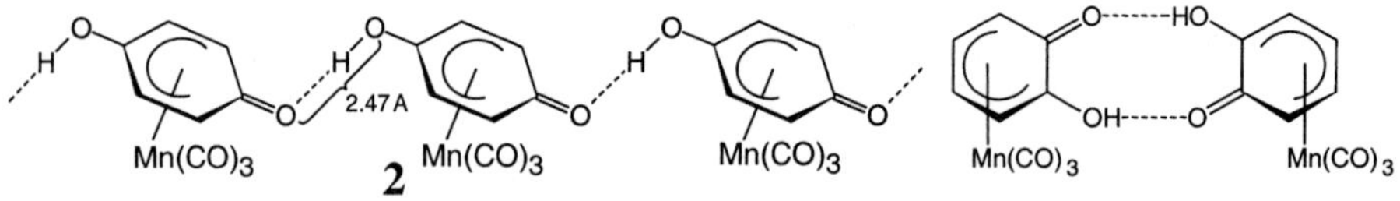

Quinone Organometalloligands

The anionic *para*- and *ortho*-benzoquinone manganese tricarbonyl complexes obtained from double deprotonation of the corresponding hydroquinone and catechol can function as bifunctional ligands ("organometalloligands") in the presence of appropriate metal ions by σ-bonding through the oxygen atoms. These ligands are called *p*-QMTC and *o*-QMTC, respectively. The chemistry of the former is described in some detail in the next section, while that of the latter is briefly summarized in this section. The ligand *o*-QMTC readily reacts with divalent metal ions to afford monomeric complexes **4** (M = Mn, Cd, Co; L = neutral ligand). Complexes containing deprotonated catechol sigma-bonded to metals via the oxygens have long been known[7,12] to exhibit redox or valence tautomerization due to facile electron transfer between the metal and the catechol ligand. There is evidence that this effect is enhanced in the π-bonded complex **4**, with the $Mn(CO)_3$ moiety acting as an electron sink. The particular tautomer that prevails - the η^4-quinone (**4**), η^5-semiquinone (**5**), or even the η^6-catecholate - appears to be a sensitive function of the axial ligand L. This conclusion is based on ν_{CO} bands that are *unusually* dependent on the electron donating properties of the axial ligand. It is possible that the facile electron transfer behavior shown by π-bonded quinone complexes such as **4** may have useful applications in catalysis. This would be especially true when the substrate(s) can partake of the self-adjusting electronic environment by binding directly to the σ-bonded metal.

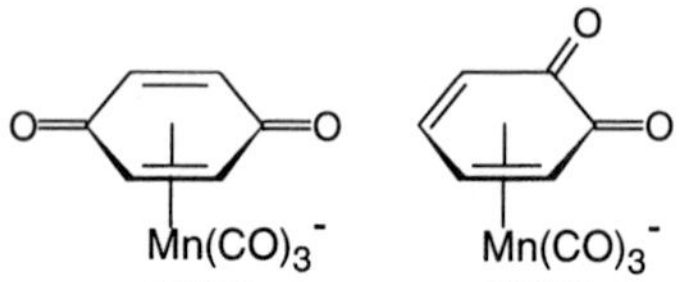

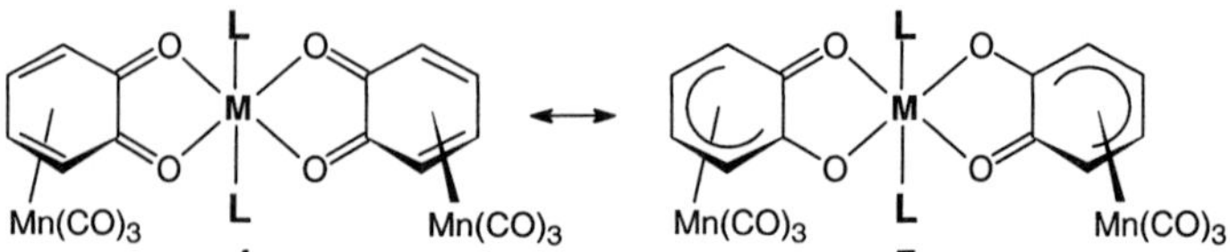

Coordination networks can be made from **4** by using organic spacers as axial ligands, but this chemistry is yet to be developed. The reaction of *o*-QMTC with divalent metal ions in the presence of 2,2-bipyridine affords stable complexes M(*o*-QMTC)$_2$(L-L) (M = Mn, Co, Cd). In the solid state these complexes self-assemble into two-dimensional supramolecular networks, the structure of which are determined by π-π stacking and interdigitation of the bipyridine ligands and by a pairwise π-π stacking of one of the two *o*-benzoquinone ligands in each monomeric unit to generate dimeric units.[13]

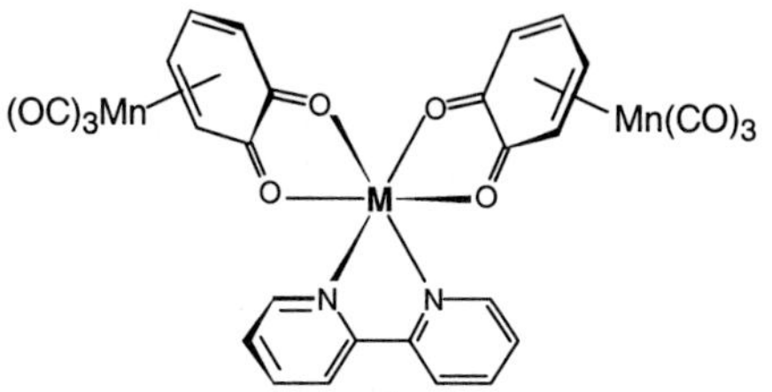

(OC)$_3$Mn Mn(CO)$_3$ O O O M O N N

6

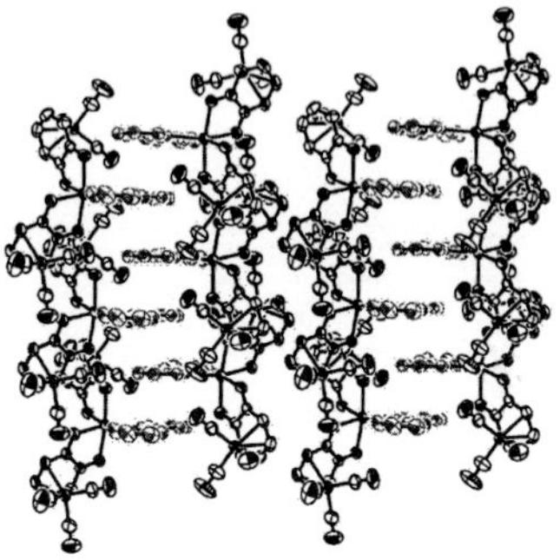

Quinonoid Metal-Organometallic Coordination Networks

The geometry of the *p*-QMTC complex suggests that it could function as a spacer in the formation of coordnation networks. Indeed, it was found[14] that *p*-QMTC readily binds to metal ions through the oxygen atoms. The key to the synthesis of MOMNs having *p*-QMTC spacers was finding the right solvent and temperature conditions for the self-assembly process to occur without premature precipitation of oligomers. Warm DMSO turned out to be an excellent medium for the high yield synthesis of *crystalline* polymers. Initially, it was found[14] that *p*-QMTC combined with the metal cations to form the 1-dimensional "string" MOMN **7** (M = Mn, Co, Ni, Cd). The crystal structure of the neutral Co quinonoid polymer with pyridine axial ligands is illustrated. It was subsequently found that the *p*-QMTC organometalloligand can be used for the synthesis of an diverse array of 1- 2- and 3-dimensional polymers, with exact architecture obtained depending on the geometrical requirements of the added metal ion, the solvent, and on the presence of added organic ligands that function as additional spacers.

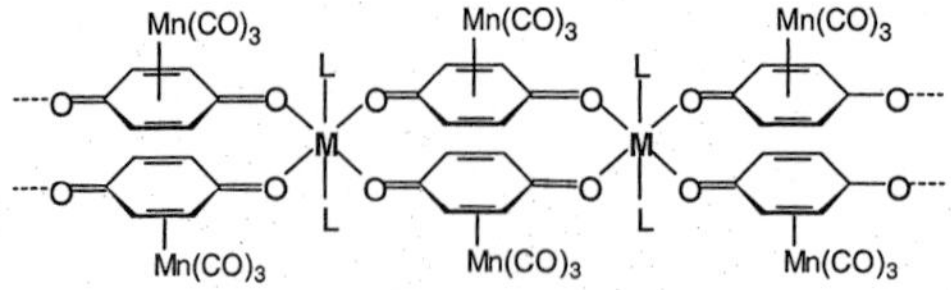

1-Dimensional MOMN 7

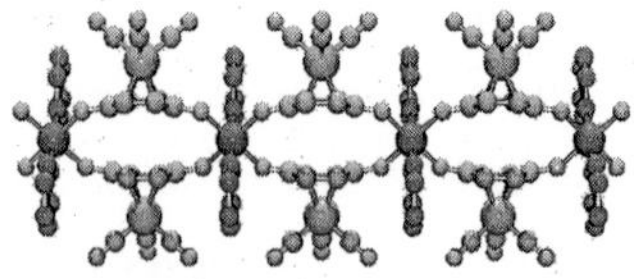

M = cobalt and L = pyridine

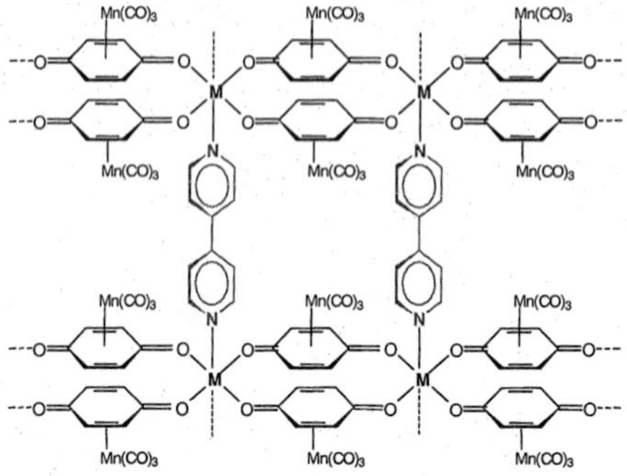

2-Dimensional MOMN 8

M = Mn(II) or Ni(II)

For example, replacement of the unidentate axial ligand in **7** with 4,4'-bipyridine generates the 2-D polymer **8**. The rectangular grids in **8** are filled with solvent DMSO, which can be easily removed thermally *without* losing crystallinity.[15] Isomeric with the 1-D string motif formed by combination of metal ion nodes and *p*-QMTC spacers is the 2-D grid structure shown in Figure. 2. It was found[15] that the prevailing isomer, 1-D or 2-D can be *predetermined* by controlling the metal ion concentration. At higher concentrations, the 2-D pseudo-planar quinonoid network is favored because *p*-QMTC binding sites are "trapped" before assembly to the 1-D string can occur. When 4,4'-bipyridine is present, the 2-D grids then link together to generate 3-D MOMNs (**9**) containing rectangular pores (12 x 6 Å) filled with DMSO solvent. Polymerization of *p*-QMTC in the presence of a bridging metal that prefers tetrahedral rather than octahedral geometry would be expected to give a nonlinear product quite different from **7-9**. This possibility was explored using Zn(II), and the result obtained from DMSO solvent was the 3-D polymer **10.** The geometry around the zinc is indeed tetrahedral and the solid state structure consists of two interpenetrating diamondoid networks, as illustrated in Figure 3.[14]

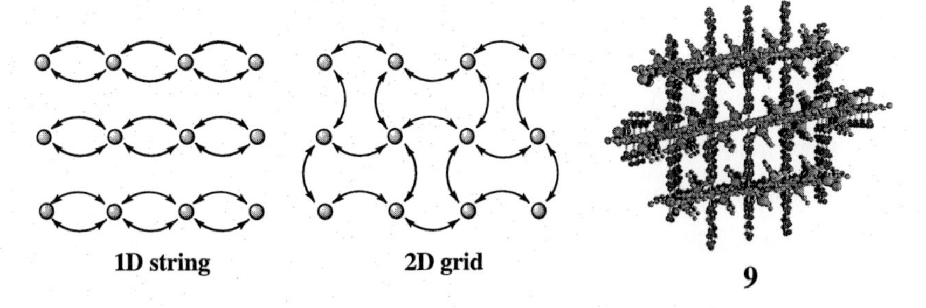

Figure 2. Supramolecular isomers with *p*-QMTC spacers (double headed arrows).

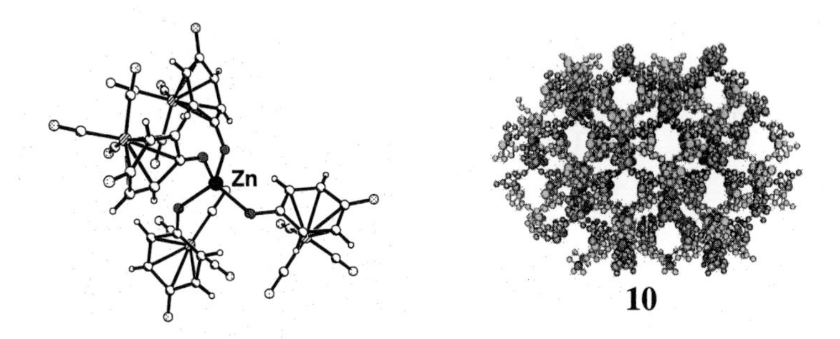

Figure 3. Diamondoid MOMN **10** with tetrahedral Zn(II) nodes.

In the reaction of *p*-QMTC with Mn^{2+} and Co^{2+}, it was found that changing from DMSO to the less coordinating solvent MeOH leads to the formation of 3-D diamondoid structures in which the metal ion nodes adopt tetrahedral rather than octahedral[16] coordination. The polymers $[Co(QMTC)_2]_\infty$ and $[Mn(QMTC)_2]_\infty$, obtained from MeOH solvent, possess overall structures virtually identical to that found for $[Zn(QMTC)_2]_\infty$ shown in Figure 3. It is concluded that with Mn^{2+} and Co^{2+}, switching the solvent from DMSO to MeOH results in a fundamental change in architecture that is triggered by a change from octahedral to tetrahedral coordination at the divalent metal ion node. This change in geometry, which is likely due to the generally weaker coordinating ability of MeOH compared to DMSO, suggests that solvent variation may be generally useful for controlling coordination network architecture. The introduction of methyl groups at the 2- and 3-positions of the benzoquinone ring was found to greatly influence the manner in which the quinone oxygen lone pairs bind to the metal nodes. Crystal structures of coordination polymers $[M(2,3\text{-}Me_2QMTC)_2]_\infty$ ($M = Mn^{2+}$, Zn^{2+}) revealed that it is the lone pairs projecting away from the methyls that bind to the nodes, resulting in the *cis*-arrangement illustrated in Figure 4.[16] Presumably, this bonding pattern is dictated by the steric requirements of the methyl groups. The stereochemical switch from the *trans* to *cis* results in a concomitant change in the polymer architecture from a two-fold interpenetrated 3-D diamondoid structure to a noninterpenetrated ruffled 2-D rhombohedral grid. In spite of the *trans* to *cis* and diamondoid to rhombohedral structural changes, the metal nodes in $[M(2,3\text{-}Me_2QMTC)_2]_\infty$ remain tetrahedrally bonded to the quinone oxygens.

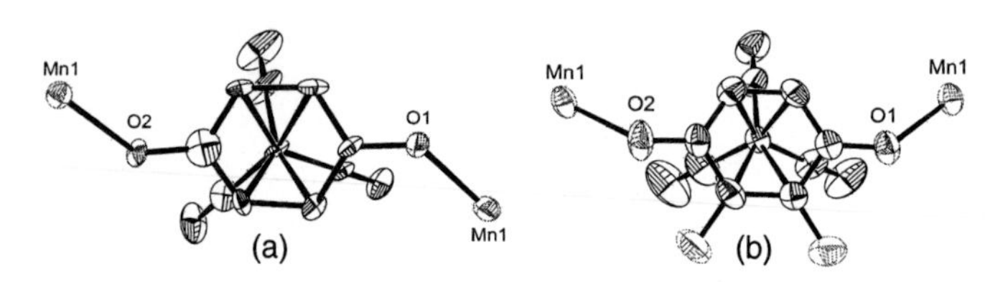

Figure 4. Bonding of the quinone oxygens to Mn(II) nodes in (a) $[Mn(p\text{-}QMTC)_2]_\infty$ and (b) $[Mn(2,3\text{-}Me_2QMTC)_2]_\infty$.

The reaction of *p*-QMTC with Mn^{2+} in a DMSO/MeOH solvent mixture produced a 3-D diamonoid polymer containing dimanganese secondary building units.[17] Mixing *p*-QMTC and $Cu(OAc)_2$ in MeOH generated a MOMN possessing an unprecedented *3-dimensional* "brick wall" structure.[18] Figure 5 illustrates how a 3-D brick wall can be generated starting with "T" shaped nodes that connect to form the familiar 2-D brick wall.[19] Extension of the brick wall from 2-D (Figure 5a) to 3-D (Figure 5b) involves converting the T-shaped node to a square pyramidal node possessing fivefold connectivity. The MOMN formed from *p*-QMTC and $Cu(OAc)_2$ consists of bimetallic SBUs of formula $Cu_2(\mu\text{-}CH_3CO_2)^{3+}$, which are linked via *p*-QMTC spacers with the square pyramidal geometry shown in Figure 5c. The resultant 3-D structure, which has the molecular formula $[Cu_2(p\text{-}QMTC)_3(\mu\text{-}CH_3CO_2)]_\infty$, constitutes a previously unknown extended (3-D) brick wall architecture.

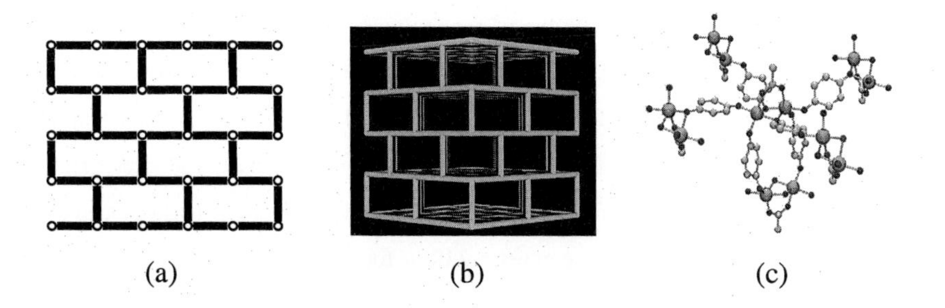

Figure 5. (a) 2-D and (b) 3-D brick wall motifs and (c) square pyramidal connectivity in MOMN $[Cu_2(QMTC)_3(\mu\text{-}CH_3CO_2)]_\infty$.

When *p*-QMTC and M^{2+} (M = Cd, Mn) were reacted in the presence of 2,2'-bipyridine, a most interesting MOMN was isolated.[20] Because both nitrogens in 2,2'-bipyridine are constrained to bind to the same metal node, two structures can result - a nanotube, which is yet to be observed and a 1-D zigzag polymer (Scheme 2). The crystal structure of the zigzag MOMNs revealed that the individual 1-D polymer units interdigitate via π-π stacking of the 2,2'-bipyridine ligands. While π-π stacking is a well known phenonomen,[21] the structure in Scheme 2 is particularly interesting for two reasons. First, the interstrand metal nodes are linked via π-π stacking of the bipyridine rings and this is likely to influence the temperature at which magnetic ordering occurs when the nodes are paramagnetic (e.g., Mn^{2+}).[22] The second interesting feature of the π-stacked polymer is the inclusion of two "free" bipyridine molecules between each pair of coordinated bipyridines, resulting in a continuous π-stacking along the entire length of the polymer. It is possible that the "π-pocket" in this or related MOMNs can be used to bind π-molecules of appropriate size other than free bipyridine, and thereby have a number of potential applications.

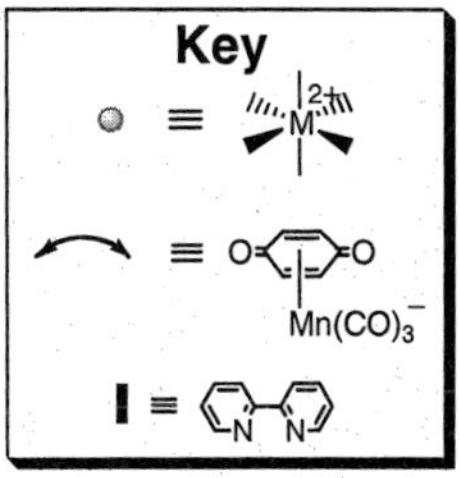

Scheme 2. Formation of π-stacked MOMNs.

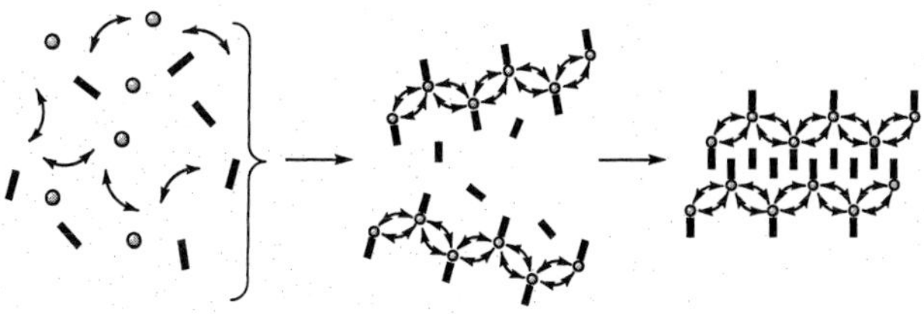

When Mn^{2+} in MeOH solvent reacts with the η^5-semiquinone complex **2** instead of *p*-QMTC, a 1-D zigzag polymer is generated that contains two semiquinone organometalloligands bound in a *cis* manner to each Mn^{2+} node.[20] The resultant MONN (**11**) contains semiquinone moieties protruding from the central core, rather like antennae, that can be used for interstrand π-π stacking or binding to external electrophilic sites. An equally interesting MOMN containing dicadmium SBUs with protruding quinone moieties (**12**) is formed from **2** and Cd^{2+} in MeOH.

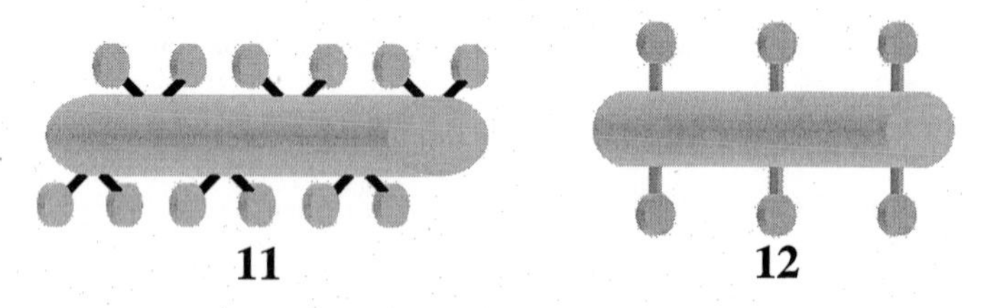

Figure 6. Cartoon representations of $\{Mn(p\text{-QMTC})_2[(\eta^5\text{-semiquinone})Mn(CO)_3]_2\}_\infty$ (**11**) and $[Cd_2(p\text{-QMTC})_4(MeOH)_4]_\infty$ (**12**).

Conclusion

The use of organometallics in coordination directed self-assembly, holds much promise for the synthesis of functional materials with useful applications. We have demonstrated the construction of supramolecular metal-organometallic networks (MOMNs) using π-bonded quinonoid complexes of manganese as "organometalloligands". Preliminary results indicate that the redox-active quinone-based complexes permit the construction of an impressive range of architectures. Further, it appears that the available architectures can be *rationally designed* based upon the coordination number, oxidation state, and geometrical requirements of the metallic nodes that link the organometalloligands.

Acknowledgment

Acknowledgment is made to donors of the Petroleum Research Fund, administered by the ACS, for support of this research.

[1] J. M. Lehn, "Supramolecular Chemistry: Concepts and Perspectives", VCH, Weinheim, **1995**; J. W. Steed, J. L. Atwood, "Supramolecular Chemistry", Wiley, Chichester, **2000**.
[2] D. Philp, J. F. Stoddart, *Angew. Chem. Int. Ed.* **1996**, *35*, 1154; S. Leininger, B. Olenyuk, P. J. Stang, *Chem. Rev.* **2000**, *100*, 853; B. J. Holliday, C. A. Mirkin, *Angew. Chem. Int. Ed.* **2001**, *40*, 2022; M. Fujita, *Chem. Soc. Rev.* **1998**, *27*, 417; L. R. MacGillivray, R. H. Groeneman, J. L. Atwood, *J. Am. Chem. Soc. 1998*, **120**, 2676; Y. L. Cho, H. Uh, S.-Y. Chang, H.-Y. Chang, M.-G. Choi, I. Shin and K.-S. Jeong, *J. Am. Chem. Soc.* **2001**, *123*, 1258.
[3] S. R. Batten, R. Robson, *Angew. Chem. Int. Ed.* **1998**, *37*, 1460; R. Robson, *J. Chem. Soc., Dalton Trans.* **2000**, 3735; M. Eddaoudi, D. B. Moler, H. Li, B. Chen, T. M. Reineke, M. O'Keeffe, O. M. Yaghi, *Acc. Chem. Res.* **2001**, *34*, 319; O. R. Evans, W. Lin, *Acc. Chem. Res.* **2002**, *35*, 511.
[4] B. Moulton, M. J. Zaworotko, *Chem. Rev.* **2001**, *101*, 1629.
[5] S. A. Bourne, J. Lu, A. Mondal, B. Moulton, M. J. Zaworotko, *Angew. Chem. Int. Ed.* **2001**, *40*, 2111; H. Abourahma, B. Moulton, V. Kravtsov, M. J. Zaworotko, *J. Am. Chem. Soc.* **2002**, *124*, 9990; Y.-B. Dong, R. C. Layland, N. G. Pschirer, M. D. Smith, U. H. F. Bunz, H.-C. zur Loye, *Chem. Mater.* **1999**, *11*, 1413; H. J. Choi, M. P. Suh, *J. Am. Chem. Soc.* **1998**, *120*, 10622; K. Biradha, M. Fujita, *J. Chem. Soc., Dalton Trans.* **2000**, 3805; K. Biradha, M. Fujita, *Angew. Chem. Int. Ed.* **2002**, *41*, 3392; Y. Cui, O. R. Evans, H. L. Ngo, P. S. White and W. Lin, *Angew. Chem. Int. Ed.* **2002**, *41*, 1159; O. R. Evans, W. Lin, *Chem. Mater.* **2001**, *13*, 3009; N. L. Rosi, M. Eddaoudi, J. Kim, M. O'Keeffe, O. M. Yaghi, *Angew. Chem. Int. Ed.* **2002**, *41*, 284.
[6] J. Kim, B. Chen, T. M. Reineke, H. Li, M. Eddaoudi, D. B. Moler, M. O'Keeffe, O. M. Yaghi, *J. Am. Chem. Soc.* **2001**, *123*, 8239; M. Eddaoudi, H. Li, O. M. Yaghi, *J. Am. Chem. Soc.* **2000**, *122*, 1391, B. Chen, M. Eddaoudi, S. T. Hyde, M. O'Keeffe, O. M. Yaghi, *Science* **2001**, *291*, 1021.
[7] C. G. Pierpont, *Coord. Chem. Rev.* **2001**, *216*, 99; A. Caneschi, A. Dei, D. Gatteschi, L. Sorace, K. Vostrikova, *Angew. Chem.* **2000**, *39*, 246; C. G. Pierpont, C. W. Langi, *Prog. Inorg. Chem.* **1994**, *41*, 331; C. G. Pierpont, *Coord. Chem. Rev.* **2001**, *216-217*, 99.
[8] Y.-S. Huang, S. Sabo-Etienne, X.-D. He, B. Chaudret, *Organometallics* **1992**, *11*, 3031; J. Le Bras, H. Amouri, J. Vaissermann, *Organometallics*, **1998**, *17*, 1116; H. Schumann, A. M. Arif, T. G. Richmond, *Polyhedron* **1990**, *9*, 1677; G. Fairhurst, C. White, *J. Chem. Soc., Dalton Trans.* **1979**, 1531; H. Amouri, J. Le Bras, *Acc. Chem. Res.* **2002**, *35*, 501.
[9] S. Sun, G. B. Carpenter, D. A. Sweigart, *J. Organomet. Chem.* **1996**, *512*, 257.
[10] M. Oh, G. B. Carpenter, D. A. Sweigart, *Organometallics* **2002**, *21*, 1290.
[11] Y.-B. Dong, M. D. Smith, R. C. Layland, H.-C. zur Loye, *J. Chem. Soc., Dalton Trans.* **2000**, 775; D. Braga, F. Grepioni, G. R. Desiraju, *Chem. Rev.* **1998**, *98*, 1375; D. Braga, F. Grepioni, *Coord. Chem. Rev.* **1999**, *183*, 19; D. Braga, L. Maini, F. Grepioni, C. Elschenbroich, F. Paganelli, O. Schiemann, *Organometallics* **2001**, *20*, 1875; C. A. S. Fraser, H. A. Jenkins, M. C. Jennings, R. J. Puddephatt, *Organometallics* **2000**, *19*, 1635, T. Steiner, *Angew. Chem. Int. Ed.* **2002**, *41*, 48.
[12] A. S. Attia, C. G. Pierpont, *Inorg. Chem.* **1998**, *37*, 3051; O.-S. Jung, D. H. Jo, Y.-A. Lee, Y. S. Sohn, C. G. Pierpont, *Inorg. Chem.* **1998**, *37*, 5875.
[13] M. Oh, G. B. Carpenter, D. A. Sweigart, *Organometallics*, in press.
[14] M. Oh, G. B. Carpenter, D. A. Sweigart, *Angew. Chem. Int. Ed.*, **2001**, *40*, 3191.
[15] M. Oh, G. B. Carpenter, D. A. Sweigart, *Angew. Chem. Int. Ed.*, **2002**, *41*, 3650.
[16] M. Oh, G. B. Carpenter, D. A. Sweigart, *J. Am. Chem. Soc.*, submitted.
[17] M. Oh, G. B. Carpenter, D. A. Sweigart, *Chem. Commun.* **2002**, 2168.
[18] M. Oh, G. B. Carpenter, D. A. Sweigart, *Angew. Chem. Int. Ed.*, in press.
[19] M. Fujita, Y. J. Kwon, O. Sasaki, K. Yamaguchi, K. Ogura, *J. Am. Chem. Soc.* **1998**, *120*, 10622; L. Carlucci,

G. Ciani, D. M. Proserpio, *New J. Chem.* **1998**, 1319; L. Carlucci, G. Ciani, D. M. Proserpio, *J. Chem. Soc., Dalton Trans.* **1999**, 1799. (f) M. Kondo, M. Shimamura, S. Noro, S. Minakoshi, A. Asami, K. Seki, S. Kitagawa, *Chem. Mater.* **2000**, *12*, 1288

[20] M. Oh, G. B. Carpenter, D. A. Sweigart, unpublished results.

[21] O. M. Yaghi, H. Li, T. L. Groy, *Inorg. Chem.* **1997**, *36*, 4292; K. V. Domasevitch, G. D. Enright, B. Moulton, M. J. Zaworotko, *J. Sol. State Chem.* **2000**, *152*, 280; C. A. Hunter, *Chem. Soc. Rev.* **1994**, 101; C. A. Hunter, X.-J. Lu, *J. Mol. Biol.* **1997**, *265*, 603; C. A. Hunter, K. R. Lawson, J. Perkins, C. J. Urch, *J. Chem. Soc., Perkin Trans.2*, **2001**, 651; L. J. Childs, N. W. Alcock, M. J. Hannon, *Angew. Chem. Int. Ed.* **2001**, *40*, 1079; A. Hori, A. Akasaka, K. Biradha, S. Sakamoto, K. Yamaguchi, M. Fujita, *Angew. Chem. Int. Ed.* **2002**, *41*, 3269.

[22] O. Kahn, *Acc. Chem. Res.* **2000**, *33*, 647; K. Barthelet, J. Marrot, D. Riou, G. Ferey, *Angew. Chem. Int. Ed.* **2002**, *41*, 281; E. Coronado, J. R. Galan-Mascaros, C. J. Gomez-Garcia, V. Laukhin, *Nature* **2000**, *408*, 447.

Macromol. Symp. **196**, *113–123 (2003)*

Electrochemical Investigations of Oligomers and Polymers Containing Ruthenium- and Iron-Arene Complexes

Christine R. de Denus,[1] Philip Baker,[1] Jaclyn Toner,[1] Sheila McKevitt,[1]*
Erin K. Todd,[2] Alaa S. Abd-El-Aziz[2]

[1]Hobart and William Smith Colleges, Department of Chemistry, Geneva, NY, 14456, USA
[2]The University of Winnipeg, Department of Chemistry, Winnipeg, MB, R3B 2E9, Canada
E-mail: dedenus@hws.edu

Summary: Cyclic voltammetry was employed to investigate the electrochemical behavior of numerous cyclopentadienyliron ($CpFe^+$) and pentamethyl-cyclopentadienylruthenium (Cp^*Ru^+) coordinated oligomers and polymers. The electrochemical behavior of the iron systems indicated the cyclopentadienyliron complexes had isolated redox centers and that changes in the reversibility of the redox couple occurred with changes in solvent and temperature. In contrast, the monometallic ruthenium systems showed large peak separations that suggested slow kinetics on the CV timescale. The cyclic voltammograms of the larger ruthenium-containing oligomers and polymers showed multiple redox steps indicating complex electrochemical behavior.

Keywords: arene complexes; electrochemistry; metal-polymer complexes; oligomers; polyethers

Introduction

Over the past decade, there has been considerable interest in the electrochemical properties and applications of organometallic oligomers and polymers.[1-8] Particular attention has been given to systems that contain arenes π-coordinated to cyclopentadienyliron moieties. The redox behavior of this class of materials has led to a thorough investigation of the electron-transfer reactions that these complexes undergo.[9-13] Much of this interest stems from the ability of these systems to exhibit mixed-valency[14, 15] and conductivity,[16] to act as electron reservoirs which initiate electron-transfer chain catalysis,[2] and to behave like molecular batteries[17] and metalloreceptors for anion recognition.[18]

Numerous electrochemical studies have been undertaken to explore the redox behavior of η^6-arene-η^5-cyclopentadienyliron complexes. The iron centers in these complexes undergo two

© 2003 WILEY-VCH Verlag GmbH & KGaA, Weinheim CCC 1022-1360/00/$ 17,50+.50/0

successive one-electron reductions (18/19 e^- and 19/20 e^-), with the reversibility of these couples being dependent on the solvent, working electrode, and temperature.[3, 9-11, 18, 19] Solodovnikov *et al* have also reported the formation of an extremely unstable 21e^- dianionic cyclopentadienyliron complex of naphthalene;[20] however, it is the neutral arene systems (19e^-) that are the most important electron reservoirs. This class of electron reservoir can be used as catalysts in the reduction of other species existing in solution via homogeneous charge transfer reactions.[1, 2, 19, 21] It has also been reported that the decoordination of unstable nineteen-electron complexes may take place by polyhapto ligand replacement in the presence of donor ligands such as $P(OMe)_3$.[22-24]

In addition to monometallic species, there has been considerable interest in the electrochemistry of bi- and poly-metallic complexes due to possible communication of the metal centers through coordinating ligands.[6-12, 25] The type of metal and bridging ligands present often dictates whether the metal centers are interacting with, or isolated from one another. For instance, Bard and co-workers have shown that for a bimetallic phenanthroline complex, the planar polycyclic ligand allows for electrochemical communication between the metal centers, with the electrochemical reversibility being dependent on the scan rate.[26] At low scan rates, two electrochemically irreversible waves were detected, whereas at much higher scan rates (20 V/s), complete reversibility was shown. For interacting systems of this nature, the degree of communication is often measured by the separation of formal potentials (ΔE^o), obtained by cyclic voltammetry.[27] In contrast to communicating systems, the cyclic voltammograms of complexes containing multiple isolated redox centers are the same as for a molecule with a single redox center.[28] Controlled potential coulometry is often employed to determine the number of electrons involved in these types of electron transfer processes.[29-31] Astruc has shown that only one electrochemically reversible wave is observed for tentacled cationic arene cyclopentadienyliron sandwich complexes with multiple isolated redox centers.[25] It was also reported that a complex containing six terminal ferrocene ($FeCp_2$) units and a core consisting of an arene cyclopentadienyliron complex exhibited two separate redox waves.[25] The neutral ferrocene complexes were oxidized while the cationic cyclopentadienyliron complex was reduced. For this material, the central $[FeCp(arene)]^+$ unit was used as a standard to determine the number of electrons involved in the ferrocene oxidations. More recently, Astruc has reported the preparation of a metallodendrimer containing sixty-four redox active cationic Fe(II) sites attached to the periphery of a polypropyleneimine dendrimer. Cyclic voltammetric studies indicated the presence of a single reversible redox couple due to the isolated centers. Sixty-four equivalents of a nineteen-electron permethylated arene complex were used to reduce the polycationic dendrimer, which resulted in the isolation of its neutral analog.[17]

Abd-El-Aziz and coworkers have also reported the electrochemical behavior of a variety of cyclopentadienyliron-coordinated mono- and poly-metallic complexes.[10, 11, 32-36] Again, it was found that the first electron transfer was chemically reversible in all complexes while the second processes were dependent on the experimental conditions. In addition, organoiron polymers that contain ferrocenyl spacers in their backbones and cyclopentadienyliron cations pendent to their backbones have also been synthesized and studied.[32] For these systems, the neutral iron centers underwent reversible oxidation and the cationic iron centers underwent reversible reduction processes.

Electrochemical investigations of arene complexes with the Cp^*Ru^+ moiety are decidedly rare, and much less successful than the analogous $CpFe^+$ systems.[37, 38] There have only been a few studies on the electrochemical behavior of pentamethylcyclopentadienylruthenium-coordinated complexes.[39-41] The electrochemical reduction of mono- and bi-metallic (Cp*)(polyarene)ruthenium cations was reported in 1990 by Koelle (Figure 1).[40] Reduction of the ruthenium complexes was observed in CH_2Cl_2 and propylene carbonate, using either platinum or vitreous carbon working electrodes. The monocations exhibited varying degrees of reversibility, which were highly dependent on the scan rates used. The pyrene complex, $\mathbf{3}^+$, was chemically reversible at fairly low scan rates (1 V/s), while the phenanthrene and anthracene complexes required much higher scan rates (20 V/s) in order to exhibit reversible reduction processes. The dications ($\mathbf{5}^{2+}$ and $\mathbf{6}^{2+}$) behaved as interacting redox centers and reduced in two discrete steps with the degree of separation being dependent on the nature of the arene.[40] It was postulated that the lower aromaticity of the polyarenes allowed for electrochemical reduction to occur within the potential window, unlike monoarene cations which have been reported to show no electrochemical activity up to -2.5 V.[37, 42] The increased stability of the neutral bimetallic complexes over the corresponding monometallics was attributed to a redistribution of electrons which may occur for $\mathbf{5}^{2+}$ and $\mathbf{6}^{2+}$.

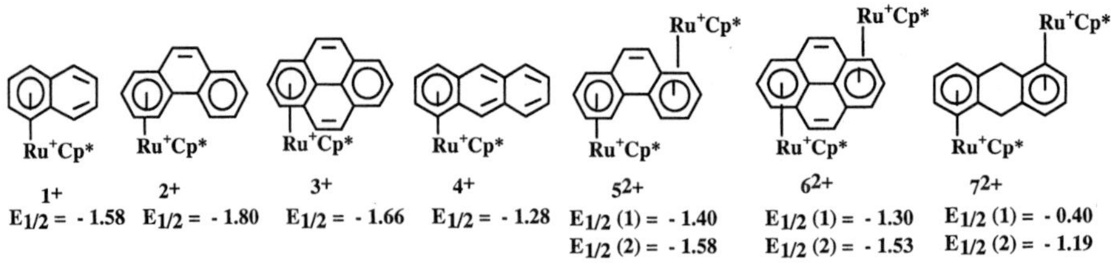

Ru⁺Cp* Ru⁺Cp* Ru⁺Cp* Ru⁺Cp* Ru⁺Cp* Ru⁺Cp* Ru⁺Cp* Ru⁺Cp* Ru⁺Cp* Ru⁺Cp*

1^+	2^+	3^+	4^+	5^{2+}	6^{2+}	7^{2+}
$E_{1/2}$ = - 1.58	$E_{1/2}$ = - 1.80	$E_{1/2}$ = - 1.66	$E_{1/2}$ = - 1.28	$E_{1/2}$ (1) = - 1.40	$E_{1/2}$ (1) = - 1.30	$E_{1/2}$ (1) = - 0.40
				$E_{1/2}$ (2) = - 1.58	$E_{1/2}$ (2) = - 1.53	$E_{1/2}$ (2) = - 1.19

Figure 1. Half-wave potentials for (polyarene)Ru^+Cp* complexes.

This article describes the electrochemical behavior of oligomers and polymers containing pendent cyclopentadienyliron and pentamethylcyclopentadienylruthenium cations. The effects of temperature, scan rate and the nature of the iron and ruthenium complexes are also examined.

Results and Discussion

Oligomeric Cp Fe^+-Coordinated Complexes

As mentioned previously, studies of monoiron cyclopentadienyliron arene complexes have shown that these materials undergo two successive one-electron reductions.[10, 11] We have also conducted cyclic voltammetric studies on oligomeric ether complexes that contain between two and fifteen cationic cyclopentadienyliron moieties pendent to alternating aromatic rings.[39-41] All of these materials exhibited two separate reduction steps, with the electrochemical reversibility being dependent on the nature of the complex, working electrode, temperature and solvent. In all oligomers, the presence of only one electrochemically reversible wave ($E_{1/2}$ between –0.99 and –1.89 V vs Fc/Fc^+), corresponding to the generation of the neutral nineteen-electron species, indicated that the iron centers were isolated from one another. For bimetallic complexes, this was also verified using controlled potential coulometry.[40] The mechanism for electron transfer in these complexes follows the general form shown in Scheme 1.

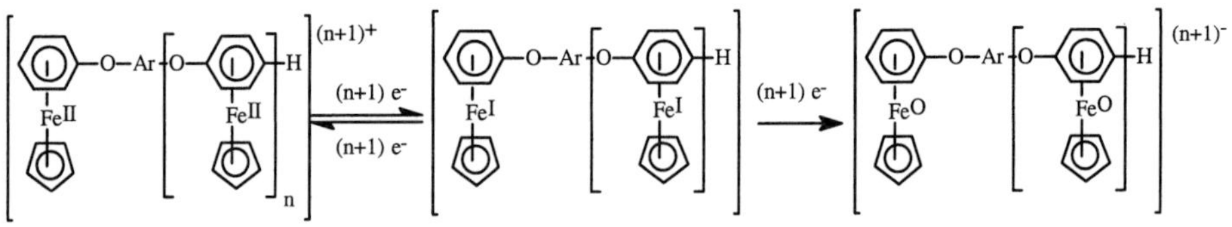

(n+1)+ —O—Ar—O— —H Fe^{II} Fe^{II} n (n+1) e⁻ (n+1) e⁻ Fe^{I} Fe^{I} (n+1) e⁻ Fe^{0} Fe^{0} (n+1)⁻

Scheme 1. Mechanism of electron transfer in polyether complexes of $CpFe^+$.

We have also prepared oligomeric complexes that contain mixed ether/sulfide or ether/sulfone backbones and these complexes showed the presence of more than one redox couple due to the different environments of the iron centers.[34] In addition, we have found that it is possible to detect subtle differences in the environment of the iron centers in star-shaped polyaromatic ethers coordinated to cyclopentadienyliron cations as shown below.[35] The cyclic voltammogram of the complex shown in Figure 2 reveals the presence of two distinct redox processes. These two redox steps can be attributed to the inner three and outer three iron centers. The $E_{1/2}$ values for these two redox processes occurred at –1.20 and –1.30 V, respectively.

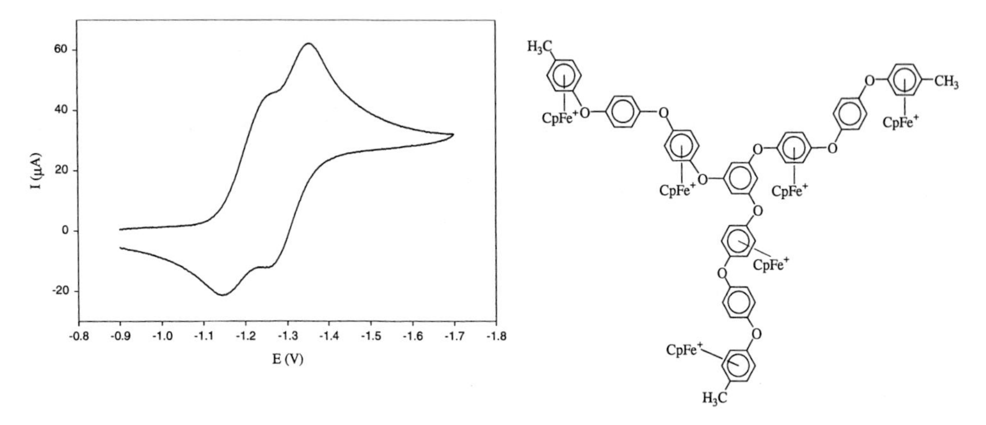

Figure 2. Cyclic voltammogram at glassy carbon of 2 mM complex in 0.1 M TBAP in DMF, ν = 0.1 V/s, T = 233K (Reproduced from "Aromatic Ether Star Oligomers Coordinated to Redox-active Cyclopentadienyliron Moieties", A.S. Abd-El-Aziz, E.K. Todd and T. Afifi, *Macromol. Rapid Commun.*, **23**, 113 (2002) Copyright © [2002] Wiley Periodicals, Inc.).

Our studies have also indicated that the coordinating ability of the solvent plays a role in the electrochemical behavior of Fe(I) complexes. In general, at low temperature (233 K) the rate of arene decomplexation was much higher in acetonitrile than in solvents such as DMSO, DMF, and acetone. This behavior is attributed to the fact that acetonitrile is a stronger coordinating solvent. The mechanism for decoordination of a bimetallic complex is shown is Scheme 1 but this can be extrapolated to the larger oligomers as well.

$$[\{(C_6H_5O)Fe^{II}Cp)\}_2C_6H_4]^{2+} + 2e^- \rightleftharpoons [\{(C_6H_5O)Fe^{I}Cp)\}_2C_6H_4]$$

$$[\{(C_6H_5O)Fe^{I}Cp)\}_2C_6H_4] + nS \xrightarrow{k} 2\,[S_n(Fe^{I}Cp)] + [(C_6H_5O)_2C_6H_4]$$

$$2\,[S_n(Fe^{I}Cp)]_2 \xrightarrow{fast} Fe^{II}(Cp)_2 + Fe^0 + 2nS$$

Scheme 1. Mechanism of electrochemical decoordination.

Oligomeric Ru$^+$Cp* and Heterometallic Ru$^+$Cp*-Fe$^+$Cp Complexes

The synthesis of a number of aromatic ether and thioether complexes containing pendent pentamethylcyclopentadienylruthenium moieties has been described.[43] Cyclic voltammetric studies of the complexes shown in Figure 3 have been performed in various solvents, at different temperatures, and with different working electrodes. Studies of the monometallic complexes performed in DMF using either a platinum disk or glassy carbon working electrode have indicated that the monometallic complexes **4.1-4.3** undergo quasi-reversible redox processes with a large ΔEp at all temperatures (213 K - 363 K) and electrodes studied.[3, 14] It was noted that the ΔEp value decreased with an increase in temperature which suggests that the kinetics of the processes were slow on the CV timescale.

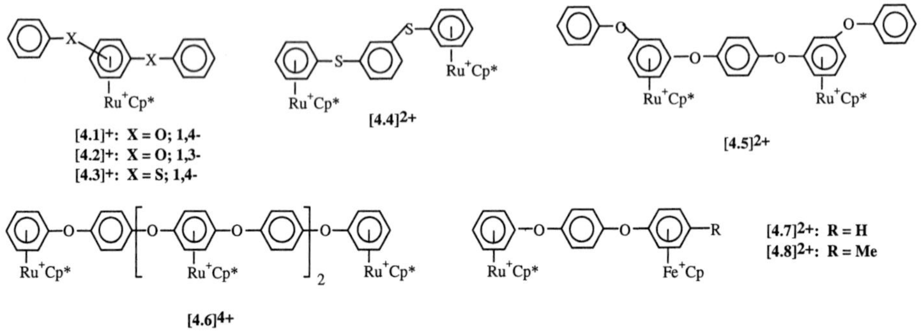

[4.1]+: X = O; 1,4-
[4.2]+: X = O; 1,3-
[4.3]+: X = S; 1,4-
[4.4]2+
[4.5]2+
[4.6]4+
[4.7]2+: R = H
[4.8]2+: R = Me

Figure 3. Structures of arene complexes of Cp^*Ru^+ and complexes containing Cp^*Ru^+ and $CpFe^+$ moieties.

Figure 4 shows typical cyclic voltammograms obtained for complex $[\textbf{4.3}]^{+}$ at different temperatures. The E_{pc} and E_{pa} values for this complex range from -2.58 V to -2.17 V and -0.81 V to -1.50V (vs Fc/Fc^{+}) as the temperature increases from 213 K to 333 K. Upon moving from the monometallic species to the bimetallic and tetrametallic complexes, the cyclic voltammograms of these materials became more complex. For the bimetallic ruthenium complexes, two or three reduction peaks were observed depending on the working electrode used, but only one oxidation peak was present. However, as in the case of the monometallic species, the peak separation did decrease with an increase in temperature.

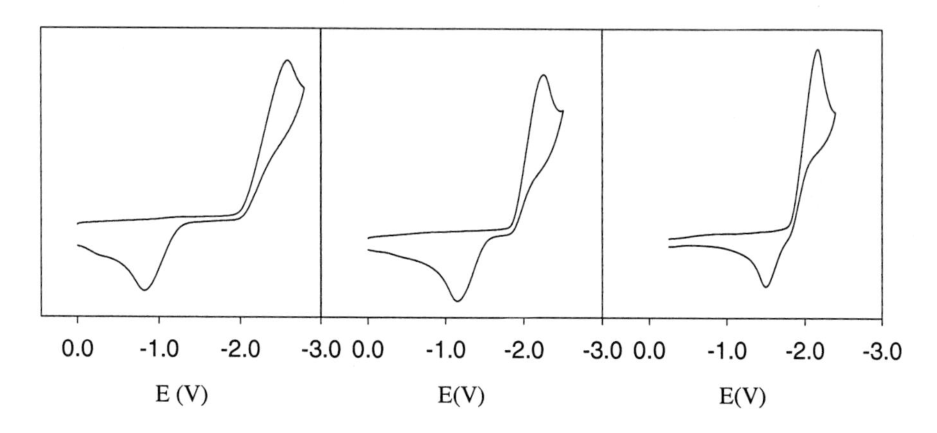

Figure 4. Cyclic voltammograms at Pt of 2.0 mM $[\textbf{4.2}]^{+}$ in DMF containing 0.1 M $TBAPF_6$; $v = 0.5$ V/s at (a) 213 K, (b) 273 K, and (c) 333 K.

Studies of the tetrametallic complex ($[\textbf{4.6}]^{4+}$) showed even more complex behavior that varied with both solvent and working electrode. For complexes $[\textbf{4.7}]^{2+}$ and $[\textbf{4.8}]^{2+}$, it was expected that a reversible one-electron reduction corresponding to the iron center, and a quasi-reversible reduction corresponding to the ruthenium center would be present. It was indeed observed that a reversible one-electron redox couple corresponding to the iron center and a quasi-reversible process attributed to the reduction of the ruthenium center was present when scanning from -0.7 V to -3.2 V. It was also found that the second redox couple corresponding to the ruthenium complex became more prominent as the temperature of the electrochemical experiment was increased (Figure 5). No second redox couple was observed for the iron center.

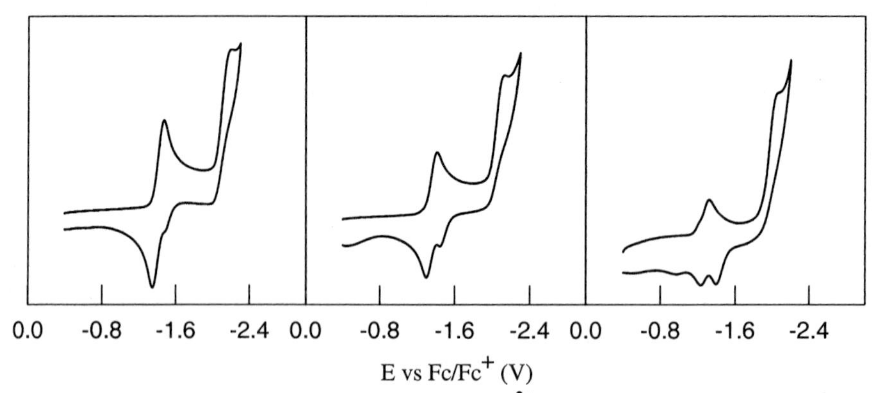

Figure 5. Cyclic voltammograms at Pt of 2.0 mM $[\mathbf{4.7}]^{2+}$ in DMF containing 0.1 M TBAP; ν = 0.2 V/s at (a) 233 K, (b) 253 K, and (c) 273 K.

Polymeric Complexes

Homo- and hetero-metallic polymers containing pendent $CpFe^+$ and Cp^*Ru^+ moieties have been prepared by a number of synthetic routes including nucleophilic aromatic substitution, polycondensation, radical polymerization and ring-opening metathesis polymerization.[36, 43-49] Typically, the cyclopentadienyliron-coordinated polymers exhibited reversible redox couples similar to the smaller oligomers with $E_{1/2}$ values between -0.99 V and -1.45 V. It was important that the cyclic voltammetric behavior of the polymers be compared to their monomeric and oligomeric analogs. For example, Figure 6 shows the cyclic voltammograms of an aryl ether complex and polymer obtained in a DMF solutions containing TBAP as the supporting electrolyte. These CVs were measured at –20 °C with a scan rate of 1 V/s. It was found that the reduction processes for this diiron complex displayed good reversibility and had a cathodic peak potential (E_{pc}) of -1.28 V, and an anodic peak potential (E_{pa}) of -1.08 V. The half-wave potential ($E_{1/2}$) for this reduction process was at -1.18 V. The cyclic voltammograms of the polymeric analog showed poorer reversibility relative to the diiron complex, in particular at low scan rates. The CV of the organoiron polymer had an E_{pc} = -1.61 V, E_{pa} = -1.18 V and $E_{1/2}$ = -1.40 V.

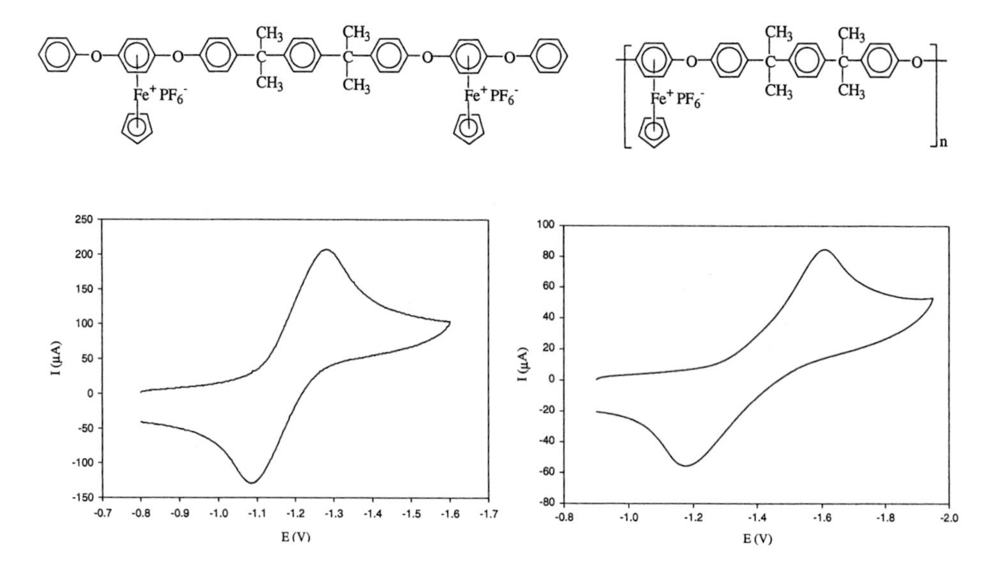

Figure 6. Cyclic voltammogram of a bimetallic complex and polymer containing aryl ether linkages.

For analogous pentamethylcyclopentadienylruthenium-coordinated polymers (Figure 7), it was found that the shape of the cyclic voltammogram depended on the scan rate and temperature. At 213 K, the cyclic voltammograms showed multiple reduction peaks at low scan rates (0.05, 0.2, 0.5 V/s) that started to coalesce as the scan rate was increased to 5 V/s. It was observed that as the scan rate increased, the voltammogram started to resemble the shape of the analogous monometallic $Cp^{*}Ru^{+}$ systems. It is also important to note that the heterometallic polymers only possessed one redox couple for the iron center with no electrochemical activity of the ruthenium center being observed. This may be a consequence of the multiple ruthenium centers becoming very difficult to reduce, and therefore no electrochemical activity is seen within the potential window of the experiment.

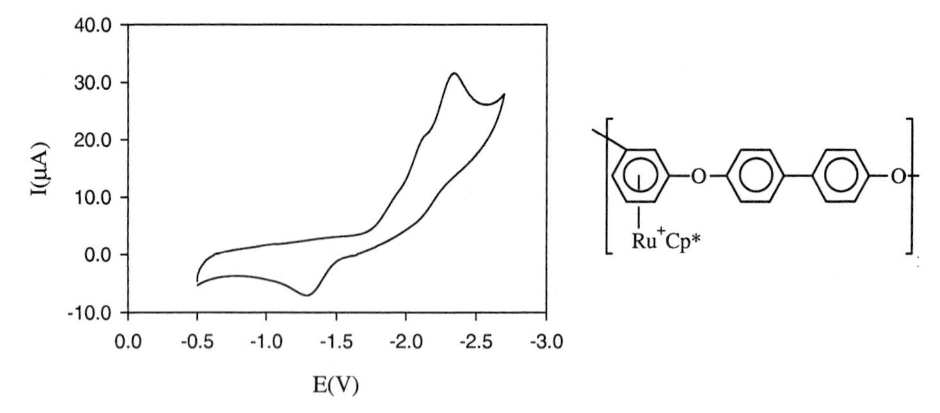

Figure 7. Cyclic voltammogram at Pt of 2.0 mM ruthenium polymer in DMF containing 0.1 M TBAP; ν = 1.0 V/s at 213 K.

Conclusion

Extensive electrochemical studies of oligomeric and polymeric arene complexes with pendent cyclopentadienyliron and pentamethylcyclopentadienylruthenium moieties have been conducted. For the cyclopentadienyliron-coordinated systems it was found that the iron centers behaved as isolated redox centers. The reversibility of the couples were found to be dependent on different factors (solvent, temperature). In contrast, it was found that the ruthenium systems exhibited quite complex electrochemical behavior regardless of the experimental conditions employed.

[1] *"Electron Transfer in Chemistry"*, Vol. II, V. Balzani, Ed. Wiley, Weinheim, 2001.
[2] D. Astruc (Editor), *"Modern Arene Chemistry: Concepts, Synthesis, and Applications"*, Wiley, 2002.
[3] D. Astruc, *"Electron Transfer and Radical Processes in Transition-Metal Chemistry"*, VCH, New York, 1995.
[4] D. H. Evans, *Chem. Rev.* **1990**, *90*, 739.
[5] A. S. Abd-El-Aziz, *Macromol. Rapid Commun.* **2002**, *23*, 995.
[6] A. S. Abd-El-Aziz, *Coord. Chem. Rev.* **2002**, *233-234*, 177.
[7] P. Nguyen, P. Gomez-Elipe, I. Manners, *Chem. Rev.* **1999**, *99*, 1515.
[8] H. Nishihara, in *"Handbook of Organic Conductive Molecules and Polymers"*, H. S. Nalwa, Ed., John Wiley & Sons, New York, 1997; Vol. 2, p. 799.
[9] R. Q. Bligh, R. Moulton, A. J. Bard, A. Piorko, R. G. Sutherland, *Inorg.Chem.* **1989**, *28*, 2652.
[10] A. S. Abd-El-Aziz, K. Winkler, A. S. Baranski, *Inorg. Chim. Acta* **1992**, *194*, 207.
[11] A. S. Abd-El-Aziz, A. S. Baranski, A. Piorko, R. G. Sutherland, *Inorg. Chim. Acta* **1988**, *147*, 77.

[12] I. Manners, *Adv. Mater.* **1994**, *6*, 68.
[13] D. A. Foucher, C. H. Honeyman, J. M. Nelson, B. Z. Tang, I Manners, *Angew. Chem. Int. Ed. Engl.* **1993**, *32*, 1709.
[14] W. E. Geiger, *Prog. Inorg. Chem.* **1985**, *33*, 275.
[15] W. E. Geiger, N. G. Connelly, *Adv. Organomet. Chem.* **1985**, *24*, 87.
[16] V. Guerchais, D. Astruc, *J. Organomet. Chem.* **1986**, *316*, 335.
[17] J. Ruiz, C. Pradet, F. Varret, D. Astruc, *Chem. Commun.* **2002**, 1108.
[18] A. N. Nesmeyanov, N. A. Vol'kenau, L. S. Shilovtseva, V. A. Petrokova, *J. Organomet. Chem.* **1973**, *61*, 329.
[19] E. El Murr, *J. Chem. Soc., Chem. Commun.* **1981**, 251.
[20] S. P. Solodovnikov, N. A. Vol'Kenau, L. C. Shilovtseva, *Izv. Akad. Nauk SSSR, Ser Khim* **1985**, *8*, 1733.
[21] A. N. Nesmeyanov, L. I. Denisovich, S. P. Gubin, N. A. Vol'kenau, E. I. Sirotkina, I. N. Bolesova, *J. Organomet. Chem.* **1969**, *20*, 169.
[22] A. Darchen, *J. Chem. Soc., Chem. Commun.* **1983**, 768.
[23] A. Darchen, *J. Organomet. Chem.* **1986**, *302*, 389.
[24] A. S. Abd-El-Aziz, A. Piorko, A. S. Baranski, R. G. Sutherland, *Synth. Commun.* **1989**, *19*, 1865.
[25] J.-L. Fillaut, D. Astruc, *New. J. Chem.* **1996**, *20*, 945.
[26] R. Moulton, T. W. Weidman, K. P. C. Vollhardt, A. J. Bard, *Inorg. Chem.* **1986**, *25*, 1846.
[27] D. E. Richardson, H. Taube, *Inorg. Chem.* **1981**, *20*, 1278.
[28] J. B. Flanagan, S. Margel, A. J. Bard, F. C. Anson, *J. Am. Chem. Soc.* **1978**, *100*, 4248.
[29] J. M. Mevs, T. Gennett, W. E. Geiger, *Organometallics* **1991**, *10*, 1229.
[30] A. J. Bard, L. R. Faulkner, "*Electrochemical Methods: Fundamentals and Applications*", John Wiley and Sons, New York, 1980.
[31] D. T. Sawyer, J. L. Roberts, "*Experimental Electrochemistry for Chemists*", John Wiley and Sons, New York, 1974.
[32] A. S. Abd-El-Aziz, E. K. Todd, R. M. Okasha, T. E. Wood, *Macromol. Rapid Commun.* **2002**, *23*, 1.
[33] R. M. Moriarty, U. S. Gill, Y. Y. Ku, *J. Organomet. Chem.* **1988**, *350*, 157.
[34] I. W. Robertson, T. A. Stephenson, D. A. Tocher, *J. Organomet. Chem.* **1982**, *228*, 171.
[35] P. J. Fagan, M. S. Ward, J. V. Caspar, J. C. Calabrese, P. J. Krusic, P. J. *J. Am. Chem. Soc.* **1988**, *110*, 2981.
[36] U. Koelle, M. H. Wang, *Organometallics* **1990**, *9*, 195.
[37] O. V. Gusev, M. A. Ievlev, M. G. Peterleitner, S. M. Peregudova, L. I. Denisovich, P. V. Petrovskii, N. A. Ustynyuk, *J. Organomet. Chem.* **1997**, *534*, 57.
[38] E. Roman, D. Astruc, *Inorg. Chim. Acta* **1979**, *37*, L465.
[39] A. S. Abd-El-Aziz, D. C. Schriemer, *Inorg. Chim. Acta* **1992**, *202*, 123.
[40] A. S. Abd-El-Aziz, C. R. de Denus, K. M. Epp, S. Smith, R. J. Jaeger, D. T. Pierce, *Can. J. Chem.* **1996**, *74*, 650.
[41] A. S. Abd-El-Aziz, E. K. Todd, T. H. Afifi, *Macromol. Rapid Commun.* **2002**, *23*, 113.
[42] C. R. de Denus, L. M. Hoffa, A. S. Abd-El-Aziz, E. K. Todd. *J. Inorg. Organomet. Polym.* **2000**, *10*, 189.
[43] A. A. Dembek, P. J. Fagan, M. Marsi, *Macromolecules* **1993**, *26*, 2292.
[44] A. S. Abd-El-Aziz, L. J. May, J. A. Hurd, R. M. Okasha, *J. Polym. Sci., Part A: Polym. Chem.* **2001**, *39*, 2716.
[45] C. R. de Denus, P. Baker, J. Toner, S. McKevitt, unpublished results.
[46] A. S. Abd-El-Aziz, E. K. Todd, G. Z. Ma, *J. Polym. Sci., Part A: Polym. Chem.* **2001**, *39*, 1216.
[47] A. S. Abd-El-Aziz, E. K. Todd, G. Z. Ma, J. DiMartino, *J. Inorg. Organomet. Polym.* **2000**, *10*, 265.
[48] A. S. Abd-El-Aziz, T. H Afifi. W. R. Budakowski, K. J. Friesen, E. K. Todd, *Macromolecules* **2002**, *35*, 8929.
[49] A. S. Abd-El-Aziz, R. M. Okasha, T. H. Afifi, E. K. Todd, *Macromol. Chem. Phys.* **2003**, *204*, 555.

Synthesis and Thermal Properties of Diblock Copolymers Utilizing Non-Covalent Interactions

Bas G. G. Lohmeijer,[1] *Helmut Schlaad,*[2] *Ulrich S. Schubert**[1]

[1]Laboratory of Macromolecular Chemistry and Nanoscience, Eindhoven University of Technology and Dutch Polymer Institute (DPI), P. O. Box 513, 5600 MB Eindhoven, The Netherlands
E-mail: u.s.schubert@tue.nl

[2]Max-Planck-Institut für Kolloid- und Grenzflächenforschung, Am Mühlenberg, 14476 Golm, Germany

Summary: Diblock copolymers of poly(styrene) and poly(ethylene oxide) were prepared utilizing a *bis*terpyridine ruthenium complex as non-covalent interaction for the connection of the two blocks. Apart from the synthesis and characterization of four metallo-supramolecular block copolymers, first studies on the thermal properties of the block copolymers have been performed. A complex crystallization behavior was observed and is described in a qualitative fashion. The influence of the metal complex on the thermal stability of the metallo-supramolecular block copolymers remains a question for further investigation.

Keywords: block copolymers; differential scanning calorimetry; supramolecular chemistry; synthesis; terpyridine

Introduction

The merging of supramolecular chemistry and polymer chemistry has become an area of increasing interest in the past decade.[1-3] In supramolecular chemistry, non-covalent interactions are employed to selectively build-up well-defined architectures. The implementation of supramolecular binding sites into polymer chains gives rise to materials that reveal additional features due to the generally weaker nature of the non-covalent interaction: external stimuli can have a great influence on the material properties by switching the non-covalent interactions reversibly and simultaneously from on to off (in the case of highly dynamic supramolecular systems). Important examples of non-covalent interactions are hydrogen bonding and metal-ligand interactions.[2-4] Supramolecular binding sites may be introduced at the chain ends of

© 2003 WILEY-VCH Verlag GmbH & KGaA, Weinheim CCC 1022-1360/00/$ 17,50+.50/0

polymers or along the polymer backbone, allowing the formation of 'classical' polymer architectures such as high molecular weight chain extended polymers, block copolymers, graft copolymers or cross-linked/gel-forming systems, among others.[5]

Our interest lies in supramolecular block copolymers that eventually enable manipulation of the morphologies formed by such block copolymers. This requires a strong non-covalent interaction that is able to withstand the opposing van-der-Waals forces described by the Flory-Huggins interaction parameter upon microphase separation of block copolymers. Moreover, it should not be highly dynamic in order to prevent the formation of a macrophase separated system or compatibalized blends. Metal-ligand complexes that are by nature inert are therefore the most suitable candidates. In particular, the possibility of selectively constructing heteroleptic complexes utilizing the same type of ligands is a major advantage to design and prepare a wide variety of block copolymer architectures. In our experience *bis*terpyridine ruthenium complexes fulfill all these requirements.[6-8] In this contribution we describe in detail the synthetic aspects of metallo-supramolecular block copolymers bearing a *bis*terpyridine ruthenium complex at the junction point between the two constituting polymer blocks (Figure 1). Last but not least, we will briefly reflect upon the thermal behavior of this new class of block copolymers.

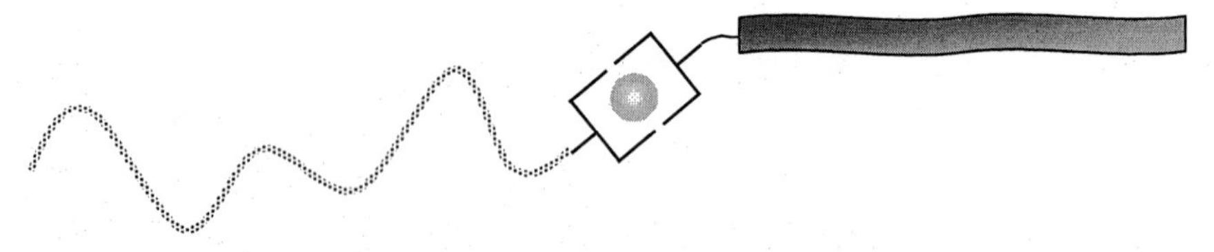

Figure 1. Schematic representation of an AB diblock copolymer bearing a metal complex at the junction between the two constituting blocks.

Experimental Part

Synthesis of terpyridine end-functionalized poly(ethylene oxide): see ref. 6.

General procedure for the synthesis of terpyridine end-functionalized poly(styrene): Hydroxy end-functionalized poly(styrene) and *t*-BuOK (3 eq.) were refluxed for 15 min in dry THF under argon. Via an addition funnel a 2-fold excess of 4'-chloro-2,2':6',2''-terpyridine was added to the solution and was refluxed for additional 4 hours. Hereafter the solution was directly precipitated in methanol (1:10 v/v). The white precipitate was re-precipitated from THF into methanol and washed with hexane. Yields varied from 76% to 85%. Selected analytical data of poly(styrene) with DP = 100: ^{1}H-NMR: 400 MHz, $CDCl_3$, 25°C): signals are broadened $\delta = 8.67$ (bm, 2H, H6;H6''), 8.60 (bm, 2H, H3;H3''), 7.91 (s, 2H, H3';H5'), 7.83 (bm, 2H, H4; H4''), 7.31-6.39 (C_6H_5 PS backbone; H5;H5''), 4.11-3.96 (bm, 2H, tpyOCH_2), 2.66, 2.18-1.10, 0.78 (PS backbone). UV/Vis (CH_2Cl_2): λ_{max} (nm) = 278, 243. MALDI-TOF MS: M_n = 9487 g/mol. GPC (RI): M_n = 9800, PDI = 1.13.

General procedure for the synthesis of $RuCl_3$ *mono*complexes of poly(styrene) and poly(ethylene oxide): A suspension of anhydrous $RuCl_3$ in dry degassed DMF was heated under argon at 130°C. Not until the color of the suspension had turned dark brown, a 1/3-fold equivalent of the polymer in dry degassed DMF was added slowly. The reaction mixture was stirred overnight at 130°C, after which it was allowed to cool to room temperature and partitioned between water and CH_2Cl_2. The organic layer was quickly separated and dried over Na_2SO_4 and removed *in vacuo*. The resulting solid was precipitated from CH_2Cl_2 in a non-solvent, methanol and diethyl ether for poly(styrene) and poly(ethylene oxide) respectively. Yields varied from 89% to 98%. Selected analytical data: PS_{20}:UV/vis (CH_2Cl_2): λ_{max} (nm): 400, 312, 277, 270, 259. PEO_{375}: UV/vis (CH_2Cl_2): λ_{max} (nm): 395, 309, 276. In ^{1}H-NMR the terpyridine signals have completely vanished due to the paramagnetic nature of the compound. Only the polymer backbones were visible.

General procedure for the synthesis of A_x-[Ru]-B_y block copolymers: see ref. 6-8. The following block copolymers have been prepared: PS_{20}-[Ru]-PEO_{70} (**1**), PS_{20}-[Ru]-PEO_{375} (**2**),

PS_{100}-[Ru]-PEO_{70} (**3**) and PS_{100}-[Ru]-PEO_{375} (**4**). Selected analytical data: ^{1}H-NMR (400 MHz, $CDCl_3$, 25°C): δ = 8.36 (td, 2H, H3;H3'', PEO), 8.26 (s, 2H, H3';H5', PS), 8.17 (bm, 2H, H3;H3'', PS), 7.88-7.73 (bm, 6H), 7.33-6.32 (bm, x = 110 H for **1** and **2**, x = 520 H for **3** and **4** C_6H_5 PS backbone and terpyridine signals), 4.74 (t, 2H, tpyOCH_2, PEO), 4.28-4.04 (m, 4H, tpyOCH$_2$CH_2 (PEO), tpyOCH_2 (PS)), 3.92-3.42 (m, x = 290 H for **1** and **3**, 1550 H for **2** and **4**, PEO backbone), 3.38 (s, 3H, OCH_3), 1.72-0.60 (m, x = 50 H for 1 and 2, x = 204 H for **3** and **4**, CH_2, CH PS backbone). UV/vis (CH_2Cl_2): λ_{max} (nm): 487, 305, 269, 262 (in case of **3** and **4**, styrene), 244.

Results and Discussion

Terpyridine is a well-known tridentate chelating ligand that forms stable octahedral complexes with a wide variety of transition metal ions.[9] In order to create diblock copolymers this ligand needs to be introduced at one chain end of two different polymers using a 'blocking onto'-procedure. We have selected poly(ethylene oxide) and poly(styrene) as constituting polymer blocks. The important difference is the solubility of these polymers and this requires a different approach for the introduction of the terpyridine ligand. In both cases the starting polymers have been prepared by living anionic polymerization techniques. Intrinsically, polymerization of ethylene oxide renders a hydroxyl group at the chain end of poly(ethylene oxide), whereas for poly(styrene) this same end group was introduced by addition of the living polymer reaction mixture to a solution of ethylene oxide in THF. This hydroxy end group was subjected in both cases to an addition-elimination reaction using a base to deprotonate the polymer and 4'-chloro-2,2':6',2''-terpyridine as the other reactant (Figure 2).

Hydrophilic poly(ethylene oxide) of various molecular weights was easily converted to the terpyridine functionalized polymer using a suspension of KOH in DMSO. The synthesis and characterization have been described elsewhere.[6,10] For poly(styrene) this approach did not work for solubility reasons. We have already used an approach in which we applied 18-crown-6 as a phase transfer catalyst for KOH in toluene-solution and this was rather successful, although the reaction times were quite long (typically 48 hours).[6]

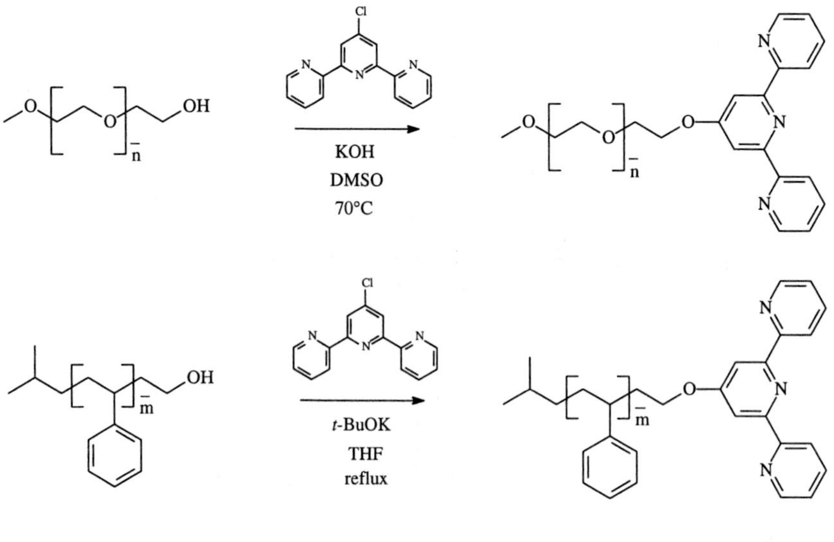

Figure 2. Synthetic scheme for the preparation of terpyridine end-functionalized poly(ethylene oxide) with n = 70 or n = 375 and poly(styrene) with m = 20 or m = 100.

We therefore applied a different method using *t*-BuOK in dry THF for the deprotonation of the polymer and then added the 4'-chloroterpyridine in excess. The reaction has gone to completion within 4 hours, requires less starting materials and represents therefore an important improvement to the route developed before. We have used poly(styrene) of two different molecular weights (~2 and 10 kDa) and the obtained results were comparable. The excess 4'-chloroterpyridine was removed by a double precipitation in methanol and subsequent washing with hexanes. Alternatively, a preparative size exclusion chromatography could be carried out. However, this was not necessary in the present case. The terpyridine functionalized poly(styrene) was analyzed by ^{1}H-NMR, size exclusion chromatography (GPC), FT-IR and MALDI-TOF MS. In IR, characteristic vibrations at 1600, 1582 and 1563 cm^{-1} of the terpyridine ligand can be observed. In ^{1}H-NMR the corresponding signals for the terpyridine unit arise and can be integrated to the backbone. This was in excellent agreement with results obtained from GPC using poly(styrene) standards and MALDI-TOF MS measurements. Figure 3 shows the MALDI-TOF MS spectra of the hydroxy end-functionalized poly(styrene) (10 kDa) and the resulting terpyridine functionalized polymer having a mass difference of 232 Da corresponding to the mass of a terpyridine ligand.

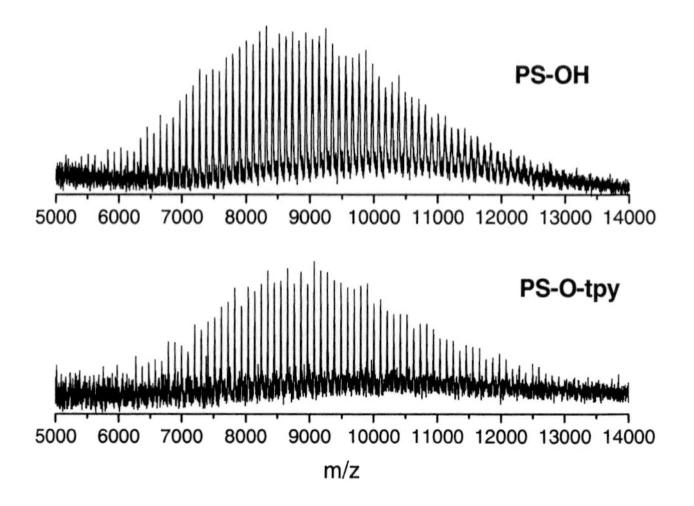

Figure 3. MALDI-TOF mass spectra of poly(styrene) of DP = 100 bearing a hydroxyl end group (top) and having a terpyridine end group (bottom).

To construct heteroleptic *bis*terpyridine metal complexes, ruthenium was employed as transition metal ion. Ruthenium(III)chloride forms a stable *mono*complex with a terpyridine ligand, which can be isolated. When using the right conditions the chlorides block the three other coordination sites, preventing the formation of *bis*complexes. This *mono*complex is paramagnetic from the d^5 configuration and therefore in ^{1}H-NMR the terpyridine signals disappear upon complexation. UV/Vis-spectroscopy revealed a typical metal-to-ligand-charge-transfer band (MLCT) at ~400 nm, corresponding to the *mono*-complex (Figure 4). The versatility of the $RuCl_3$ *mono*complex formation is demonstrated through the general applicability to both poly(styrene)s and poly(ethylene oxide)s. This means that either the *mono*complex of poly(styrene) or of poly(ethylene oxide) may be used for further reaction. This enhances the applicability of preparative size exclusion chromatography for purification in later steps: the lower molecular weight polymer with a free terpyridine ligand can be applied in excess and later easily removed.

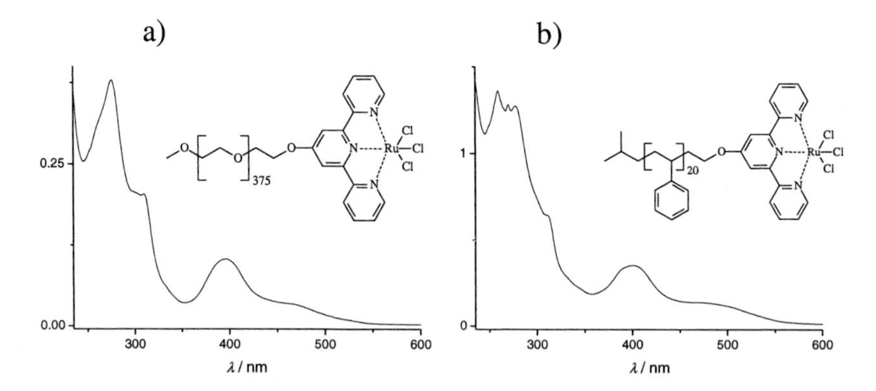

Figure 4. UV/Vis-spectra showing the MLCT band at ~ 400 nm for the $RuCl_3$ *mono*-complexes of a) poly(ethylene oxide) of 17 kDa and b) poly(styrene) of 2 kDa (both measured in CH_2Cl_2).

Metallo-supramolecular block copolymers were prepared upon addition of a slight excess of free terpyridine end-functionalized polymer to the *mono*complex of poly(styrene) or poly(ethylene oxide) respectively. The metallo-supramolecular polymers that have been prepared are described by the following acronyms: PS_{20}-[Ru]-PEO_{70} (**1**), PS_{20}-[Ru]-PEO_{375} (**2**), PS_{100}-[Ru]-PEO_{70} (**3**) and PS_{100}-[Ru]-PEO_{375} (**4**) (see Figure 5).

PS_m-[Ru]-PEO_n

1 : m = 20 n = 70
2 : m = 20 n = 375
3 : m = 100 n = 70
4 : m = 100 n = 375

Figure 5. Metallo-supramolecular block copolymers based on poly(styrene) and poly(ethylene oxide).

The block copolymer formation was carried out in a mixture of chloroform and methanol: the reduction of Ru(III) to Ru(II) is accompanied by an oxidation of methanol and is catalyzed by *N*-ethylmorpholine. The reactions were followed by UV/Vis spectroscopy, where a shift of the MLCT band had occurred from 400 nm for the Ru(III) *mono*complex to 490 nm for the Ru(II) *bis*complex (Figure 6). Purification was carried out by preparative size exclusion chromatography. Figure 6 also shows a photograph of such a size exclusion of the PS_{20}-[Ru]-PEO_{375} on a column packed with BioBeads SX-1. Three bands are visible: the lower (red) band represents the product, the top band can be assigned to the starting material and in between an intense purple band can be observed, which is not understood yet due to analytical difficulties. Nevertheless, the isolated amount of this fraction was very little (~5%).

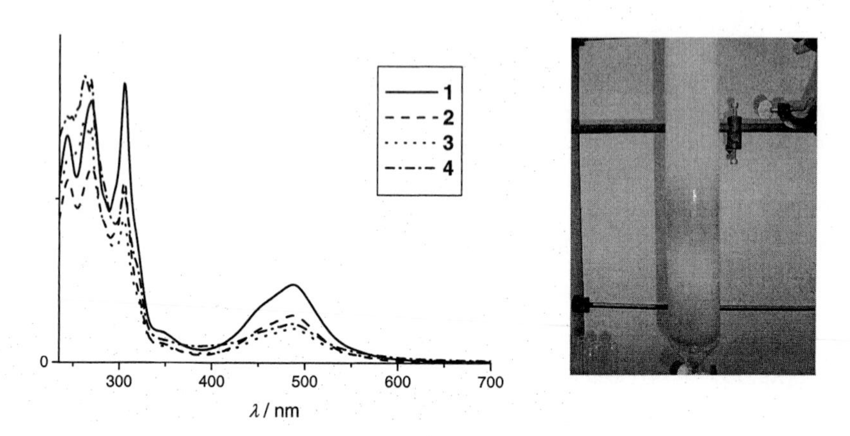

Figure 6 (left): UV/vis spectra of the four block copolymers, clearly showing the MLCT band at 490 nm for all compounds. Right: Photograph of preparative BioBeads size exclusion chromatography column applied on PS_{20}-[Ru]-PEO_{375}.

All the metallo-supramolecular block copolymers were purified using preparative size exclusion chromatography. Characterization was carried out by UV/Vis (Figure 6), FT-IR, ^{1}H-NMR and where possible GPC and MALDI-TOF MS. In the latter technique a breakage of metal-ligand interaction could be observed depending on the laser intensity applied. GPC showed severe problems owing to column interactions and shear forces on the column. We are currently trying to find a reliable set-up. From ^{1}H-NMR we could obtain an idea of the molecular weight of the block copolymers by integrating the terpyridine-signals to the polymer backbones. Figure 7 shows a ^{1}H-NMR-spectrum of PS_{20}-[Ru]-PEO_{70}, where the peaks have been assigned.

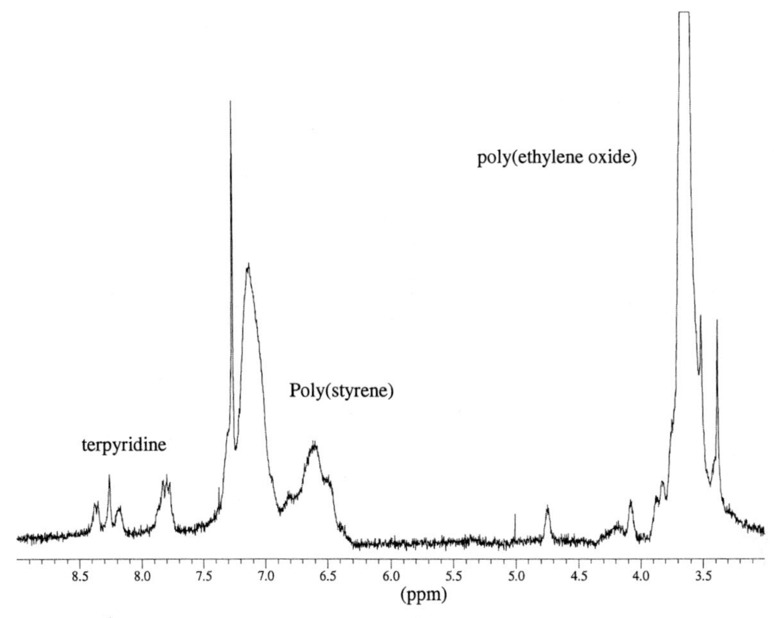

Figure 7. ^{1}H-NMR spectrum of PS_{20}-[Ru]-PEO_{70} in $CDCl_3$. Indicated are the signals coming from the metal complex, poly(styrene) and poly(ethylene oxide). The sharp signals at 4.7 and 4.1 ppm originate from the methylene protons next to the terpyridine ligand on the PEO-backbone, whereas the broad peak at 4.2 ppm is caused by the methylene protons next to the terpyridine on the PS-backbone.

As a first step towards the understanding of the thermal properties of the metallo-supramolecular block copolymers, DSC-measurements have been performed on all samples. Figure 8 shows the respective heating and cooling runs for the four block copolymers. Compounds 2 and 4 show clear melting and crystallization temperatures. They contain a high volume fraction of the long PEO-block. It is worth noting that in both cases the same T_m of 59°C was found for the corresponding uncomplexed poly(ethylene oxide). Interestingly, in the cooling runs 2 and 4 show different crystallization temperatures of 19°C and 34°C, respectively. This may be attributed to compatibilization of the short PS-block in 2, leading to retarded crystallization effects in the PEO-block.

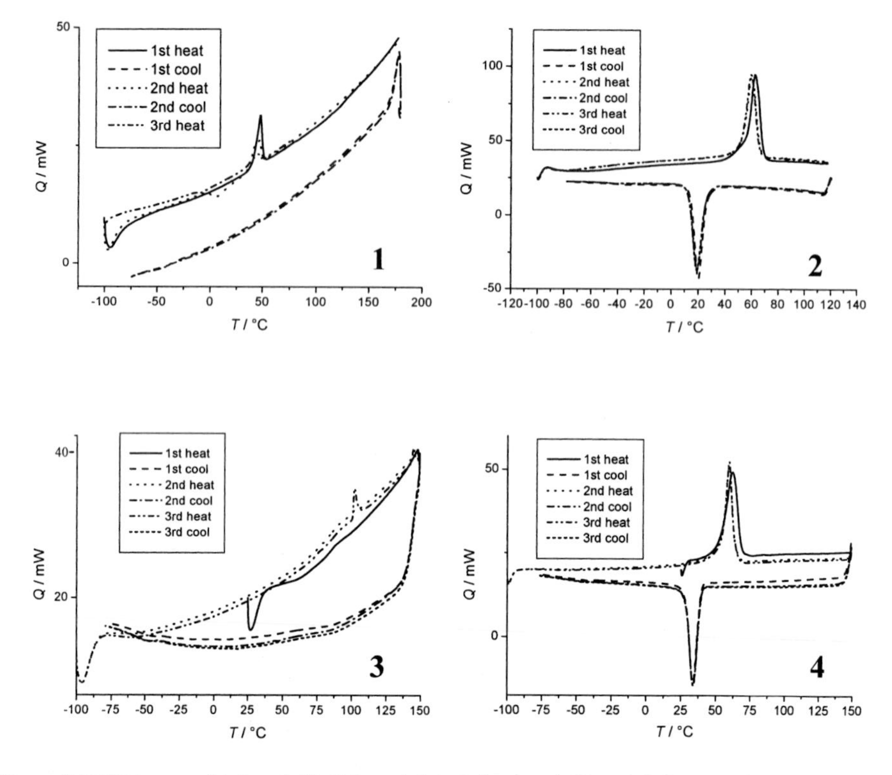

Figure 8. DSC-traces of 1 (top left), 2 (top right), 3 (bottom left) and 4 (bottom right) at a heating rate of 40°C/min.

In 1 and 3 a crystallization peak is absent and the heat required for melting 1 is much smaller compared to the corresponding uncomplexed poly(ethylene oxide). Therefore, the poly(styrene)-block must prevent the crystallization of the poly(ethylene oxide) block. The complete absence of a melting peak for compound 3, (with the highest PS-content) also supports this hypothesis. Moreover, at 102°C the T_g of the PS can be observed. Similar effects have been reported for covalent PS-*b*-PEO analogues.[11-13]

Conclusions and Outlook

Block copolymers in which the constituting blocks of poly(styrene) and poly(ethylene oxide) are held together via a metal complex have been prepared and characterized. Efficient and high-yield strategies have been developed for the various steps. The thermal behavior of the new block copolymers was studied by DSC. It is difficult to draw conclusions about the influence of the metal complex on the thermal properties of the block copolymers. The block copolymers in itself show, in analogy with their covalent counterparts, interesting behavior regarding crystallization retardation and compatibilization effects. In the near future, WAXS and SAXS experiments will be carried out in order to elucidate the crystallization phenomena as well as the role of the metal complex. More block copolymers having different PS and PEO contents will be prepared for this purpose, leading to a library of PS_x-[Ru]-PEO_y block copolymers. Morphological studies will be undertaken as well. Finally, the opening of the metal complex and the manipulation of defined nano-structures on surfaces will be addressed.

Acknowledgements

The authors thank Dipl.-Chem. Michal A. R. Meier for MALDI-TOF MS-measurements as well as the Dutch Polymer Institute and Fonds der Chemischen Industrie for funding.

[1] J.-M. Lehn, *Makromol. Chem. Macromol. Symp.* **1993**, *69*, 1.
[2] L. Brunsveld, B. J. B. Folmer, R. P. Sijbesma, E. W. Meijer, *Chem. Rev.* **2001**, *101*, 4071.
[3] U. S. Schubert, C. Eschbaumer, *Angew. Chem. Int. Ed.* **2002**, *41*, 2892.
[4] J.-M. Lehn, "*Supramolecular Chemistry - concepts and perspectives*", VCH, Weinheim, Germany, **1995**.
[5] B. G. G. Lohmeijer, U. S. Schubert, *J. Polym. Sci.: Part A: Polym. Chem.* **2003**, in press.
[6] B. G. G. Lohmeijer, U. S. Schubert, *Angew. Chem. Int. Ed.* **2002**, *41*, 3825.
[7] J.-F. Gohy, B. G. G. Lohmeijer, U. S. Schubert, *Macromolecules* **2002**, *35*, 4560.
[8] J.-F. Gohy, B. G. G. Lohmeijer, S. K. Varshney, B. Decamps, E. Leroy, S. Boileau, U. S. Schubert, *Macromolecules* **2002**, *35*, 9748.
[9] E. C. Constable, *Adv. Inorg. Chem. Radiochem.* **1986**, 69.
[10] U. S. Schubert, C. Eschbaumer, *Macromol. Symp.* **2001**, *163*, 177.
[11] Y. Shimura, T. Hatakeyama, *J. Polym. Sci.: Part B: Polym. Phys.* **1975**, 13, 653.
[12] L. Zhu, Y. Chen, A. Zhang, B. H. Calhoun, M. Chun, R. P. Quirk, S. Z. D. Cheng, B. S. Hsiao, F. Yeh, T. Hashimoto, *Phys. Rev. B* **1999**, *60*, 10022.
[13] C. Tsitsilianis, G. Staikos, P. Lutz, P. Rempp, *Polymer* **1992**, *33*, 3369.

Precious Metal Polymers

Richard J. Puddephatt

Department of Chemistry, University of Western Ontario, London, Canada N6A 5B7
E-mail: pudd@uwo.ca

Summary: The synthesis and properties of polymers containing precious metal centres in the backbone are described. Polymers with labile transition metal centres can be prepared by ring-opening polymerization of macrocycles, and examples are given with silver, gold and palladium as the metal centres. In some cases, the polymers can be further organized by using ligands with hydrogen bonding substituents, and self-assembly into sheet or network structures can then occur. Secondary bonding between inorganic centres can also lead to ordered self-assembly. Bicyclic precursors can ring open to form either chains or sheets.

Keywords: bis(pyridine) ligands; diphosphine; gold; palladium; silver

Introduction

There is intense interest in the synthesis and properties of metal-containing polymers.[1] Such polymers may be broadly classified as those having metallic units appended to the polymer backbone or those with metal atoms incorporated into the backbone. Polymers with peripheral metal complexes are most commonly synthesised by functionalization of an organic polymer or by polymerization of an organic monomer with a peripheral metal-containing unit. Two examples are shown in Figures 1 and 2. Figure 1 shows how poly(4-vinylpyridine) can be functionalized by the addition of diphoshinedigold(I) fragments.[2] Short chain diphosphines such as $Ph_2PCH_2CH_2PPh_2$ tend to form intrachain links whereas longer chain diphosphines give a greater degree of interchain crosslinks. With shorter chain diphosphines, the gold(I) centers may be close enough together to form a secondary aurophilic attraction as indicated by the broken bond in Figure 1. Figure 2 shows how a vinyl monomer containing an organoplatinum(IV) unit can be polymerized to give a polymer with pendant organometallic units.[3] A polymer having the same functionality can be prepared by adding the platinum unit by oxidative addition after polymerization as shown in Figure 2.

© 2003 WILEY-VCH Verlag GmbH & KGaA, Weinheim CCC 1022-1360/00/$ 17,50+.50/0

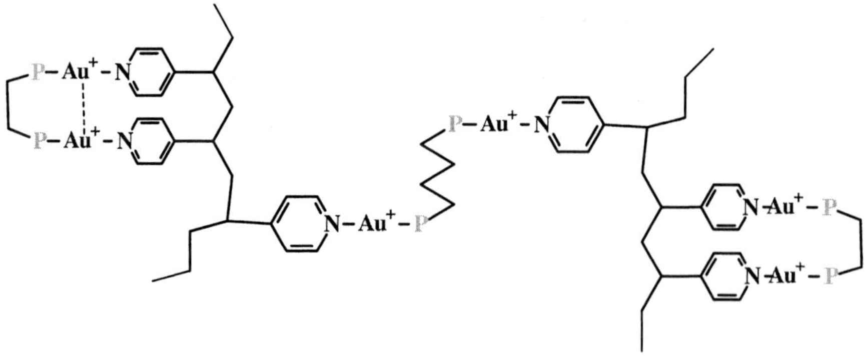

Figure 1. Poly-4-vinylpyridine functionalized with diphosphinedigold(I) units.

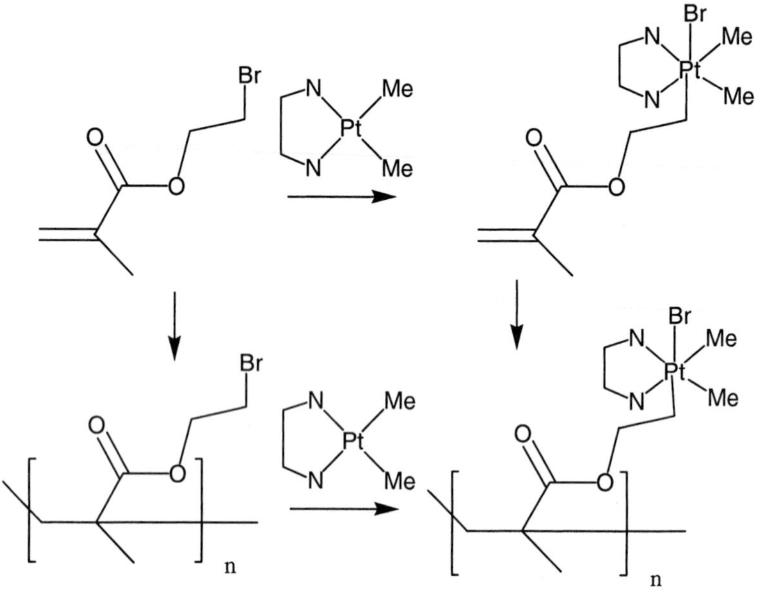

Figure 2. Two routes to synthesis of organoplatinum polymers (NN = 2,2'-bipyridine).

This article is concerned with the synthesis of polymers with transition metal atoms in the backbone. Three mechanisms of polymerization have been studied as detailed below.

Addition Polymerization

The first method involves addition polymerization at a metal center. An example involving oxidative addition as the propagation step is shown in equation (1).[4]

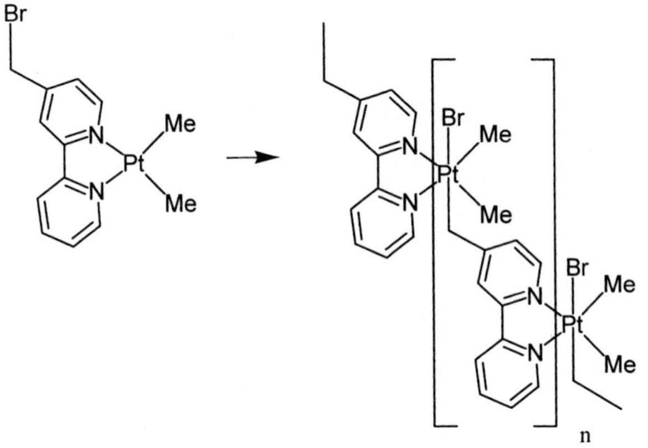

(1)

Condensation Polymerization

Several examples of condensation oligomerization or polymerization have been found.[5,6] A recent example is shown in equation (2). The product polymer can be crystallized and forms a double-stranded polymer. The polymer strands are held together by aurophilic attractions in the solid state as shown in Figure 3.

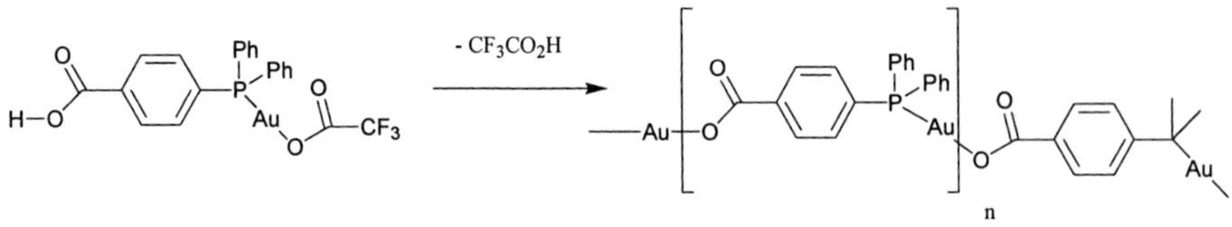

$- CF_3CO_2H$

(1)

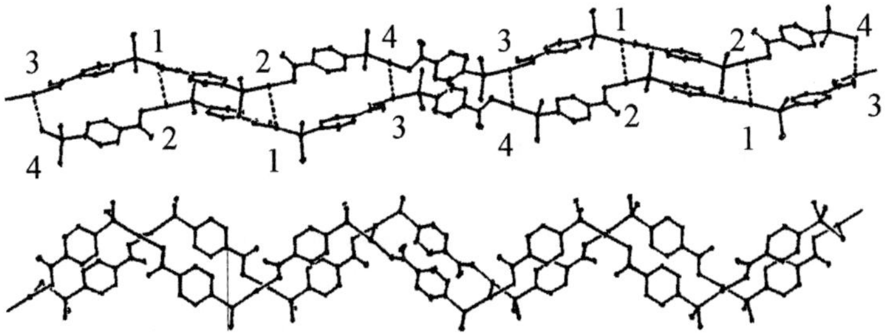

Figure 3. Two views of the double stranded polymer of equation 1, with only the *ipso* carbon atoms of the phenyl groups shown. The aurophilic bonds are shown as dashed lines.

Ring-Opening Polymerization

Recent advances have been made in designing strained ring complexes that can easily undergo ring-opening polymerization, in which ring strain is relieved. Several examples have been discovered in gold(I) chemistry. For example, the dithiolate complex with bridging *trans*-bis(diphenylphosphino)ethylene exists as a macrocycle in solution but crystallizes as the ring-opened polymer, whose structure is shown in Figure 4.[7] The individual chains are associated through weak interchain attractions between AuS groups as shown in Figure 4.

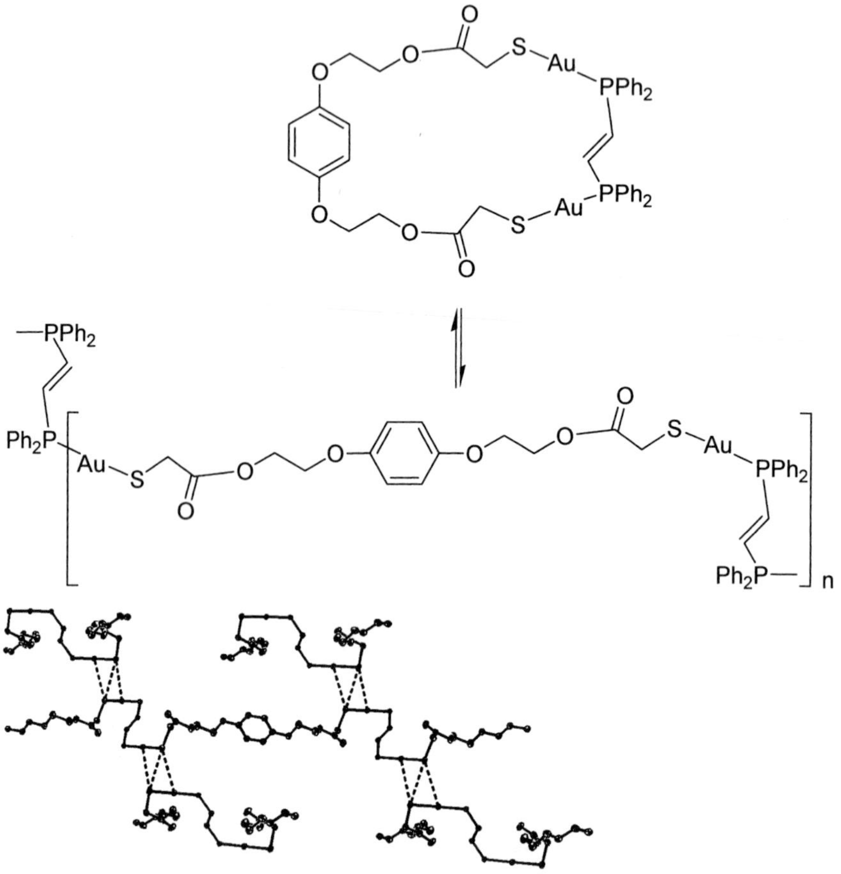

Figure 4. The structure of the dithiolate(diphosphine)digold(I) polymer. The dashed lines indicate weak interchain bonds and the phenyl groups of the diphosphine are omitted for clarity.

An example of a bis(pyridine) complex of gold(I) thatcan ring open is shown in Figure 5. When the diphosphine is $Ph_2PCH_2CH_2PPh_2$, the complex exists in solution as a mixture of the macrocycle and the ring-opened oligomer. Crystallization then yields the unusual polymer which contains alternating left and right handed helical turns along the polymer chain. The helicity is induced by the amide links in the bis(pyridine) ligand.[8]

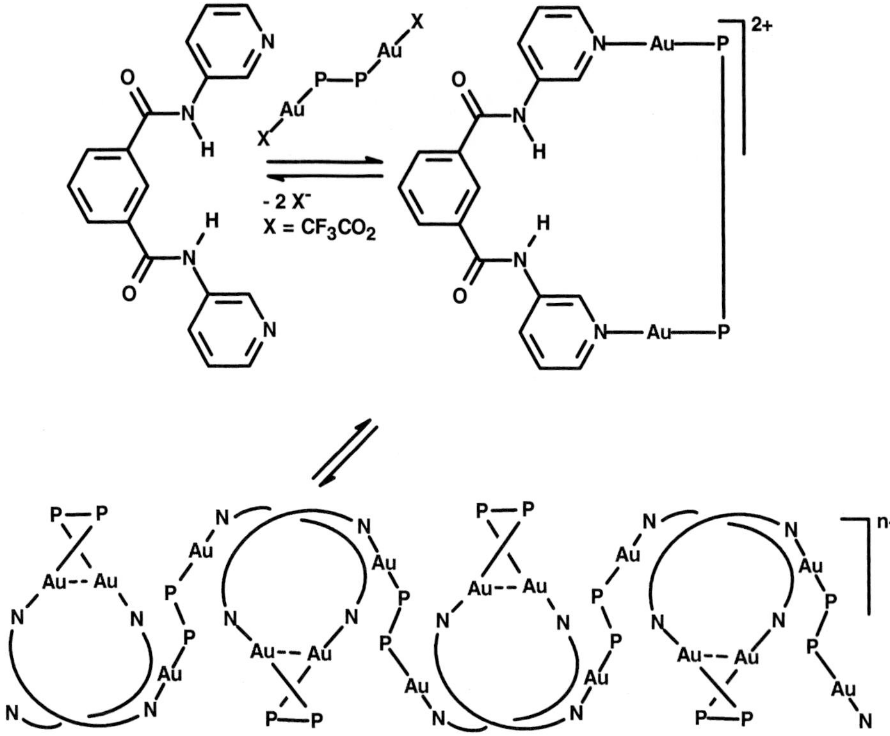

Figure 5. Ring-opening polymerizationof a cationic bis(pyridine)digold(I) macrocycle.

In this example the amide groups of the ligand are not involved in interchain hydrogen bonding but, in other cases, such bonding can occur and can lead to supromolecular association of the polymer chains to give duplexes, or sheet or network structures. An example is shown in Figure 6, in which the supramolecular structure formed through interchain hydrogen bonding is a network structure.[8]

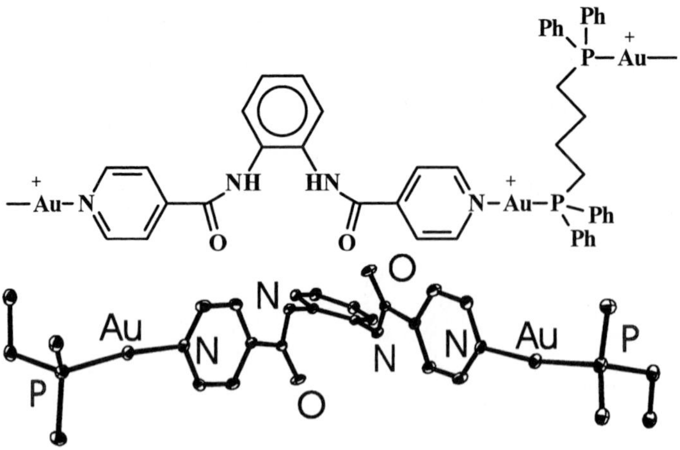

Figure 6. The repeat unit in the polymer chain of a cationic gold(I) polymer. The chains are crosslinked through interchain NH..O=C hydrogen bonding to give a supramolecular network structure.

Double Ring-Opening Polymerization

A polycyclic transition metal complex can undergo multiple ring-opening and this can also lead to formation of unusual structures. The first ROP gives a chain structure and the second gives a sheet.

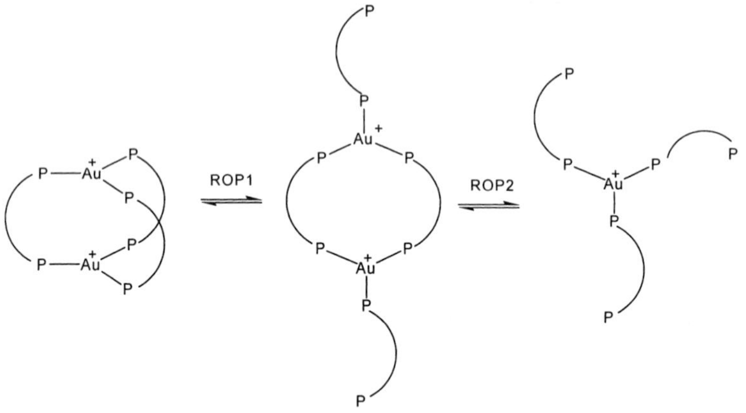

The structures of two such reactions are shown in Figures 7 and 8. The one-dimensional chain compound is formed when the diphosphine PP = *trans*-$Ph_2PCH{=}CHPPh_2$ and is shown in Figure 7. It contains alternating ring and chain sections, like a chain-link fence.[10]

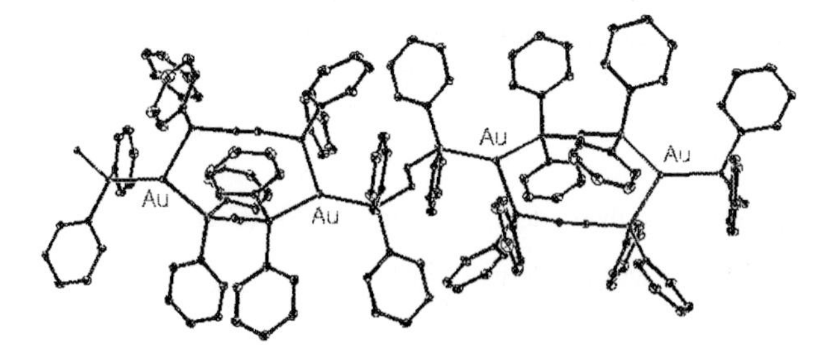

Figure 7. The chain link fence structure formed by single ring-opening polymerization (trifluoroacetate anions are not shown).

The double ring-opening polymerization occurs when the the diphosphine PP = $Ph_2P(CH_2)_4PPh_2$ and the structure is shown in Figure 8 (left).[10] The anions in this case are $[Au(CN)_2]^-$ and two of them are present in each giant ring. This is a honeycomb structure with the gold atoms having trigonal planar stereochemistry. A similar double ring opening is observed with silver(I) when the the diphosphine PP = $Ph_2P(CH_2)_6PPh_2$ and the structure is shown in Figure 8 (right).[11] In this case, the trifluoacetate anions also coordinate to silver(I), which therefore has distorted tetrahedral stereochemistry. In this case, the structure is a puckered sheet with alternate anions coordinated on either side of the plane of the sheet structure.

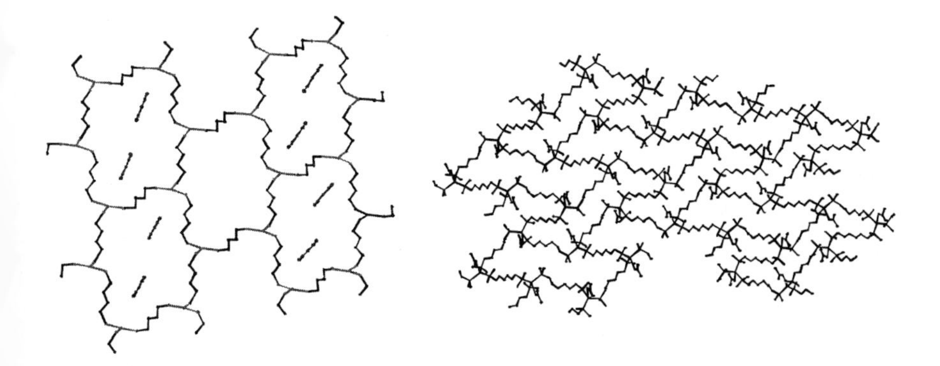

Figure 8. The honeycomb (left) and puckered sheet (right) structures (phenyl groups omitted).

Conclusion

Ring-opening polymerization is shown to be a very powerful method for the synthesis of polymeric, sheet and network structures. Further structural complexity and ordering may be obtained by using interchain aurophilic attractions or hydrogen bonding. These synthetic advances should prove useful in the design of functional molecular materials.

Acknowledgments

I thank the excellent students listed below for their ideas and efforts, NSERC and EMK for funding, and the Government of Canada for a Canada Research Chair.

[1] R.J. Puddephatt, *Coord. Chem. Rev.*, **2001**, *216-217*, 313.
[2] M.-C. Brandys, R.J. Puddephatt, unpublished work
[3] S. Achar, J.D. Scott, R.J. Puddephatt, *Organometallics*, **1992**, *11*, 2325.
[4] S. Achar, J.D. Scott, J.J. Vittal, R.J. Puddephatt, *Organometallics*, **1993**, *12*, 4592.
[5] G. Jia, R.J. Puddephatt, J.J. Vittal, N.C. Payne, *Organometallics*, **1993**, *12*, 263.
[6] F. Mohr, R.J. Puddephatt, unpublished work.
[7] W.J. Hunks, M.C. Jennings, R.J. Puddephatt, *Chem. Comm.*, **2002**, 1834.
[8] Z. Qin, M.C. Jennings, R.J. Puddephatt, *Chem. Eur. J.*, **2002**, *8*, 735.
[9] T. Burchell, R.J. Puddephatt, unpublished work.
[10] M.-C. Brandys, R.J. Puddephatt, *J. Am. Chem. Soc.,* **2001**, *123*, 4839.
[11] M.-C. Brandys, R.J. Puddephatt, *J. Am. Chem. Soc.,* **2002**, *124*, 3946.

Dendritic Polymers: From Efficient Catalysis to Drug Delivery

Ashok K. Kakkar

Department of Chemistry, McGill University, 801 Sherbrooke St. West, Montreal, Quebec H3A 2K6 Canada E-mail: ashok.kakkar@mcgill.ca

Summary: Dendritic polymers constitute an intriguing class of macromolecules that offer tremendous potential in designing new materaisl for applications in areas such as catalysis and small molecule loading and delivery. Synthesis of a variety of dendritic polymers using a simple and highly versatile synthetic methodology has enabled us to carry out a detailed investigation of dendritic effects in transition metal catalyzed organic transformations. Small dye molecules such as *p*-nitroaninline and DR1 could be loaded into the intrinsic cavities of the backbone of 3,5-dihydroxybenzyl alcohol based dendrimers, leading to a change in physical properties of both the dye and the dendrimer. We are also exploring the use of dendrimers as templates to prepare network carriers containing cavities of predetermined size and disposition.

Keywords: catalysis; dendritic networks; olefin hydrogenation; organometallic dendrimers; small molecule loading

Introduction

Dendrimers constitute an intriguing class of macromolecules with a unique set of properties which can be related to their monodisperse nature and a well-defined three-dimensional structure. The potential offered by these systems of nanoscopic dimensions for applications in significant areas such as recoverable catalysts and medicinal chemistry, continues to provoke intense interest in their syntheses and a detailed understanding of their physical characterisitics.[1] Divergent (inside out) and convergent (outside in) are two common methodologies that have been extensively used to prepare a variety of dendrimers with varied backbone architecture. In principle, divergent build-up of dendrimers could lead to inherent defects in generation-by-generation build-up of dendrimers. However, it is an easy synthetic route to prepare large macromolecules with an overall high structural integrity.

Metallodendrimers, a much more recent addition to the dendrimer family, provide a good platform to address key issues related to recoverable catalysts.[2] The latter class of dendrimers

© 2003 WILEY-VCH Verlag GmbH & KGaA, Weinheim CCC 1022-1360/00/$ 17,50+.50/0

can be divided into two categories, the more common organic backbone dendrimers that are periphery functionalized with organometallic fragments, and those in which organometallic moieties are distributed throughout the backbone. A large variety of organometallic dendrimers have now been prepared,[3] however, key issues related to the effect of dendritic shape and the location of functional groups in the dendritic architecture in catalysis, still remain. In addition, the advantages of using multistep build-up of dendrimers which is an arduous task, compared with their hyperbranched analogs prepared using simple one-pot reaction chemistry, are not clearly defined. It is understood that dendrimers have a perfect structure compared to hyperbranched polymers, but the role of defects in altering catalytic efficiency, needs to be elaborated. A detailed investigation of the behavior of similar organometallic dendritic polymers under catalytic conditions will address these important concerns.

Efficient loading and delivery of small molecules are two noteworthy goals in drug design.[4] To achieve these, one needs to construct backbone structures with tailorable nanosized cavities and well-defined structures. There is intense research activity in this area and several strategies to develop such nanostructures have been proposed. These include conjugated proteins,[5] liposomes,[6] antibodies,[7] and foldamers.[8] More recently there has been much interest in exploiting the intriguing properties of dendrimers including a well-defined molecular architecture for applications in drug delivery.[9-11] Despite extensive efforts in this area, synthesizing materials with holes of tunable sizes and distribution has alluded us. Dendrimers offer another unexplored venue in developing nanomaterials with pre-determined physical properties i.e., size and distribution of nanocavities. We are exploring a novel strategy that involves dendrimer templated sol-gel synthesis. The dendrimer generation helps control the size of the sphere from which sol-gel process is affected, leading to a well-defined architecture with precisely defined distribution of dendrimers in the matrix. The dendrimer fractions could then be removed from the backbone using simple hydrolysis, yielding desired nanomaterials.

Results and Discussion

Dendrimers in Catalysis: The synthetic methodology to construct dendrimers is based on a general acid-base hydrolysis strategy developed in our laboratory to prepare a variety of new materials ranging from simple monolayers to complex functionalized polymers and networks.[12] In order to evaluate dendritic effects in catalysis, a detailed study should involve the construction of dendrimers i) in which the transition metal centers are distributed throughout the backbone; ii) the transition metal fragments are attached at the periphery; iii) dendrimers in which the overall shape of the macromolecule is non-spherical; and iv) synthesis of hyperbranched polymers containing transition metals in the backbone or at the periphery. In order to prepare dendrimers containing transition metal centers distributed throughout the backbone, we developed a new divergent synthetic methodology to phosphorus containing dendrimers.[13] It involved the reaction of a trifunctional phosphine ($[HO(CH_2)_3]_3P$) with bis(dimethylamino)silane in an iterative sequence as shown in Scheme **1**. Using this methodology, dendrimers upto fourth generation containing phosphorus donor centers distributed throughout the backbone were prepared. The latter could be easily functionalized with Rh(I) fragments via a bridge splitting reaction with Rh(I) dimer $[Rh(COD)Cl]_2$. In order to ascertain that the metal centers are bound to each phosphine unit in the dendrimers, phosphine monomer ($[HO(CH_2)_3]_3P$) was first ligated with rhodium before being employed in the build-up of the dendrimer.

Characterization of the dendrimer structure is an important step in understanding dendritic effects in catalysis. In addition to ^{1}H and ^{13}C NMR spectra, MALDI-TOF mass spec and elemental analyses, ^{31}P NMR spectra provided useful information about the evolving structure of these dendrimers. The phosphorus centers in each generation were found to be chemical shift distinct, and the appearance of a new singlet upon each addition of a new layer was observed.[14] Upon Rh(I) complexation, coupling of phosphorus and rhodium nuclei led a doublet in the ^{31}P NMR spectra with a coupling constat of ~145Hz (Table **1**).

These Rh(I) organometallic dendrimers were found to be active in the hydrogenation of decene. In a typical run, the organic substrate to metal in a ratio of 200:1 were mixed in benzene and allowed to react at room temperature for 30 minutes under 20 bars of hydrogen pressure.

Scheme **1**

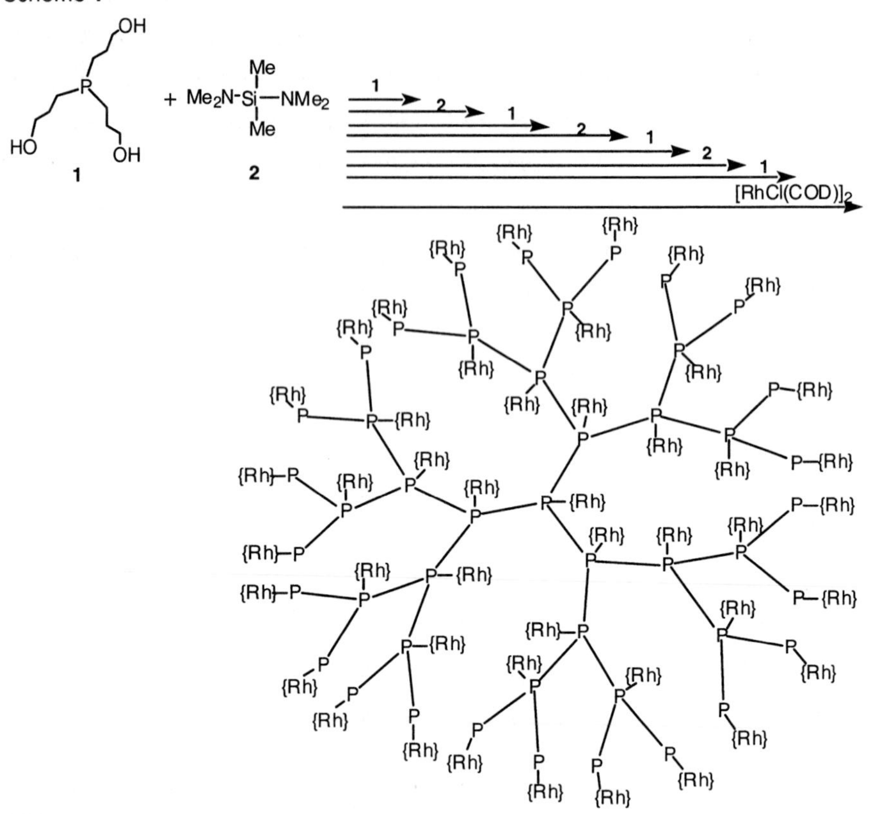

Table **1**

Compound	$^{31}P\{^{1}H\}$ NMR δ ppm	Conversion %	TOF (h^{-1})
RhP_1, $CODRhClP[(CH_2)_3OH]_3$	11.26 (d)	97.3	413
Rh_4P_4	15.4 (d), 18.7 (d)	100	399
$Rh_{10}P_{10}$	8.62 (d), 14.53 (d), 18.4 (d)	98.7	401
$Rh_{22}P_{22}$	8.73 (d), 10.89 (d), 14.55 (d), 18.39 (d)	99.2	400
$Rh_{46}P_{46}$	9.36 (d), 11.18 (d), 12.04 (d), 14.62 (d), 18.36 (d)	98.3 93.5 (Recycle)	393 374 (Recycle)

A conversion efficiency of 97-100% was observed with generations 0-4 (Table **1**). The catalyst was easily separated from the reaction mixture by crystallization from THF/hexanes mixture, and upon recycling showed efficiency that was only slighly lowered from the original run.[14]

The synthesis of dendrimers in which the peripheries could be functionalized with organometallic fragments was achieved by reacting 3,5-dihydroxybenzylalcohol with bis(dimethylamino)silane in a step-by-step build-up of generations 1-3.[15] These dendrimers contain 6-24 terminal OH groups which were used to anchor phosphine donor ligands by reacting them with bis(dimethylamino)silane and $Ph_2P(CH_2)_3OH$ in seqeunce. The latter were then reacted with Rh(I) dimer $[Rh(COD)Cl]_2$ to yield dendrimers containing 6-24 organometallic centers at the periphery (Scheme **2**).[15]

Scheme **2**

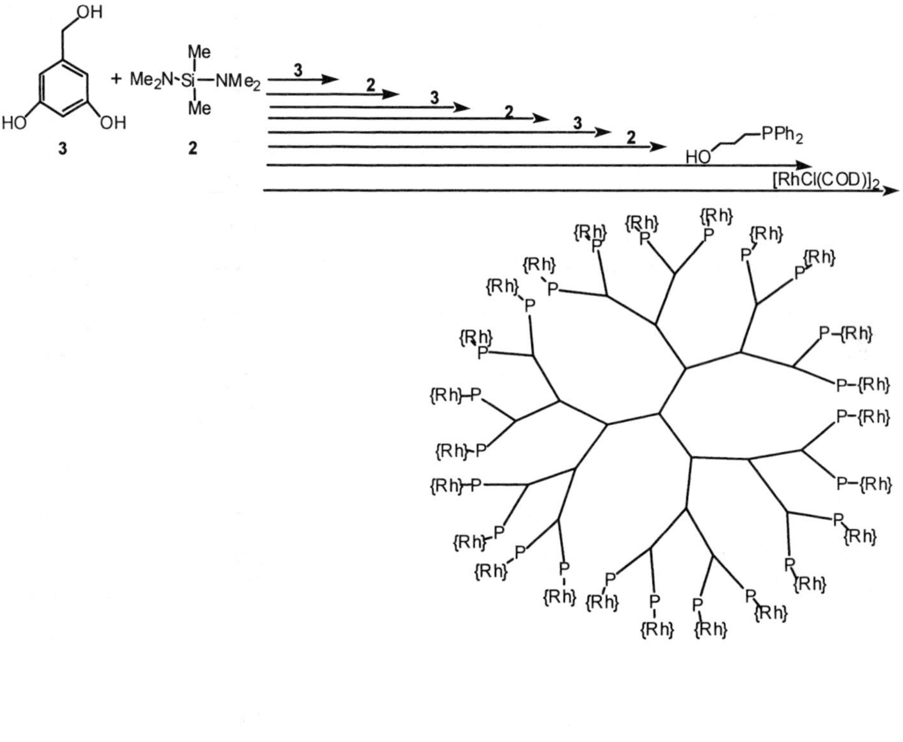

OH

+ Me_2N-Si(Me)(Me)-NMe_2

HO OH

3 2

3, 2, 3, 2, 3, 2, HO~~~PPh2, [RhCl(COD)]2

{Rh} P

The catalytic efficiency of these dendrimers was examined under similar conditions as those used above (200:1 decene to metal ratio; 20 bars H_2 pressure, room temperature). The hydrogenation efficiency was found to be dependent on dendrimer generation and reaction time.[15] It increased upon an increase in generation number and reaction time. Dendrimer generation 3 functionalized with Rh(I) showed maximum conversion in 2h while generations 1-2 showed maximum conversion only after 5 h of reaction time. The dendrimer supported catalysts could be easily recovered and recycled with retention of catalyst activity.

In order to examine the role of dendrimer shape on catalyst efficiency, we then prepared organometallic dendrimers in which the overall dendrimer structure was dictated by the complexation environments of phosphorus ligands.[16] Tripodal polyphosphorus ligand $O{=}P[O(CH_2)_3PPh_2]_3$ in which the phosphorus at the core is unable to coordinate to the transition metals, carries out typical bridge-splitting reaction of the terminal phosphine units with the rhodium dimer $[(\mu\text{-}Cl)(1,5\text{-}C_8H_{12})Rh]_2$. The latter upon further reaction in sequence first with $PhP[O(CH_2)_3PPh_2]_3$ and then the rhodium dimer yields, typical spherical multimetallic complexes (Scheme **3**).[16,17] On the other hand, treatment of tripodal polyphosphorus ligands such as $PhP[O(CH_2)_3PPh_2]_2$ and $P[O(CH_2)_3PPh_2]_3$ with the rhodium dimer $[(\mu\text{-}Cl)(1,5\text{-}C_8H_{12})Rh]_2$ leads to complexation with rhodium at the bridgehead phosphite group first, and then replacement of the 1,5-cyclooctadiene ligand with two terminal phosphines. Continuation of the dendrimer build-up process using $PhP[O(CH_2)_3PPh_2]_2$ or $P[O(CH_2)_3PPh_2]_3$ and the rhodium dimer $[(\mu\text{-}Cl)(1,5\text{-}C_8H_{12})Rh]_2$ in sequence, leads to organometallic dendrimers with atypical spherical structures.[16,17] The structures of the multimetallic compounds were also controlled by using phosphines with longer alkyl arms. For example, in contrast to the phosphines, $PhP[O(CH_2)_3PPh_2]_2$ or $P[O(CH_2)_3PPh_2]_3$, $PhP[O(CH_2)_{10}PPh_2]_2$ and $P[O(CH_2)_{10}PPh_2]_3$ react with the rhodium dimer in a typical bridge splitting reaction without displacement of the 1,5-cyclooctadiene ligand.[16] The catalytic activity of the multimetallic compounds with a typical spherical structures for hydrogenation of decene (1:200 metal to substrate ratio, 20 bar H_2 pressure, room temperature) was found to be lower than similar non-spherical compounds under similar conditions.

Scheme 3

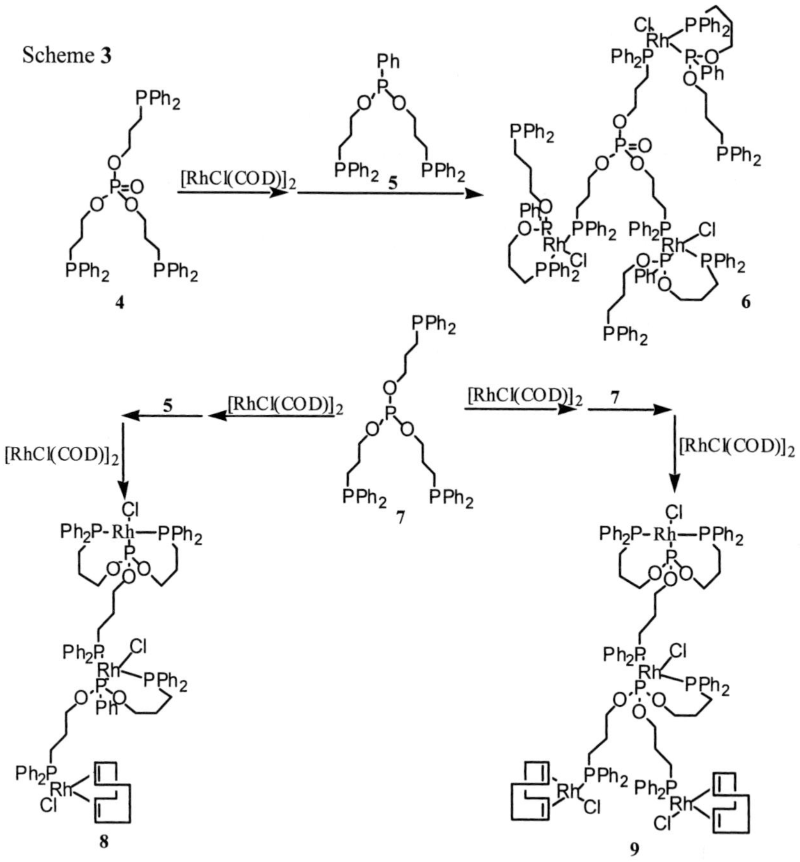

Dendrimers as Hosts: Small molecule loading and its efficient delivery consitute a topical area of research, and dendrimers offer a unique opportunity to achieve these goals. Due to a layer-by-layer build-up that includes adding branching units in each generation, dendrimers contain a densely packed periphery but a hollow interior. Suitable functional groups ideally located at the periphery can also lead to aggregation using weak intermolecular interactions such as hydrogen bonding. 3,5-dihydroxybenzyl alcohol based dendrimers containing terminal OH groups have been demonstrated to yield aggregates of 130-180 nm in diameter depending on the concentration and dendrimer generation.[15] The aggregation phenomenon is reversible, and offers potential in encapsulating small molecules of interest in hollow interiors of these dendrimers at higher dilutions, followed by their assembly at above critical micellar concentrations. We have studied this behavior of dendrimers with a 3,5-dihydroxybenzyl alcohol backbone using the commercially available reagents such as *p*-nitroaniline and the dye DR1.[18] Encapsulation of DR1 (Scheme **4**) leads to changes in the physical properties of the dendrimer and the dye. For example, the solubility of the third-generation dendrimer is significantly enhanced upon loading with DR1. The UV-Vis absorption spectrum of DR1 is blue shifted upon trapping in the dendrimer interior. Upon loading of small molecules, the dendrimer could be easily zipped by reacting the peripheral OH groups withMe_3SiNEt_2 that leads to complete entrapment of the dye molecules.

Dendrimers as Templates: Some of the key issues in the design of carriers include, control on the size of the internal cavities into which designed molecules could be trapped, and their regular distribution in the backbone. We are studying a novel technique to achieve these goals by creating a merger of dendrimer chemistry with our newly developed sol-gel methodology.[19] The globular macromolecules of nanodimensions that have reactive groups at the periphery can be used to build network structures from which dendrimer fractions are finally removed. The terminal OH groups of different generations of 3,5-dihydroxybenzylalcohol based dendrimers are first reacted with $Si(NMe_2)_4$, followed by slow hydrolysis and condensation.[20] The dendrimer fractions are then removed from the network material by acid hydrolysis. Such an approach provides the desired control in tailoring the size and distribution of cavities that can be fine tuned to the specific needs of the host molecules. The latter can be physically trapped in the hollow spaces or chemically bound using residual Si-OH groups.

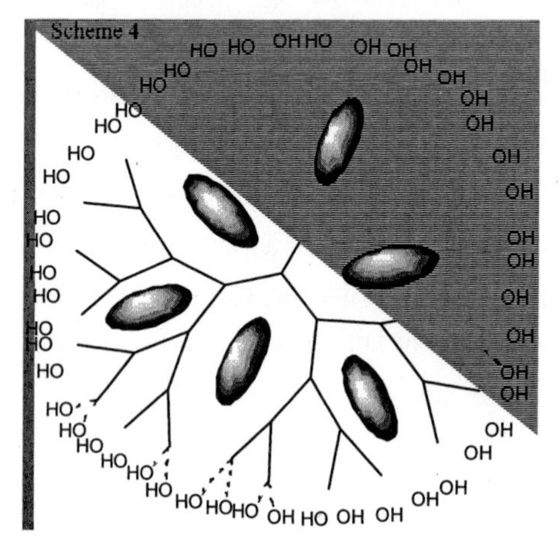

Scheme 4

Hyperbranched Polymers: We have also used our acid-base hydrolytic chemistry route to prepare hyperbranched polymers in which the organometallic fragments can be added throughout the backbone or are attached at the periphery.[3,15] To prepare the former type, uncontrolled reaction of ($[HO(CH_2)_3]_3P$) with bis(dimethylamino)silane was carried out in a one pot reaction of these building blocks in a 2:3 molar ratio. It led to the formation of a white viscous liquid that showed broad peaks in its ^1H NMR spectra, and three singlets in its $^{31}P\{^1H\}$ NMR spectra. The MALDI-TOF mass spectrum of the polymer showed a maximum mass corresponding to that of a second generation dendrimer. The data confirmed the polymeric and hyperbranched nature of these compounds. The phosphine backboned polymer was easily metallated upon reacting with $[Rh(COD)Cl]_2$ with the appearance of a doublet in the $^{31}P\{^1H\}$ NMR spectra ($J_{Rh\text{-}P}$ = 140 Hz) with no residual free phosphine.[14]

To prepare hyperbranched polymers containing phosphine donors at the periphery, 3,5-dihydroxybenzyl alcohol was reacted with bis(dimethylamino)silane in a manner typical of dendrimer construction (stepwise addition) but without any control on reaction conditions. It led to the formation of a hyperbranched polymer with a dendritic type growth but with more defects, as expected.[15] The peripheries of these polymers could be easily functionalized with donor phosphine ligands by reacting them first with $Me_2Si(NMe_2)_2$ and then with $Ph_2P(CH_2)_3OH$.[21] The latter were then reacted with $[Rh(COD)Cl]_2$ that yielded hyperbranched polymers functionalized with organometallic fragments at the periphery. A detailed investigation of the catalytic properties of these hyperbranched polymers containing active metallic sites in the backbone or at the periphery is currently in progress.

Conclusions

Acid-base hydrolytic chemistry of molecules containing multisite OH groups with aminosilanes is a useful approach to construct a variety of dendritic structures. The position of donor groups in these globular structures can be tailored to include them throughout the backbone or exclusively at the periphery. Upon metallation with rhodium, the resulting organometallic dendrimers are found to be efficient and recyclable catalysts for olefin hydrogenation. The dendrimers containing rhodium centers distributed throughout the cascade structures are more efficient in catalysis while the efficiency of periphery functionalized dendrimers is dependent on the generation number and temperature. We have also developed a methodology to build organometallic dendrimers in which their overall shape can be tailored by controlling the type of polyphosphorus ligands at the core. The latter is helping us evaluate the effect of dendrimer shape in catalysis. The hollow interiors of these dendrimers can be used as hosts for small molecules as evidenced by the successful entrapment of *p*-nitroaniline and the dye DR1. Using dendrimers as templates for building network materials using mild sol-gel processing affords structures with cavities of tailorable size and distribution. A detailed investigation of the latter materials is currently being pursued.

Acknowledgment

We would like to thank NSERC (Canada) and FCAR (Quebec, Canada) for financial support.

[1] A.W. Bosman, H.M. Janssen, E.W. Meijer, *Chem. Rev.* **1999**, *99*, 1665.
[2] R. van Heerbeek, P.C.J. Kamer, P.W.N.M. van Leeuwen, J.N.H. Reek, *Chem. Rev.* **2002**, *102*, 3717.
[3] M. Dasgupta, M.B. Peori, A.K. Kakkar, *Coord. Chem. Rev.* **2002**, *233-234*, 223.
[4] M.V. Backer, R. Aloise, K. Przekop, K. Stoletov, J.M. Backer, *Bioconjugate Chem.* **2002**, *13*, 462.
[5] E.K. Gaidamakova, M.V. Backer, J.M. Backer, *J. Controlled Release*, **2001**, *74*, 341.
[6] G.M. Dubowchik, M.A. Walker, *Pharmacol. Ther.* **1999**, *83*, 67.
[7] J. Kriangkum, B. Xu, L.P. Nagata, R.E. Fulton, M.R. Suresh, *Biomol. Eng.* **2001**, *18*, 31.
[8] D.J. Hill, M.J. Mio, R.B. Prince, T.S. Hughes, J.S. Moore, *Chem. Rev.* **2001**, *101*, 3893.
[9] H.R. Ihre, O.L. Padilla De Jesus, F.C. Szoka Jr., J.M.J. Frechet, *Bioconjugate Chem.* **2002**, *13*, 443.
[10] S-E. Stitiba, H. Frey, R. Haag, *Angew. Chem. Int. Ed.* **2002**, *41*(8), 1329.
[11] R. Goller, J-P. Vors, A-M. Caminade, J-P. Majoral, *Tet. Lett.* **2001**, 3587.
[12] (a) C.M. Yam, A.J. Dickie, A.K. Kakkar, *Langmuir*, **2002**, *18*, 8481. (b) A.J. Dickie, A.K. Kakkar, M.A. Whitehead, *Langmuir*, **2002**, *18*, 5657. (c) A.K. Kakkar, *Chem. Rev.* **2002**, *102*, 3579. (d) F. Chaumel, H. Jiang, A.K. Kakkar, *Chem. Mater.* **2001**, *13*, 3389. (e) H. Jiang, A.K. Kakkar, *Adv. Mater.* **1998**, *10*, 1093. (f) H. Jiang, A.K. Kakkar, *Macromolecules*, **1998**, *31*, 4170.
[13] M.G.L. Petrucci, V. Guillemette, M. Dasgupta, A.K. Kakkar, *J. Am. Chem. Soc.* **1999**, *121*, 1968.
[14] M.G.L. Petrucci, Ph.D. Thesis, McGill University, 1999.
[15] O. Bourrier, A.K. Kakakr, *J. Mater. Chem.* **2003**, in press.
[16] M. Brad Peori, A.K. Kakkar, *Organometallics*, **2002**, *21*, 3860.
[17] M. Brad Peori, M.Sc. Thesis, McGill University, 2002.
[18] J. Butlin, A.K. Kakkar, unpublished results.
[19] H. Jiang, A.K. Kakkar, *J. Am. Chem. Soc.* **1999**, *121*, 3657.
[20] A. Tsiodras, A.K. Kakkar, unpublished results.
[21] S. Dufresne, A.K. Kakkar, unpublished results.

Synthesis of Novel Phenylazomethine Derivatives as a Multi-Dentate Ligand for Advanced Metal-Organic Hybride Nano-Materials

Kimihisa Yamamoto

Department of Chemistry, Keio University, Yokohama 223-8522, Japan
E-mail: yamamoto@chem.keio.ac.jp

Summary: Oligophenylazomethines (OPAs) and the aniline-capped OPAs were synthesized via dehydration of 4-aminobenzophenone in the presence of $TiCl_4$ as a Lewis acid. $TiCl_4$ acts as an excellent dehydration agent for formation of the C=N bond. OPAs have some isomers, which are based on the *E*/*Z* conformation of the azomethine moieties. Novel cyclic molecules, the cyclic phenylazomethine trimer (CPA_3-ab) derivatives, were synthesized via the dehydration of the 4-aminobenzophenone. The yields of the macrocycles were enhanced to over 90% by induction of the bulky α-substituent to the substrate. The highly preferential formation of novel aabb-type phenylazo-methine macrocycles was also achieved. The resulting CPA_6-aabb has unique structure based on the *E*/*Z* conformation of the azomethine bonds, the extremely regular molecular-packing state. The novel dendritic phenylazomethines(DPAs), were synthesized by the convergent method. The DPAs have a high solubility unlike the conventional linear polyphenylazomethines. The tin chloride complexed to the imine groups of DPAs in a stepwise fashion. The complexation occurs in the order of 1st, 2nd, 3rd, and 4th layers of DPA G4.

Keywords: dendrimer; macrocycle metal-organic hybride; multi-dentate ligand; nano-materials; phenyazomethine

Introduction

In the past several years, much interest has been devoted to the study of rigid and/or π-conjugated chain polymers with metal ion, π-conjugated macromolecule-metal complexes, mainly because of their superior mechanical, electronic , photonic, and magnetic properties.[1-4] These malecules will provide novel advanced organic-metal hybrid nanomaterials for the next generation.[5,6] Azo-methine polymers which have a long history, should be carefully noted in order to find novel properties and applications becase the polymers act as useful multi-dentate ligands for metal assembling. [7-14] The phenylazomethine polymers exhibit an attractive thermal stability, good

© 2003 WILEY-VCH Verlag GmbH & KGaA, Weinheim CCC 1022-1360/00/$ 17,50+.50/0

mechanical strength, meltability, fiber-forming properties, electronic conductivity, redox activity, photoluminescence, non-linear optical properties and so on. [15-17] We focused on the novel azomethine polymers[18-24] with a supramolecular structure such as cyclic and dendritic topology. Dendrimers[25-33] are highly branched organic macromolecules with successive layers or `*generations*' of branch units surrounding a central core with a beautiful sphere-like shape. The rigid dendrimers, especially dendrimers with a π-conjugated backbone, such as the phenylazomethine dendrimers are expected to expand the field of nano-marerials because dendrimers have the possibility to be regularly assembled with a packing structure on a plate. [34-37] Organic inorganic hybrid versions have also been produced by trapping metal ions or metal clusters within the voids of the dendrimers. [38-44] Here we show that the tin chroride, $SnCl_2$, complex to the imines groups of a spherical polyphenylazomethine dendrimer in a stepwise fashion[45-47] according to an electron gradient, with complexation in a more peripheral generation proceeding only after complexation in generations closer to the core has been completed. It is possible to control the number and location of metal ions incorporated into the dendrimer structures, which might are used as tailored catalysts or building blocks for advanced materials. In this review, the synthesis and properties of novel phenyl azomethine polymers including cyclic and dendritic structures were introduced in recent studies.

Synthesis of Oligophenylazomethines

For the dehydration of aldehydes/ketones with amines, the addition of an acid is sometimes necessary to enhance the electrophilicity of the carbonyl carbon. $TiCl_4$ is an effective Lewis acid for this dehydration. On the other hand, *p*-toluenesulfonic acid (PTS) is often used for the dehydration of aldehydes with amines, but ineffective in the reaction of aromatic ketones with aromatic amines because of the low electrophilicity of the carbonyl carbon having expanded π-conjugation. Oligophenylazomethines (OPAs) are synthesized via the dehydration of 4-aminobenzophenone in the presence of $TiCl_4$ and diazabicyclo[2.2.2]octane (DABCO) (Scheme 1). This dehydration is known to proceed via metathesis of benzophenone with a Ti=N compound, which is formed by the reaction of aniline with $TiCl_4$. In this oligomerization using PTS, the oligomers over the tetramer were not obtained due to the low reactivity. Aniline-capped OPAs are obtained in a one-pot synthesis. These results support the fact that $TiCl_4$ is a useful dehydration agent to form additional azomethine bonds without hydration of the already-formed azomethine bonds, because the dehydration using $TiCl_4$ is an irreversible reaction unlike PTS. ^{13}C NMR measurement revealed that the OPA3(trimer) and OPA4(tetrame) have 2 and 4 isomers, respectively, based on the *E/Z* conformation of the azomethine moiety. On the basis of the area ratio of 2 peaks attributed to the carbonyl carbon in OPA3, the formation ratio of the *E/Z* isomers in an α-phenyl-substituted azomethine bond, in principle, is 1:1. The ratio in OPA4 was determined to be 9:6:3:1. Thses results show that the formation ratio of isomers in the OPAs changes based on the steric hindrance between the intramolecular phenyl rings. The α-Ph-phenylazomethine polymers are considered to have various non-linear structures due to the *Z* conformational azomethine bonds unlike the H/Me-substituted ones.

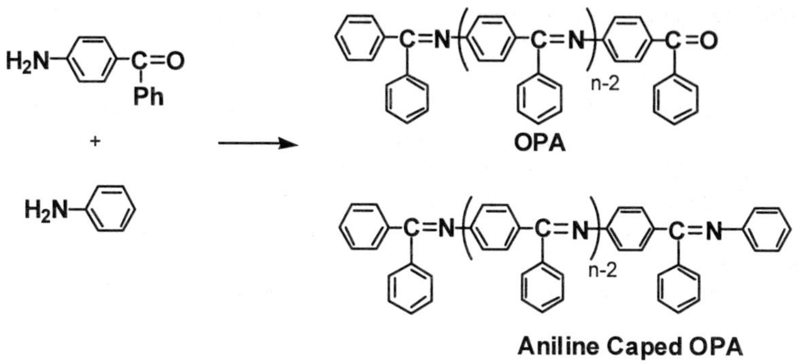

Scheme 1. Synthesis of OPAs by dehydration in the presence of $TiCl_4$.

Selective Synthesis of Macrocycles

We succeeded in the selective synthesis of cyclic tris[(α-phenyl)phenyl-azomethine] derivatives with a high yield using $TiCl_4$(Figure 1). Our results reveal that benzophenone with a bulky group at the α-position facilitates the predominant formation of the cyclic structure. The formation of the novel cyclic (α-phenyl) phenylazomethine trimer (CPA_3-ab) was carried out by dehydration of 4-aminobenzophenone using $TiCl_4$. [48,49] Using titanium(IV) tetrachloride as a Lewis acid, only the CPA_3-ab trimer was obtained with a 20% yield in spite of the non-dilute conditions. The high yield and the selectivity are caused by the steric effect of the bulky α-phenyl ring of the monomer. The selectivity and yield were emphasized by the bulkiness of the monomer. The dehydration of the 4-amino-4'-octylaminobenzophenone resulted in the formation of the corresponding cyclic trimer in 49% yield. The dehydration of 4-amino-4'-dioctylamino-benzophenone gave the corresponding trimer with a 92% isolated yield. These results indicate that bulky substituents at the α -position of the phenylazomethine facilitate the forming of the cyclic phenylazomethine trimers. The single conformation of CPA_3-ab is supported by NMR, and molecular modeling of the *E* isomer of CPA_3-ab is very difficult to build. These results support the idea that CPA_3-ab only has the *Z* conformation.

During the course of the AABB-type polycondensation of 1,4-dibenzolybenzene with 1,4-phenylenediamine, the total macrocyclization was realized by further addition of $TiCl_4$ and the diamine monomer. Only peaks attributed to the novel cyclic polyphenylazomethines (CPA_n-aabb, where n is the degree of polymerization) were confirmed in the TOF-MS spectrum of the crude products (Figure 2).

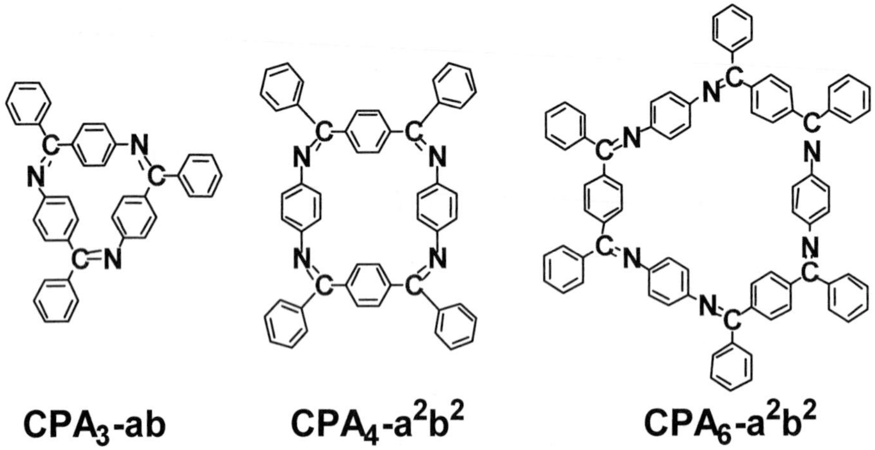

CPA_3-ab　　CPA_4-a^2b^2　　CPA_6-a^2b^2

Figure 1. Structures of phenylazomethine macrocycles.

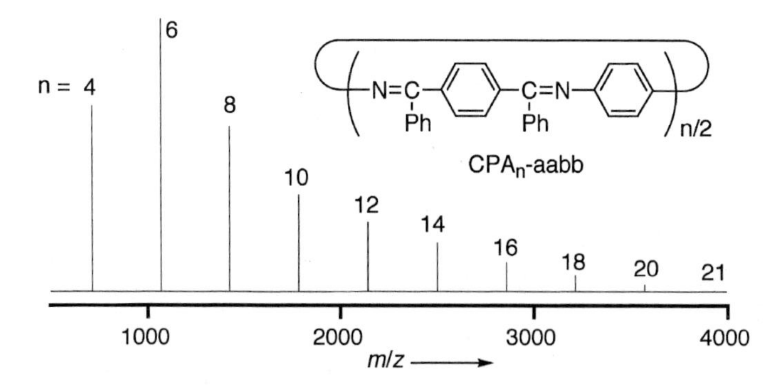

Figure 2. MALDI-TOF-MS spectrum of CPA_n-aabb.

The CPA_n-aabb products (n = 4, 6, 8, 10, 12, 14, 16, 18, and >20) were easily isolated in 13, 23, 16, 11, 8, 6, 5, 3, and 6% yields (total: 91%), respectively, by gel permeation chromatography. The ^{13}C NMR spectra of CPA_4-aabb and CPA_6-aabb are relatively simple, which support the fact that they have only one or a few isomers. One peak attributed to the azomethine carbon in the spectrum of CPA_4-aabb shows that CPA_4-aabb has a single isomer with one (*Z*) conformation of the azomethine bonds. The ^{13}C NMR spectrum of CPA_6-aabb shows that CPA_6-aabb has three isomers (Scheme 2).

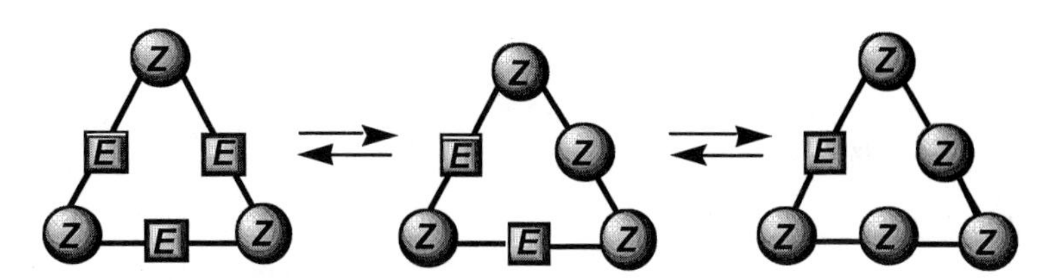

Scheme 2. *E/Z* conformations of CPA_6-aabb.

The molecular structure and packing of CPA_6-aabb were directly confirmed by X-ray crystal analysis. The triangle shape and *E/Z* conformations of the crystal molecule agreed with those of isomer I. Interestingly, the triangle molecules are two-dimensionally fine-packed (Figure 3), and aligned in a column. Such a unique and regular packing structure of the π-conjugated macrocycles has not yet been reported to the best of our knowledge. [50-54]

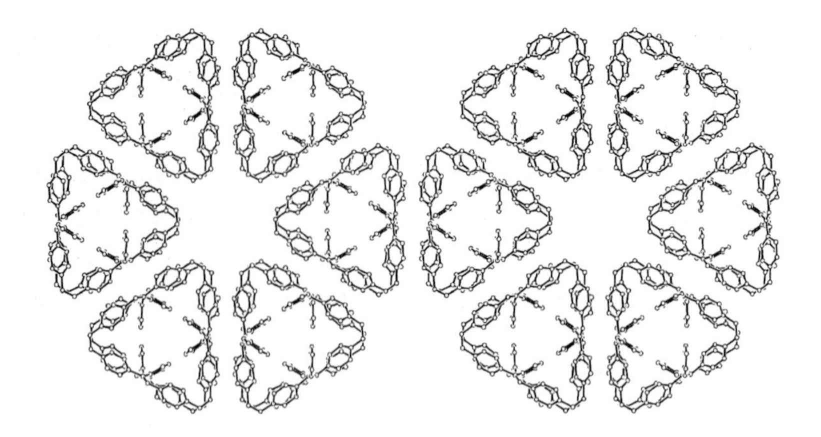

Figure 3. The packing structure(top view) of CPA_6-aabb. Crystal data, trigonal, space group *P-3c1*, a = 21.090(3) Å, b = 21.090(3) Å, c = 15.77(1) Å, U = 6075 (4) $Å^3$, Z = 4, R = 0.087, wR = 0.201.

Convergent Synthesis of Phenylazomethine Dendrimers (DPAs)

DPAs (DPA G1, G2, G3, and G4, designated as GX where X is the generation number) were synthesized by the convergent method as shown in Schemes 3. Benzophenone was allowed to react with 4,4'-diaminobenzophenone in the presence of $TiCl_4$ and DABCO. The DPA dendron G2 was synthesized via dehydration, and isolated by silica gel column chromatography with a 48% yield. DPA dendrons G3 and G4 were obtained in 64 and 20% yields by dehydration of the dendrons G2 and G3 with 4,4'-diaminobenzophenone, respectively. DPAs G1, G2, G3, and G4 were obtained by the dehydration with p-phenylenediamine, and isolated in 91, 62, 45, and 31% yields, respectively(Scheme 3). The high purity of the DPAs was confirmed by the GPC analysis, MALDI-TOF-MS, and ICP-MS.

The DPA dendrimers have a high solubility in the common solvents such as chloroform and DMSO, because their dendritic structures prevent intermolecular stacking. In each ^1H NMR spectrum of DPAs G1, G2, and G3, one singlet peak was observed at 6.6-6.3 ppm, which was attributed by COSY to the four protons of the phenyl ring in the core of the dendrimer. This singlet peak supports the symmetrical structure of the DPA. On the other hand, the singlet peak was not observed in the spectrum of DPA G4 because of the fixed conformation of the core by the bulky dendrons.

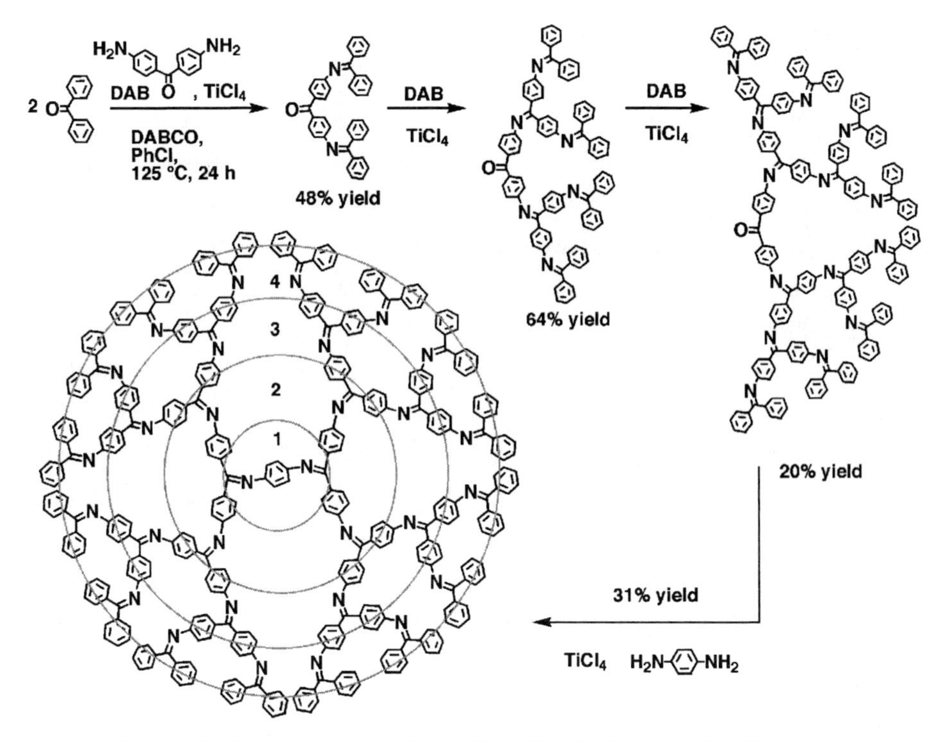

Scheme 3. The convergent synthesis of the DPA dendrons and dendrimers.

Spin-lattice (T1) relaxation measurements of the DPAs were perfomed. The NMR studies for the relaxation gave useful information about the structural density. With increasing generation, T1 of the external proton increased more than that of the internal one. This increase in T1 means a restriction in the molecular motion of the external phenyl rings; the exterior of DPA G4 is proposed to be close to the solid-state.

Stepwise Radial Complexation in DPA Dendrimers

The DPA dendrimers having many azomethine groups as the coordination site for metal ions increase their application as a novel nano-ordered film of the DPA-metal complex. The imine sites present in the DPAs strongly coordinate to various metal ions; $SnCl_2$ is a Lewis acid forming 1:1 complexes with imine compounds, and complexation of the 30 imine sites in DPA G4 should therefore yield a nano-sized spherical complex. The addition of $SnCl_2$ to a dichloromethane/acetonitrile solution of DPA G4 resulted in a color change from yellow to orange, attributed to complexation. Using UV-vis spectroscopy to monitor the titration until an equimolar amount of $SnCl_2$ had been added, we observed four changes in the position of the isosbestic point (Figure 4), indicating that the complexation proceeds not randomly, but stepwise. An isosbestic point appears when a compound is quantitatively transformed into another by complexation,[55] so the four different isosbestic points we observed suggest that four different complexes are successively formed upon the $SnCl_2$ addition. For DPA G4, four isosbestic points centered at 375, 364, 360 and 355 nm appeared when adding between 0-2, 3-6, 7-14 and 15-30 equivalents of $SnCl_2$, respectively. Overall, the isosbestic point shifted about 20 nm from 375 to 355 nm, and the number of added equivalents of $SnCl_2$ required to induce a shift was in agreement with the number of imine sites present in the different layers of DPA G4. The titration results suggest that based on our observations under equilibrium conditions, the process is proceeding in a stepwise fashion from the core imines to the terminal imines of DPA G4 as shown in Fig. 5. The metal ions are incorporated in a stepwise fashion, first filling the layers close to the dendrimer core and then progressively the more peripheral layers(Figure 5).

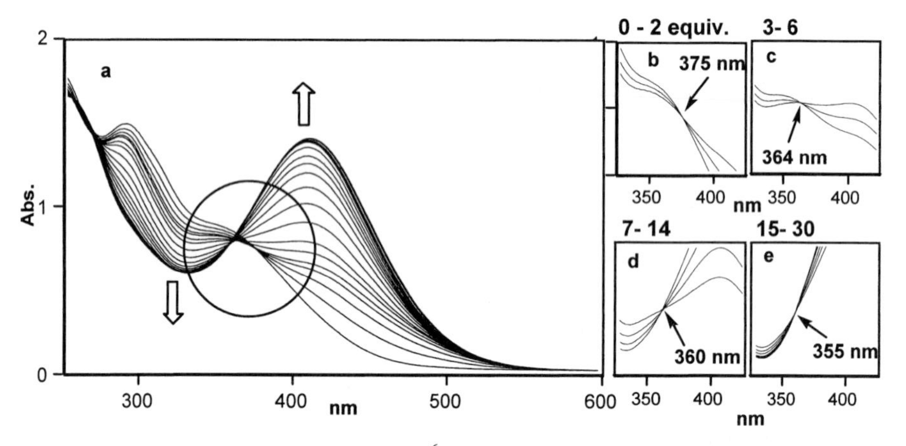

Figure 4. UV-vis spectra of DPA G4 (5 x 10^{-6} M) complexed with (a) 0-30, (b) 0-2, (c) 3-6, (d) 7-14, and (e) 15-30 equiv. of $SnCl_2$ (solv. 1:1 dichloromethane:acetonitrile).

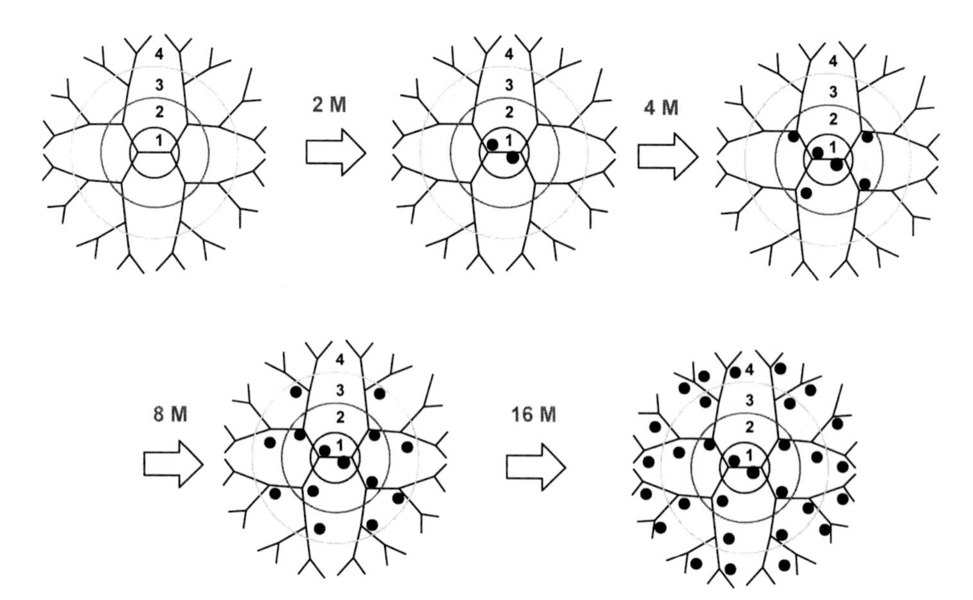

Figure 5. Stepwise radial complexation of sncl$_2$ in dpa g4 dendrimer.

Acknowledgement

This work was partially supported by a Grant-in-Aid for Scientific Research (Nos. 13022263, 13555261) and the 21st COE Program(Keio-LCC) from the Ministry of Education, Science, Culture and Sports, and a Grant-in-Aid for Evaluative Technology (12407) from the Science and Technology Agency, and Kanagawa Academy Science and Technology Research Grant(Project No. 23).

[1] *"Macromolecular-metal Complexes"*, eds. by F. Ciardelli, E. Tsuchida, D. Worle, Springer-Verlag, Berlin, 1995.
[2] *"Macromolecular Complexes"*, ed. by E. Tsuchida, VCH Publisher, New York, 1991.
[3] *"Metal Containing Polymeric Materials"*, ACS Symp. Ser. eds. by C. U. Pittman, B. M. Cullberston, J. E., Sheet, Plenum, New York, 1996.
[4] *"Handbook of Organic Conductive Molecules and Polymers"*, Vol. 1-5, ed. by H. S. Nalwa, Wiley-VCH, New York, 1997.
[5] Y. F. Lu, Y. Yang, A. Sellinger, M. C. Lu, J. M. Huang, H. Y. Fan, R. Haddad, G. Lopez, A. R. Burns, D. Y. Sasaki, J. Shelnutt, C. J. Brinker, *Nature*, **2001**, *410*, 913.
[6]R. M. Crooks, M. Zhao, L. Sun, V. Chechik, L. K. Yeung, *Accounts Chem. Res.*, **2001,** *34*, 181.
[7] R. Adams, R. E. Bullock, W. C. Wilson, *J. Am. Chem. Soc.*, **1923**, *45*, 521.
[8]C. S. Marvel, W. H. Hill, *J. Am. Chem. Soc.*, **1950**, *72*, 4891.
[9] K. Suematsu, K. Nakamura, J. Takada, *Polym. J.*, **1983,** *15*, 71.

[10]P. W. Morgan, S. L. Kwolek, T. C. Pletcher, *Macromolecules*, **1987**, *20*, 729.
[11]C. J. Yang, S. A. Jenekhe, *Macromolecules*, **1995**, *28*, 1180.
[12]O. Thomas, O. Inganäs, M. R. Andersson, *Macromolecules*, 1998, *31*, 2676.
[13] P. W. Morgan, S. L. Kwolek, T. C. Pletcher, *Macromolecules*, **1987**, *20*, 729
[14] S. Park, H. Kim, W. Zin, J. C. Jung *Macromolecules* **1993**, *26*, 1627
[15] T. Matsumoto, F. Yamada, T.Kurosaki *Macromolecules* **1997**, *30*, 3547.
[16] S.Destri, M.Pasini, C.Pelizzi, W. Porzio, G. Predieri, C. Vignali *Macromolecules* **1999**, *32*, 353.
[17]*"Recent Progress in Polycondensation"*, Ed. T. Matsumoto, Res. Signpost , K. Yamamoto, M. Higuchi, 153 (2002)
[18] M. Higuchi, K. Yamamoto, *Polym. Adv. Technol.*, **2002**, *13*, 765..
[19] K. Yamamoto, M. Higuchi, H. Kanazawa, Chem. Lett., **2002**, 692.
[20]M. Higuchi, K. Yamamoto, *Org. Lett.*, **1999**, *1*, 1881.
[21]M. Higuchi, A. Kimoto, S. Shiki, K. Yamamoto, *J. Org. Chem.*, **2000**, *65*, 5680.
[22]M. Higuchi, H. Kanazawa, M. Tsuruta, K. Yamamoto, *Macromolecules*, **2001**, *34*, 8847.
[23]M. Higuchi, H. Kanazawa, K. Yamamoto, *Org. Lett*, **2003**, *5*, 345.
[24]M. Higuchi, S. Shiki, K. Yamamoto, *Org. Lett.*, **2000**, *2*, 3079.
[25] *"Supramolecular Polymers"*, eds. by A. Ciferri, M. Dekker, New York, 2000.
[26]*"Dendrimer I, II, III"*, ed. by F. Vogtle, Springer, New York, 2001.
[27]*"Dendrimer and Dendrons"*, eds. by G. R. Newkome, C. N. Moorefield, F. Vogtle, Wiley-VCH, New York, 2001.
[28]D. A. Tomalia, H. Baker, J. R. Dewald, *Polym. J.*, **1985**, *17*, 117.
[29]G. R. Newkome, Z. Q. Yao, G. R. Baker, V. K. Gupta, *J. Org. Chem.*, **1985**, *50*, 2003.
[30]D. A. Tomalia, A. M. Naylor, W. A. Goddard III, *Angew. Chem. Int. Ed. Engl.*, **1990**, *29*, 138.
[31]A. W. Bosman, H. M. Janssen, E. W. Meijer, *Chem. Rev.*, **1999**, *99*, 1665.
[32]M. Fischer, F. Vögtle, *Angew. Chem. Int. Ed. Engl.*, **1999**, *38*, 884.
[33]D. L. Jiang, T. Aida, *Nature*, **1997**, *388*, 454.
[34] M. Higuchi, S. Shiki, K. Ariga, K. Yamamoto, *J. Am. Chem. Soc.*, **2001**, *123*, 4414.
[35]A. Hierlemann, J. K. Campbell, L. A. Baker, R. M. Crooks, A. J. Ricco, *J. Am. Chem. Soc.*, **1998**, *120*, 5323.
[36]H. Tokuhisa, M. Zhao, L. A. Baker, V. T. Phan, D. L. Dermody, M. E. Garcia, R. F. Peez, R. M. Crooks, T. M. Mayer, *J. Am. Chem. Soc.*, **1998**, *120*, 4492.
[37]J. Li, D. R. Swanson, D. Qin, H. M. Brothers, L. T. Piehler, D. Tomalia, D. T. Meier, *Langmuir*, **1999**, *15*, 7347.
[38]L. Balogh, D. A. Tomalia, *J. Am. Chem. Soc.*,**1998**, *120*, 7355.
[39]M. Zhao, L. Sun, R. M. Crooks, *J. Am. Chem. Soc.*, **1998**, *120*, 4877.
[40]M. Tominaga, J. Hosogi, K. Konishi, T. Aida, *Chem. Commun.*, **2000**, 719.
[41]D. J. Díaz, G. D. Storrier, S. Bernhard, K. Takada, H. D. Abruña, *Langmuir*, **1999**, *15*, 7351 (1999).
[42]M. Petrucci-Samija, V. Guillemette, M. Dasgupta, A. K. Kakkar, *J. Am. Chem. Soc.*, **1999**, *121*, 1968.
[43]M. Kawa, J. M. J. Fréchet, *Chem. Mater.*, **1998**, *10*, 286.
[44]K. Onitsuka, S. Takahashi, *J. Syn. Org. Chem. Jpn.*, **2000**, *58*, 988.
[45]K. Yamamoto, M. Higuchi, S. Shiki, M. Tsuruta, H. Chiba, *Nature*, **2002**, *415*, 509.
[46]C. Gorman, *Nature*, **2002**, *415*, 487.
[47]T. Imaoka, H. Horiguchi, and K. Yamamoto, *J. Am. Chem. Soc*, **2003**, *125*, 340.
[48] H. W. Boone, J. Bryce, T. Lindgren, A. B. Padias, H. K. Hall, Jr. *Macromolecules* **1997**, *30*, 2797.
[49] H. W. Boone, H. K. Hall, Jr. *ibid* **1996**, *29*, 5835.
[50] D. Zhao, J. S. Moore, *J. Org. Chem.* **2002**, *67*, 3548.
[51]S. Lahiri, J. L. Thompson, J. S. Moore, *J. Am. Chem. Soc.* **2000**, *122*, 11315.
[52]Y. Tobe, N. Utsumi, K. Kawabata, A. Nagano, K. Adachi, S. Araki, M. Sonoda, K. Hirose, K. Naemura, *J. Am. Chem. Soc.* **2002**, *124*, 5350.
[53] J. R. Nitschke, S. Zürcher, T. D. Tilley, *J. Am. Chem. Soc.* **2000**, *122*, 10345.
[54]Y. Tain, J. Tong, G. Frenzen, J. Sun, *J. Org. Chem.* **1999**, *64*, 1442.
[55] E. Schmidt, H. Zhang, C. K. Chang, G. T. Babcock, W. A. Room *J. Am. Chem. Soc.* **118**, 2954-2961 (1996).

Macromol. Symp. ***196***, *165–171 (2003)*

Electron Transport in Conjugated Metallopolymers

*Colin G. Cameron, Brian J. MacLean, Peter G. Pickup**

Department of Chemistry, Memorial University of Newfoundland, St. John's, NF A1B 3X7, Canada
E-mail: ppickup@mun.ca

Summary: Superexchange interactions between metal centers coordinated to various conjugated polymer backbones have been shown to enhance the rate of electron transport through the polymer. Results for Ru and Os bipyridine moieties complexed to polybenzimidzoles and poly(bithiophene-co-bithiazole)s are reviewed. The evidence for superexchange mediated electron transport, and the factors that influence the rate of electron hopping between metal centers, are discussed.

Keywords: charge transport; conjugated polymers; electrochemistry; metal-polymer complexes; superexchange

Introduction

Electron transport through molecular materials is of immense fundamental importance to diverse fields of science ranging from the study of redox proteins,[1,2] to the development of molecular electronic devices.[3,4] The broad field of conducting polymers is built upon this phenomenon, and photosynthesis and respiration could not occur without it. It will be crucial to the development of nano-electronics technology.

The work reviewed in this paper was designed to demonstrate and develop a new mechanism for electron transport in molecular materials, based on the well known phenomenon of electronic coupling (superexchange) between metal complexes bound to the same ligand.[5] Thus, metal complexes have been bound to conjugated polymers such that there are superexchange interactions between them through the polymer chains (Structures **1** and **2**). Electron transport rates in these materials have been measured by electrochemical impedance spectroscopy. Their dependence on variables such as the metal centre (Ru or Os), the structure of the polymer backbone, and pH, clearly demonstrate the enhancement of electron transport rates in these materials by a superexchange mechanism.

This work draws on the extensive literature on redox polymers[6-9] and conducting polymers,[10,11] and the concept of hybridization of these materials that has been developed over the past

© 2003 WILEY-VCH Verlag GmbH & KGaA, Weinheim CCC 1022-1360/00/$ 17,50+.50/0

decade.[12-14] Similar materials have been reported by Swager and coworkers,[15] and Wolf and Wrighton.[16]

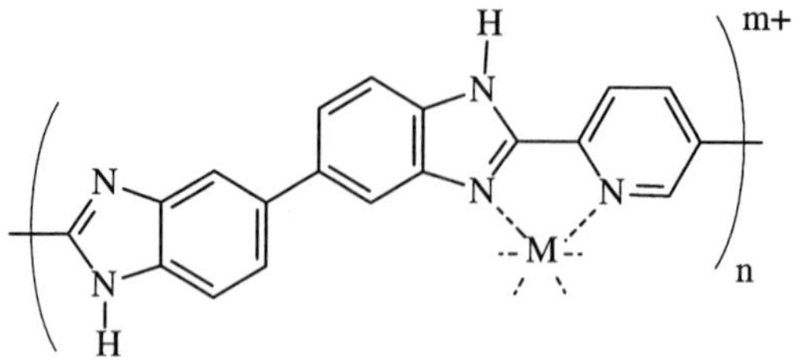

1Ru: M = $Ru(bpy)_2$; $D_eC_M^2 = 1.5 \times 10^{-14}$
1Os: M = $Os(bpy)_2$; $D_eC_M^2 = 0.71 \times 10^{-14}$

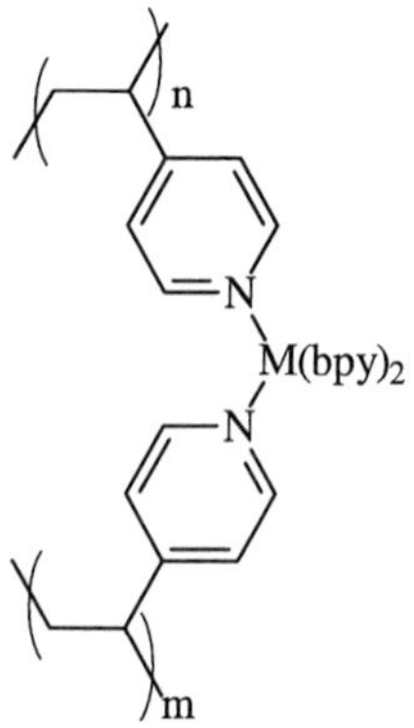

3Ru: M = $Ru(bpy)_2$; $D_eC_M^2 = 0.18 \times 10^{-14}$
3Os: M = $Os(bpy)_2$; $D_eC_M^2 = 1.2 \times 10^{-14}$

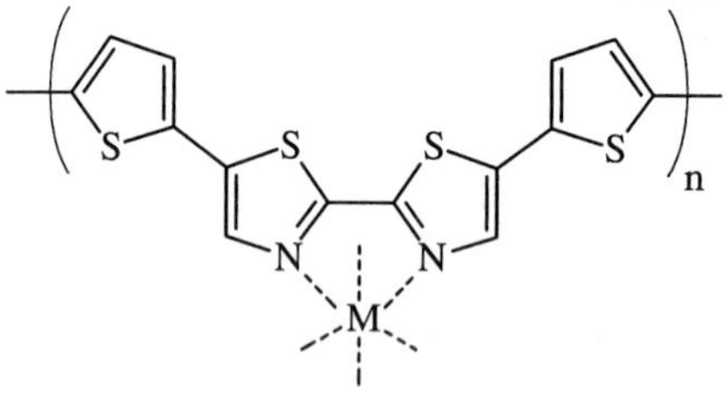

2Ru: M = $Ru(bpy)_2$; $D_eC_M^2 = 32 \times 10^{-14}$
2Os: M = $Os(bpy)_2$; $D_eC_M^2 = 5.8 \times 10^{-14}$

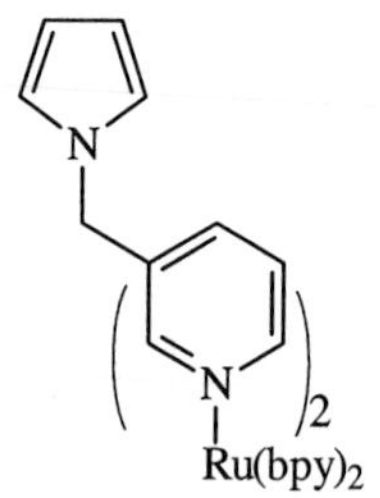

4

The synthesis and characterization of the metallopolymers discussed here as well as details of the electron transport measurements have been described elsewhere.[17-20] A variety of techniques have been used for the electron transport measurements, and these are compared in detail in ref.[17]. Results are expressed as $D_eC_M^2$ values, where D_e is the effective diffusion coefficient for electron transport and C_M is the concentration of metal centres, since it is difficult to accurately determine C_M. It is assumed in the discussion that variations in C_M between materials are small relative to the variations in D_e.[19]

Results and Discussion

Effect of the Polymer Backbone. The structures of the polymers discussed here are shown as Structures **1-3** (bpy = 2,2'-bipyridine) together with $D_eC_M^2$ values (in $mol^2\ cm^{-4}\ s^{-1}$). In polymers **1** and **2** the metal is complexed to a conjugated backbone, and so these polymers exhibit enhanced electron transport rates relative to the non-conjugated reference materials, **3**. The role of the HOMO of the conjugated backbone in enhancing electron transport is illustrated in Figure 1, where electron transport rates are plotted as a function of the estimated energy gap between the metal d orbitals and the HOMO of the polymer. These energy gaps were estimated from the difference between the M(III/II) formal potential and the formal potential for oxidation of the polymer backbone.[18,20] The latter was estimated to be ≥ 2.2 eV for **1Ru** and **1Os**.[21]

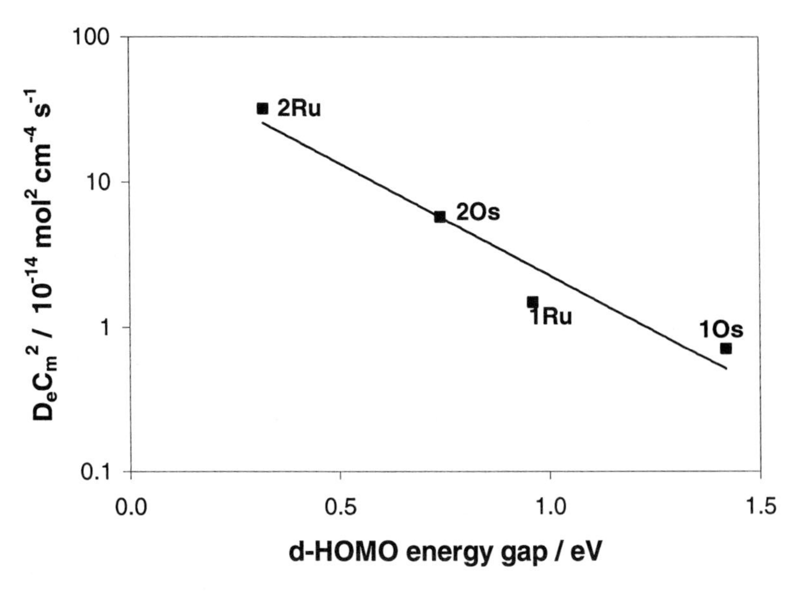

Figure 1. Electron transport rate as a function of the estimated energy gap between the metal d orbitals and the HOMO of the polymer.

The data in Figure 1 show that the electron transport rate increases greatly as the d-HOMO energy gap is decreased. Data for the two different types of polymer (polybenimidazole or poly(bithiophene-co-bithiazole)) fit an exponential relationship quite well. As illustrated in Figure

2, hole-type superexchange involves the bridging ligand's HOMO and is therefore enhanced by decreasing the d-HOMO gap. The data in Figure 1 therefore strongly implicate the involvement of a hole-type superexchange process in electron transport in these materials. There is evidence that electron-type superexchange becomes dominant for Os-benzimidazole polymers.[22]

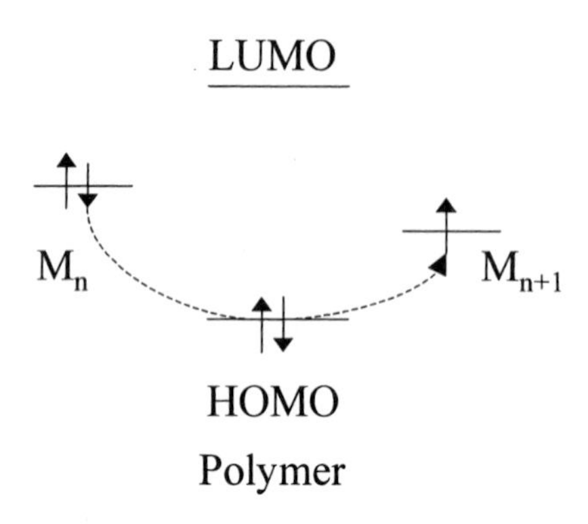

Figure 2. Schematic diagram of hole-type superexchange.

The enhanced electron transport rates observed for the conjugated polymers **1** and **2** could be explained by mediation of electron transfer between metal centres by the conjugated backbone. The distinction between this and a superexchange mechanism is subtle,[12] depending on whether there is an intermediate state in which the hole (for hole-type mediation) is located on the polymer backbone.

Arguments against the dominance of a mediated mechanism for polymers **1** and **2** have been made elsewhere.[17,19] However, there are cases in which mediation is the dominant charge transport process, and it is relevant to discuss them here.

Zotti et al[23] have shown that electron transfer rates between metal sites in polythiophenes with pendant ferrocene moieties are enhanced when a conjugated linkage is used. Ochmanska and Pickup[24] have explored the effect of copolymerization of $[Ru(2,2'\text{-bipyridine})_2(3\text{-}\{pyrrol\text{-}1\text{-}ylmethyl\}pyridine)_2]^{2+}$ (**4**) with 3-methylthiophene on its electron transport properties. The homopolymer of the complex shows no electrochemistry or conductivity due to the pyrrole moieties, while copolymerization with 3-methylthiophene generates redox waves and conductivity due to oligo-thiophene linkages. Moreover, these conjugated linkages enhance the rate of electron transfer between Ru centres by mediation. Superexchange is not possible because of the saturated linkage between the complex and the conjugated linkages. A particularly important finding from this work, in the present context, was that mediation of electron transport between Ru sites was accompanied by a distortion from the normal wave-shape for redox conduction. This therefore appears to be a useful diagnostic for mediated electron transport. The fact that it is not observed for polymers **1** and **2** (i.e. that $D_eC_M^2$ is not potential dependent) provides strong evidence that they do not exhibit significant mediated electron transport.

Effect of the Metal. It is very significant that for the conjugated polymers **1** and **2** $D_eC_M^2$ values are higher for the Ru complex than for the Os complex, while this is reversed for the non-conjugated polymer **3**. This is consistent with domination of the charge transport rate by hole-type superexchange for the conjugated polymers, and outer sphere electron transfer for **3**. The higher redox potential of Ru relative to Os causes a decrease in the d-HOMO energy gap, and therefore enhances hole-type superexchange. On the other hand, the larger Os centre enhances the rate of outer sphere electron exchange.[25]

Effect of pH. A dinuclear complex based on the structure of polymer **1Ru** has been shown to exhibit enhanced superexchange when deprotonated.[26] The imidazole of polymer **1Ru** has a pK_a of 5.5, and its deprotonation produces changes in electrochemistry and electronic absorption similar to those seen for the dinuclear complex.[17] It also leads to a significant increase in the electron transport rate of the polymer,[17] consistent with dominance of electron transport by superexchange. Deprotonation of the Os polymer, **1Os**, causes a decrease in $D_eC_M^2$, suggesting that electron-type superexchange is dominant in this case.[20]

Degradation of the Polymer Backbone. The role of a conjugated polymer backbone can be exposed by disruption of the conjugation. This can conveniently be achieved by brief exposure of the polymer to potentials that are sufficiently high to cause overoxidation, which involves nucleophilic attack of the conjugation pathway by water or other species.[27]

The electron transport rate of **1Ru** was found to decrease with time at high potentials, indicating that it was enhanced by the conjugated backbone.[17] This phenomenon was investigated in much more detail for **2Ru** and **2Os**.[19] For the Os polymer, the potential at which disruption of conjugation begins is much higher than the redox potential of the Os, and so $D_eC_M^2$ values could be compared before and following complete disruption of the conjugation. A decrease of a factor of 7 was observed, indicating that the conjugated backbone plays a major role in electron transport. For **2Ru**, the Ru oxidation and polymer overoxidation processes overlap, and so $D_eC_M^2$ decreases while it is being measured (as for **1Ru**[17]). Full disruption of conjugation resulted in a ca. 7 fold decrease, again indicating that the conjugated backbone is responsible for the extremely high electron transport rates observed for **2Ru**.

Conclusions

The combined evidence for a superexchange electron transport mechanism for polymers **1-2** is very strong. Electron transport is fastest in the Ru complexes and involves a hole-type mechanism. The highest rates are obtained by minimizing the energy gap between the Ru d orbitals and the HOMO of the polymer. An electron-type mechanism may dominate for some Os complexes (e.g. **1Os**).

Acknowledgements

This work was supported by the Natural Sciences and Engineering Research Council of Canada and Memorial University.

[1] J.R. Winkler, A.J. Di Bilio, N.A. Farrow, J.H. Richards, H.B. Gray, *Pure and Applied Chemistry* **1999**, *71*, 1753.
[2] V.L. Davidson, *Acc. Chem. Res.* **2000**, *33*, 87.
[3] R.L. Carroll, C.B. Gorman, *Angew. Chem. Int. Ed. Engl.* **2002**, *41*, 4379.
[4] A. Nitzan, *Ann. Rev. Phys. Chem.* **2001**, *52*, 681.
[5] C. Creutz, *Progr. Inorg. Chem.* **1981**, *30*, 1.
[6] G. Inzelt, in *Electroanalytical Chemistry*, vol. 18, A.J. Bard, Ed., Marcel Dekker, New York, 1994, pp. 89-241.
[7] R.W. Murray, Ed., *Molecular Design of Electrode Surfaces*, vol. XXII, Wiley, New York, 1992.
[8] N. Oyama, T. Ohsaka, in *Molecular Design of Electrode Surfaces*, R.W. Murray, Ed., Wiley, New York, 1992, pp. 333-402.
[9] R.A. Durst, A.J. Baumner, R.W. Murray, R.P. Buck, C.P. Andrieux, *Pure. Appl. Chem.* **1997**, *69*, 1317.
[10] T.A. Skotheim, R.L. Elsenbaumer, J.R. Reynolds, Eds., *Handbook of Conducting Polymers*, 2nd ed., Marcel Dekker, New York, 1998.
[11] H.S. Nalwa, Ed., *Handbook of Advanced Functional Molecules and Polymers*, vol. 1-4, Gordon & Breach,, 2001.
[12] P.G. Pickup, *J. Materials Chem.* **1999**, *8*, 1641.
[13] R.P. Kingsborough, T.M. Swager, *Progr. Inorg. Chem.* **1999**, *48*, 123.
[14] M.O. Wolf, *Advan. Mater.* **2001**, *13*, 545.
[15] S.S. Zhu, R.P. Kingsborough, T.M. Swager, *J. Mater. Chem.* **1999**, *9*, 2123.
[16] M.O. Wolf, M.S. Wrighton, *Chem. Mater.* **1994**, *6*, 1526.
[17] C.G. Cameron, P.G. Pickup, *J. Amer. Chem. Soc.* **1999**, *121*, 11773.
[18] B.J. MacLean, P.G. Pickup, *J. Materials Chem.* **2001**, *11*, 1357.
[19] B.J. MacLean, P.G. Pickup, *J. Phys. Chem. B.* **2002**, *106*, 4658.

[20] C.G. Cameron, T.J. Pittman, P.G. Pickup, *J. Phys. Chem. B.* **2001**, *105*, 8838.
[21] C.G. Cameron, *Enhanced Rates of Electron Transport in Conjugated-Redox Polymer Hybrids*, Ph.D. thesis, Memorial University of Newfoundland, 2000.
[22] C.G. Cameron, P.G. Pickup, *J. Amer. Chem. Soc.* **1999**, *121*, 7710.
[23] G. Zotti, G. Schiavon, S. Zecchin, A. Berlin, G. Pagani, A. Canavesi, *Synthet. Metal.* **1996**, *76*, 255.
[24] J. Ochmanska, P.G. Pickup, *J. Electroanal. Chem.* **1991**, *297*, 197.
[25] M.-S. Chan, A.C. Wahl, *J. Phys. Chem.* **1978**, *82*, 2542.
[26] M. Haga, T. Ano, K. Kano, S. Yamabe, *Inorg. Chem.* **1991**, *30*, 3843.
[27] A.A. Pud, *Synthet. Metal.* **1994**, *66*, 1.

Coordination/Organometallic Polymers Based on Diphosphine and Diisocyanide Ligands

Pierre D. Harvey

Département de chimie, Université de Sherbrooke, Sherbrooke, PQ, Canada, J1K 2R1
E-mail: pierre.harvey@usherbrooke.ca

Summary: This paper reviews various coordination/ organometallic polymers in which the metal atoms are incorporated in the backbone using diphosphine and diisocyanide ligands. Such ligands includes diphosphines of the type bis(diphenylphosphino)alkane where alkane is $(CH_2)_m$ with m = 1, 3-6, bis(diphenylphosphino)acetylene (dpa), and bis(dimethyl-phosphino)methane (dmpm), and diisocyanides such as 1,8-diiso-cyano-*p*-menthane (dmb) and *p*-diisocyanotetramethylbenzene (ditmb). The metal fragments are monocations such as Cu^+, Ag^+, and Au^+, dinuclear species such as $Pd_2(dmb)_2^{2+}$, $Pd_2(dppm)_2^{2+}$, $M_2(dmpm)_3^{2+}$ (M = Cu, Ag), and clusters such as $M_4(dmb)_4^{2+}$ (M = Pd, Pt).

Keywords: bridging ligands; 1-D polymers; exciton; photoconductivity; photophysics

Introduction

The syntheses and characterization of coordination and organometallic polymers in which the metal atoms are part of the macromolecule backbone, with the aim to prepare new materials finding applications in electric conductivity and photoconductivity, non-linear optical properties, sensors, liquid crystals, etc., is a still growing field. A number of challenges are associated with the fact that the fragments, or repetitive, units are held together by coordination bonds. The ligand lability directs the dissociation property of the polymer, causing two important problems. The first one is the addressibility of the structure in the solid state after dissolving the materials in different solvents or in the presence of different counter anions. Also depending on the ligand, some polymers can make rings, hence drastically changing the polymeric nature of the material. The second problem is that dissociation can be so extensive that the polymer is not present, or even oligomer in solution.

Over the years, this group has exploited two strategies to minimize the consequences of ligand lability. These are the use of two bridging ligands per metal fragment, and of more covalent M-L

© 2003 WILEY-VCH Verlag GmbH & KGaA, Weinheim CCC 1022-1360/00/$ 17,50+.50/0

bondings. In the former case, the enhanced stability gained by the use of a second anchoring ligand reminds the gained stability with chelates. In the latter case, the electrostatic interactions supplementing those between soft metal cations and ligands contribute to render these ligands less labile. The danger is that if these M-L bonds are too "strong", then the polycationic polymers may become insoluble. This situation represents the other extreme. An undissociated and soluble polymers is the ideal case, but such a situation does not always occur. The intermediate scenario is when a polymer must partially dissociate in solution to form weakly soluble oligomers, which upon precipitation reform the original polymer. In this way, the polymer structure is still reproducibly addressable, and simple chemical transformation such as counter anion exchange can be performed without altering the polymeric nature of the materials. This paper describes the chemistry between various diphosphine and diisocyanide ligands,[1] with M^+ (M = Cu, Ag, Au), Pd_2^{2+}, and M_4^{2+} (M = Pd, Pt) metallic fragments.

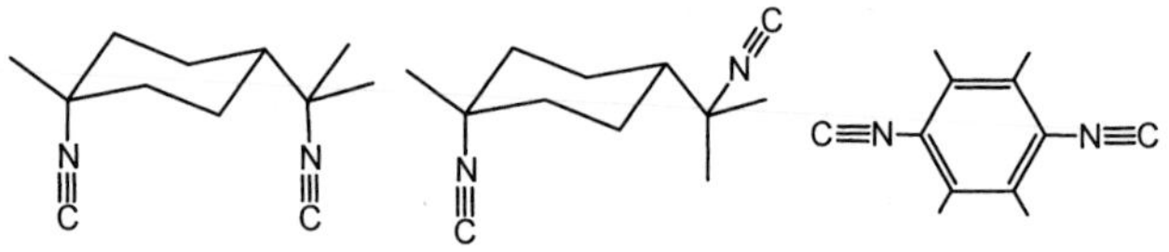

dmb, U-conformer   dmb, Z-conformer   ditmb

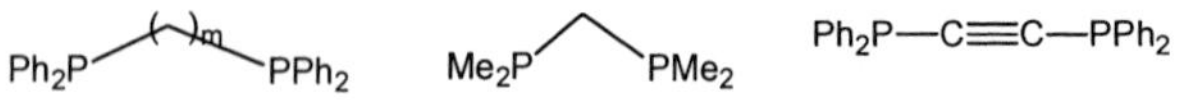

diphos; m = 1-6   dmpm   dpa

Simple $\{M(L\text{-}L)^{m+}\}_n$ Polymers

Using the diphosphines $Ph_2P(CH_2)_mPPh_2$, polymers of the type $\{M(Ph_2P(CH_2)_mPPh_2)(CN\text{-}t\text{-}Bu)_2^{2+}\}_n$ (M = Ag, Cu; m = 4, 5) have been prepared, some of which have been characterized from X-ray crystallography (M = Ag and M = Cu, the materials are amorphous). 1- and 2-D polymers have been observed. The replacement of CN-*t*-Bu by dmb leads to insoluble materials, presumably 3-D materials due to reticulation. These materials are found amorphous, and no glass transition has been depicted between 0 and 200°C. However, computer modelling shows that the replacement of two adjacent CN-*t*-Bu ligands by one dmb does not perturb greatly the polymer chain, and the 1-D chain remains intact. The poor solubility may be then associated with a great molecular dimension of these materials.

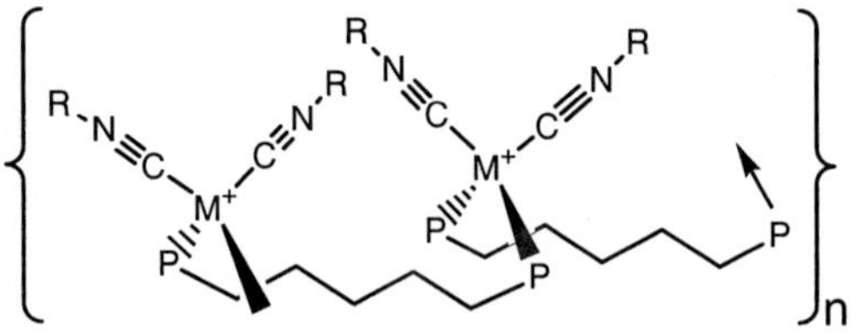

$M^+ = Cu^+, Ag^+$; R = *t*-Bu; P = PPh_2

The strongly luminescent and amorphous polymers of the type $\{Au_2(dpa)(disphos)^{2+}\}_n$ (diphos = $Ph_2P(CH_2)_mPPh_2$; m = 3-5, and Ph_2PCH_2CH**R**CH_2PPh_2; **R** = O-(CH_2)-O-naphthyl) have been prepared as well.[2] However, the mass spectra indicate ligand scrambling in the fragments, and the large FWHM (full-width-at-half-maximum) of the ^{31}P NMR signal is consistent with the presence of chemical exchange in solution. The T_g observed in the differential scanning calorimetry traces (DSC), the emission lifetimes are function of the number of methylene fragments in the diphos ligand. Despite the scrambling of the ligands within the polymeric chain, these results indicate that it is still possible to control the thermal and photophysical properties of the materials. Table 1 summarizes the DSC data as examples. The photophysical data are presented below.

Table 1. DSC data of the $\{Au_2(dpa)(disphos)^{2+}\}_n$ polymers

Polymers[a]	T_g	ΔC_p
	°C	J/(g·°C)
$\{Au_2(dpa)(Ph_2P(CH_2)_3PPh_2)^{2+}\}_n$	102.5	0.07±0.01
$\{Au_2(dpa)(Ph_2P(CH_2)_4PPh_2)^{2+}\}_n$	99.4	0.08±0.01
$\{Au_2(dpa)(Ph_2P(CH_2)_5PPh_2)^{2+}\}_n$	97.8	0.10±0.01
$\{Au_2(dpa)(Ph_2PCH_2CH\mathbf{R}CH_2PPh_2)^{2+}\}_n$	83.6	0.09±0.01

[a] **R** = O-(CH_2)-O-naphthyl

Polymers with Doubly Bridged Metal Atoms

The Ag^+ and Cu^{2+} ions, and $Cu(NCMe)_4^+$ complex react with dmb in excess to form the corresponding 1-D polymer $\{M(dmb)_2^+\}_n$ where dmb is its U-shape, and the counter anions are PF_6^-, BF_4^-, ClO_4^-, NO_3^-, $CH_3CO_2^-$ and $TCNQ^-$ (tetracyanoquinodimethane anion).[3-5]

$\{M(dmb)_2^+\}_n$ (M = Cu, Ag)

$TCNQ^-$

The structures reveal a M····M separation and M_3 angle of 5Å and 140°, respectively, in all cases. However, if the dmb is not in excess, the three latter counter anions lead to the formation of dimers,[6] while the three formers leads to dimers or trimers even in the excess of dmb.[7,8] Osmometry (M_n = 133000)[5] and light scattering (M_w = 155000)[4] establish the polymeric nature of $\{Cu(dmb)_2^+\}_n$, while these techniques fail for $\{Ag(dmb)_2^+\}_n$ (M_n < 10000). ^{13}C NMR spin-lattice (T_1) and NOE measurements (Nuclei Overhauser Enhancement) show that in fact the latter materials exist as oligomers in solution (7-8 units).[9] This property is due to the greater lability of the RNC ligand on Ag^+. This observation is further evidenced from the observation of the conformational isomers in the $\{[Ag(dmb)_2]TCNQ\}_n$ polymers, where dmb can be found in either the U- or Z-shape.[10] The polymeric isomers can be converted back and forth depending on the solvent used:

$$\{Ag(U\text{-}dmb)_2^+\}_n \rightleftharpoons \{Ag(Z\text{-}dmb)(U\text{-}dmb)^+\}_n \rightleftharpoons \{Ag(Z\text{-}dmb)_2^+\}_n \quad (1)$$

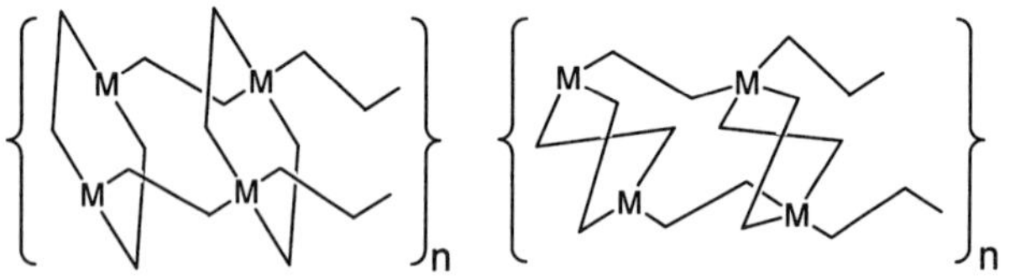

$\{Ag(U\text{-}dmb)(Z\text{-}dmb)^+\}_n$ $\{Ag(Z\text{-}dmb)_2^+\}_n$

Control over the thermal properties for these materials can be achieved from the change of counter anion and doping agent such as in the $\{[M(dmb)_2]TCNQ^{\cdot}xTCNQ^0\}_n$ polymers (M = Cu, Ag; x =1, 1.5) as exemplified in Table 2. In these materials the glass transition is not associated with changes in chain conformation due to the rigidity of the polymer backbone. T_g is related to motion of the counter-ions, so that more free the counter-ion, lower T_g is.

Table 2. DSC data for the $\{[M(dmb)_2]TCNQ^{\cdot}xTCNQ^0\}_n$ polymers

Polymers	T_g	ΔCp	Morphology
	°C	J/(g·°C)	
$\{[Ag(dmb)_2]TCNQ^{\cdot}CH_2Cl_2\}_n$	25.8	0.39	crystalline
$\{[Cu(dmb)_2]TCNQ^{\cdot}\ CH_2Cl_2\}_n$	25.0	0.37	semi-crystalline
$\{[Ag(dmb)_2]TCNQ^{\cdot}TCNQ^0\}_n$	37.7	0.19	highly crystalline
$\{[Cu(dmb)_2]TCNQ^{\cdot}TCNQ^0\}_n$	48.8	0.21	crystalline
$\{[Ag(dmb)_2]TCNQ^{\cdot}1.5TCNQ^0\}_n$	100.8	0.20	highly crystalline
$\{[Cu(dmb)_2]TCNQ^{\cdot}1.5TCNQ^0\}_n$	106.0	0.28	crystalline

Mixed-metal materials of the type $\{Cu_{1-x}Ag_x(dmb)_2^+\}_n$ can also be prepared using a mixture of the starting materials in the desired ratio.[2] Such preparations lead to the control of the average size of the oligomers in solution, and the relative morphology. A systhematic change in the unit cell parameters are noticed. The 1-D mixed-ligand polymers $\{M(dmb)(dppm)^+\}_n$ (M = Cu, Ag) have also been prepared as well.[2] Because of steric hindrance, the dmb ligand adopts the Z-conformation. These polymers are best described as polymers of $M_2(dppm)_2^{2+}$ dimers (M = Cu, Ag) bridged by two dmb`s. The $Ag^+\cdots Ag^+$ separations are 4.4 and 11.5 Å for the dmpm- and dmb-bridged units, respectively.[2] Similarly to this example, the dimer $Ag_2(dmpm)_2Br_2$ crystallizes to form a polymeric structure.[11] The $Ag\cdots Ag$ distances are 3.61 and 3.92 Å for the dmpm- and Br-bridged fragments, respectively. The dimers are weakly held together by longer Br-Ag bonds (2.95 Å), while the shorter bonds are 2.74 Å. The corresponding $Au_2(dmpm)_2Cl_2$ dimer does not form a polymer, instead the complex behaves like a salt ($[Au_2(dmpm)_2]Cl_2$). This behaviour is due to the poor tendency of the Au^+ ion to form stable 4-coordinate complexes, while the 2-coordinate form is strongly favoured.

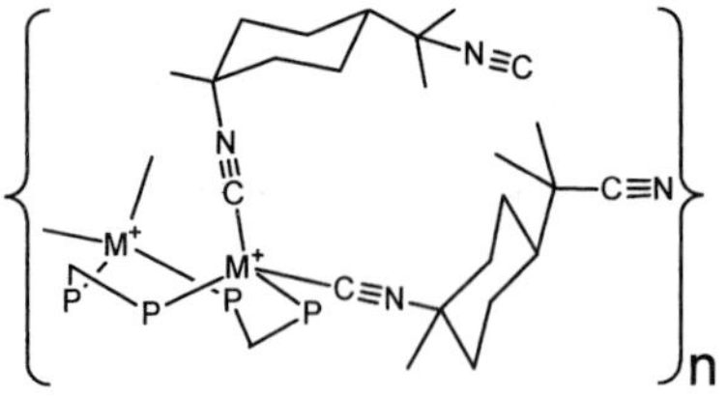

M = Cu, Ag: P-P = dppm

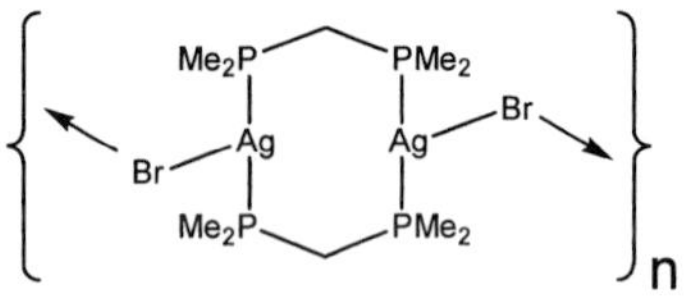

The ligand 5,11,17,23-tetraisocyano-25,26,27,28-tetrapropoxycalix[4]arene (calix) can be used to coordinate AuCl fragments.[12, 13] The coordination of Ag^+ produces a crystalline polymeric $\{M(calix)^+\}_n$ materials which is most likely related to the crystallographically characterized polymer $\{$5,11,17,23-tetracyano-25,26,27,28-tetrabenzoxycalix[4]arenesilver-(I)$\}_n$.[14] Because of the tetradentate structure of the ligand, this 1-D material falls in this polymers with doubly bridged metal atom category. Contrarily to the corresponding AuCl complex, no luminescence in the solid state, nor in solution at 77K is depicted.

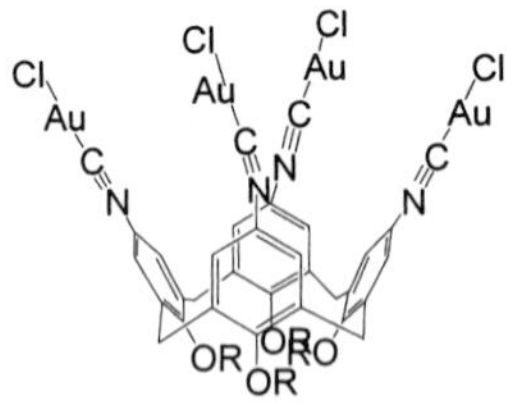

R = Pr

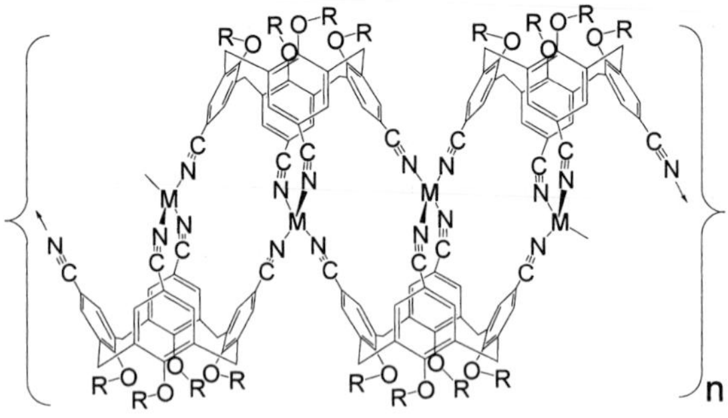

M = Ag^+; R = CH_2Ph

Polymers of Dimers and Clusters

Numerous polymers can be prepared from axially functionalizable polynuclear complexes. Examples include dimers such as $Pd_2(dppm)_2^{2+}$, $Pd_2(dmb)_2^{2+}$ and $M_2(dmpm)_3^{2+}$ (M = Cu, Ag), which are polymerized by one rigid, semi-rigid, and flexible bridging ligand such as ditmb, dmb, and diphosphines ($Ph_2P(CH_2)_mPPh_2$; m = 4-6), respectively. Evidence for their polymeric structure in the solid state can be found in the thermal gravitometry analysis (TGA: large temperature window of decomposition), DSC (presence of T_g), X-ray powder diffraction pattern (XRD: presence of semi-crystalline or amorphous materials), and the fact that polymer film can easily be made. Using T_1 measurements, these materials appear as oligomers in solution. For instance, the $\{[Pd_2(dmb)_2](Ph_2P(CH_2)_mPPh_2)^{2+}\}_n$ materials (m = 4-6) exhibit only 3-4 units, which is a dimension similar to that of $\{Ag(dmb)_2^{+}\}_n$.[9]

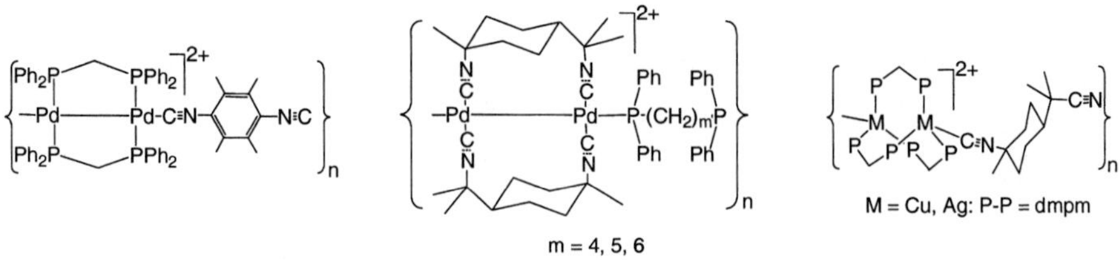

m = 4, 5, 6

M = Cu, Ag: P-P = dmpm

The polymers of clusters are also prepared from the axially functionalizable clusters, here the linear $M_4(dmb)_4^{2+}$, using dmb (M = Pd)[15] and ($Ph_2P(CH_2)_mPPh_2$) (m = 4-6).[16] The former has been characterized from X-ray crystallography, while the latters were characterized from the measurements of the intrinsic viscosity, [η], and XRD; they were found to be amorphous polymers. It is believed that the limitation of the chain length is due to the presence of residual phosphine oxyde and solubility. No glass transition was observed between 0 and 200°C. The M_4^{2+} fragment shows some sensitivity towards mild and unexpected oxidants, where Pd(II)[13] and Pt(I) species[15] have been observed as decomposition products.

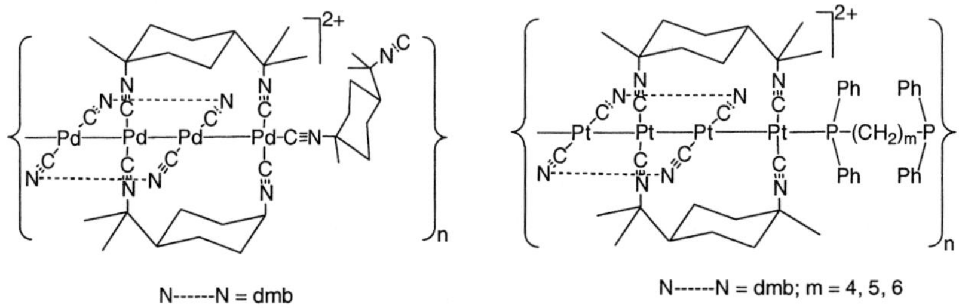

N------N = dmb

N------N = dmb; m = 4, 5, 6

Table 3. Dimension of the $\{Pt_4(dmb)_4(Ph_2P(CH_2)_mPPh_2)\}_n$ polymers

Polymers	$[\eta]$ cm³/g	M_n	numbers of units
$\{Pt_4(dmb)_4(Ph_2P(CH_2)_4PPh_2)\}_n$	3.66	203000	100
$\{Pt_4(dmb)_4(Ph_2P(CH_2)_5PPh_2)\}_n$	4.78	307000	150
$\{Pt_4(dmb)_4(Ph_2P(CH_2)_6PPh_2)\}_n$	2.06	84000	40

Semi- and Photoconductivity

The $\{[M(dmb)_2]TCNQ^{\cdot}xTCNQ^{o}\}_n$ polymers (x = 1, 1.5) are semi-conductors with resistivities (Ω) for M = Ag << than M = Cu, consistent with the higher crystallinity of the Ag-materials.[5,18] The electric conductivity occurs through a mixed-valent $\{TCNQ^{0.5-}\}_n$ or $\{TCNQ^{0.4-}\}_n$ 2-D layer (i.e. side-by-side chains placed in diagonal). An angle of about 50° is seen between the TCNQ-chain and polymer axes (Fig. 1). This result contrasts greatly with the generally encountered quasi-1-D $\{TCNQ^{x-}\}_n$ material where the counter ion is a simple alkalin or other simple cation, and where both cation and anion chains are placed parallel. This is due to the fact that the M^+ ions are placed rigidly in the polymer structure at a distance that does not match the π-stacking distance of the TCNQ`s. The observed resistivities (4-point probe technique on a pressed pellet) place the Ag-materials at the upper limit of the best semi-conducting materials. This performance is attributed to this 2D structure of the $\{TCNQ^{x-}\}_n$ layers.

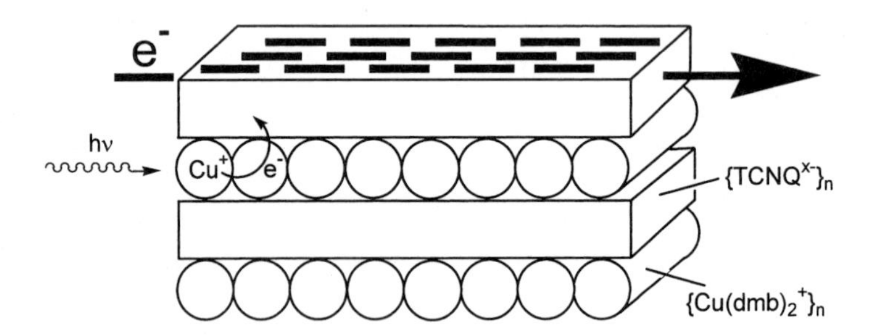

Figure 1. Scheme showing the layered structure of the $\{[M(dmb)_2]TCNQ^{\cdot}xTCNQ^{o}\}_n$ materials, and the mechanism for the photoconductivity. The fat bars represent up-right TCNQ chains.

The lesser conducting $\{[M(dmb)_2]TCNQ^{\cdot}xTCNQ^{o}\}_n$ polymers (x = 1, 1.5) are also photo-conducting. The current measured from the 4-point probe technique increases by 20% when a 200W lamp (broad band) is placed at 30cm in front of the pressed pellets. From various analyses of the photoreactions and photophysical processes using UV-vis and luminescence spectroscopy, it was possible to establish that the enhanced conductivity upon a water jacketed irradiation came from an electron transfer from the $Cu(CNR)_4^+$ chromophore to the mixed-valent $\{TCNQ^{x-}\}_n$ layer.

Luminescence Properties

The emission properties have been investigated in all cases in the solid state at room temperature. While most materials are weakly luminescent, the $\{Au_2(dpa)(diphos)^{2+}\}_n$ and $\{M_2(dmpm)_3(dmb)^{2+}\}_n$ polymers (M = Cu, Ag) are strongly luminescent. Table 4 summerizes the emission data for these strongly luminescent materials. Similarly, the $\{Pt_4(dmb)_4(Ph_2P(CH_2)_mPPh_2)\}_n$ polymers are luminescent, but somewhat less. At 77K, the intensity of luminescence increases. The $[M_2(dmpm)_3(CN\text{-}t\text{-}Bu)_2]^{2+}$ and $[Pt_4(dmb)_4(PPh_3)_2]^{2+}$ complexes are used as models for the building block of the corresponding polymers, and allow one to establish the effect of "polymerization" on the photophysical data. The key observations are that τ_e for the non M-M bonded systems decreases as Cu > Ag > Au, consistent with the heavy atom effect on the phosphorescence due to the larger spin-orbital coupling constants. For the Pt_4 systems, this heavy atom effect is more important as four heavy atom are connected together within the chromophore. The data also indicate that τ_e is affected in the polymers. The expected decrease in τ_e is associated with an increase in non-radiative pathways such as intramolecular vibrations (internal conversion into heat). In the Pt_4 systems, the reverse case is observed because the replacement of the bulker PPh_3 ligands by the diphos $Ph_2P(CH_2)_mPPh_2$ produces the reverse effect for the same reason.

Table 4. Photophysical data for some polymers

Materials[a]	λ_{max} nm	τ_e	Conditions
$[Cu_2(dmpm)_3(CN\text{-}t\text{-}Bu)_2]^{2+}$	477	290±10 μs	room temp./solid
$\{Cu_2(dmpm)_3(dmb)^{2+}\}_n$	482	259[b]	room temp./solid
$\{Ag_2(dmpm)_3(dmb)^{2+}\}_n$	448	88[b]	room temp./solid
$\{Au_2(dpa)(Ph_2P(CH_2)_3PPh_2)^{2+}\}_n$	495	8.5±0.1 μs	room temp./solid
$\{Au_2(dpa)(Ph_2P(CH_2)_4PPh_2)^{2+}\}_n$	495	8.3±0.1 μs	room temp./solid
$\{Au_2(dpa)(Ph_2P(CH_2)_5PPh_2)^{2+}\}_n$	495	5.7±0.1 μs	room temp./solid
$\{Au_2(dpa)(Ph_2PCH_2CH\mathbf{R}CH_2PPh_2)^{2+}\}_n$	none	-	room temp./solid
$[Pt_4(dmb)_4(PPh_3)_2)]^{2+}$	750	2.71±ns	77K/EtOH
$\{Pt_4(dmb)_4(Ph_2P(CH_2)_4PPh_2)\}_n$	736	4.78±ns	77K/EtOH
$\{Pt_4(dmb)_4(Ph_2P(CH_2)_5PPh_2)\}_n$	750	5.15±ns	77K/EtOH
$\{Pt_4(dmb)_4(Ph_2P(CH_2)_6PPh_2)\}_n$	755	5.17±ns	77K/EtOH

a) **R** = $O(CH_2)_6O$-naphthyl

b) The decays are polyexponential (see below), and only the maximum of the distribution is reported.

Exciton Phenomena

For the polymers where weak emissions are observed at room temperature in the solid state, an unusual property is depicted. These weak emission are unusually broad and polarized, and exhibit non-exponential decays. Time-resolved emission spectroscopy shows that at short delays times, the emission bands are blue-shifted with respect to the maxima, and at longer ones, these are red-shifted. The progression of the shift from the blue to the red is continuous with delay times with no evidence for isosbestic points. These observations are diagnostic of an exciton phenomenum; an energy delocalization within the material. This delocalization can proceeds in the bulk (intermolecular) or with the polymer itself (Fig. 2).

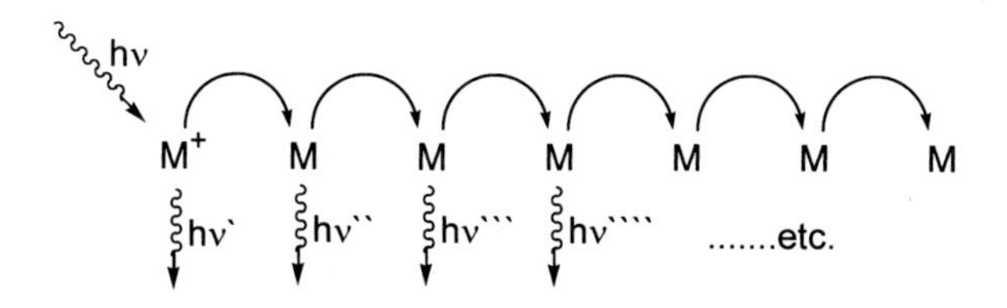

Figure 2. Scheme showing the exciton process.

For instance, the $\{M(dmb)_2^+\}_n$ polymers (M = Cu, Ag) exhibit emission decay traces that are non-exponential, but also are independent of the state of the materials (single crystal, powder, or in solution.[4] This important result indicates that energy transfer occur along the polymer chain through each $M(CN)_4^+$ chromophore. Interestingly, the emission lifetime and band position measured at the very early stage of the photophysical event are very similar to that of the corresponding isolated mononuclear complexes $M(CN\text{-}t\text{-}Bu)_4^+$ (M = Cu, Ag). This means that the absorption is followed by an spontaneous emission from this same mononuclear chromophore, an event that occurs faster than the exciton process, but competes with the latter. It is possible to analyse the decay traces with a curve distribution. This distribution of lifetimes is characterized by maxima and width. More the exciton process is important, more the width is large, and more the emission band is large as well.

Understanding the Synthons

The reaction between $Cu_2(Ph_2P(CH_2)_mPPh_2)_2(NCMe)_4^{2+}$ (m= 4, 5) with CN-*t*-Bu lead to colorless materials of the type $\{Cu(Ph_2P(CH_2)_mPPh_2)(CN\text{-}t\text{-}Bu)_2^{2+}\}_n$. these amorphous materials swell when dissolved in various organic solvents and form thick solid film or wax. Parallely, the reaction between $Pd(Ph_2P(CH_2)_mPPh_2)Cl_2$ (m = 3, 4), or $Pd_2(Ph_2P(CH_2)_m\text{-}PPh_2)_2Cl_4$, with dmb lead to completely insoluble colorless solids. The Cu^+ and $Pd2^+$ metal atoms coordinate 4 electron-donors in a tetrahedral and square planar fashion, respectively, and depending on the diphos used, monomer, dimer or polymer structures are observed CN-t-Bu and dmb complexes as shown in the next scheme.

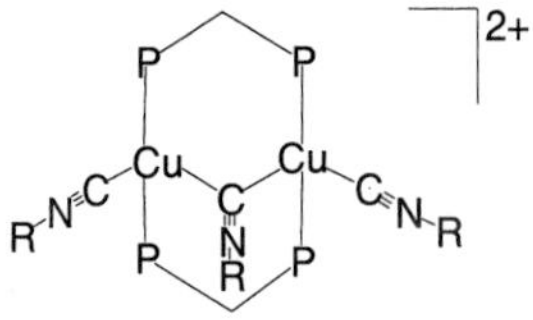

R = *t*-Bu

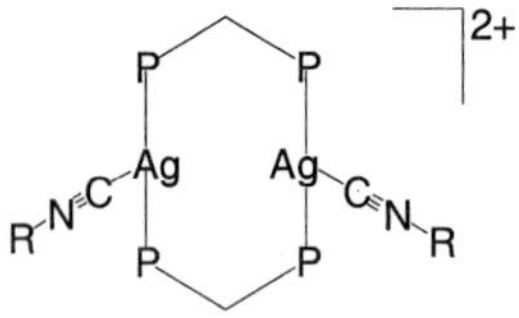

R = *t*-Bu

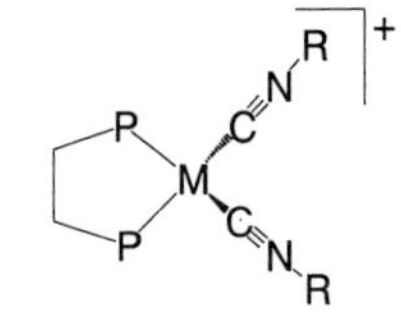

M = Cu, Ag; R = *t*-Bu

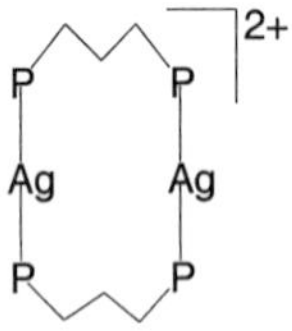

$\{Ag_2(Ph_2P(CH_2)_4PPh_2)_3(CN\text{-}t\text{-}Bu)_2^{2+}\}_n$

$\{Ag(Ph_2P(CH_2)_5PPh_2)(CN\text{-}t\text{-}Bu)_2^{2+}\}_n$

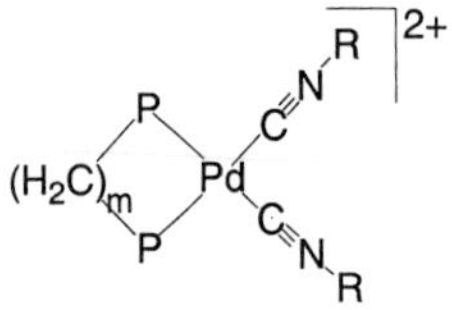

R = *t*-Bu; m = 2-4

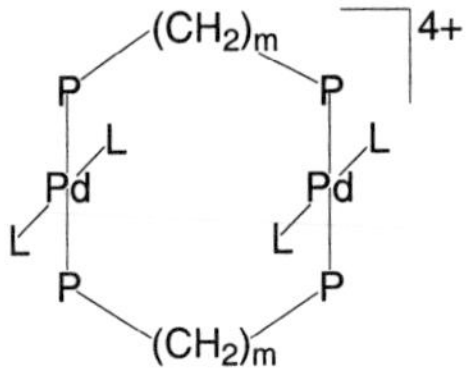

L = CN-*t*-Bu; m = 5,6

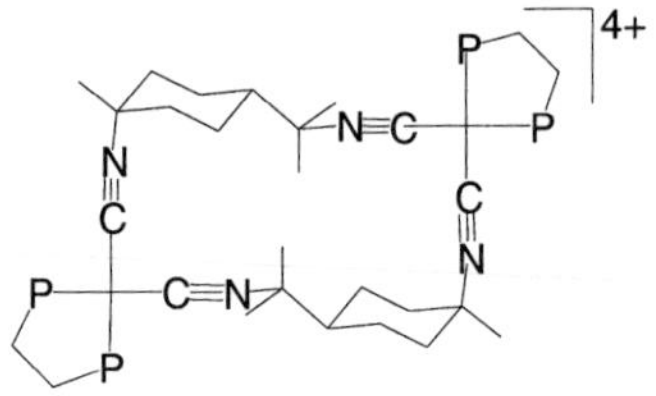

While the dppm ligand is largly (but not exclusively) an assembling ligand forming dimer species, the $Ph_2P(CH_2)_mPPh_2$ ligands form a series of complexes where the structure varies gradually from mononuclear (chelate), to dimers (bridging), to polymers (bridging) with the increase in m. The dmb ligand (both U- and Z-forms) favours either dimers or polymers. Because of steric hindrance, longer methelene chains (avoiding chelate structures), and incompatibility of the bite distances, polymers are favoured with dmb/diphos-containing materials, as those described above. The "M(diphos)(CN-*t*-Bu)" structures summerized in the Scheme above reveal some clues about what is the metal environment for these amorphous or insoluble materials is.

Acknowledgments

This research was supported by the NSERC (Natural Sciences and Engineering Research Council of Canada). PDH thanks the graduate students who did all the work; D. Perreault, D. Fortin, N. Jourdan, T. Zhang, M. Turcotte, F. Lebrun, E. Fournier, S. Sicard, P. Mongrain, and J.-F. Fortin.

[1] P.D. Harvey, *Coord. Chem. Rev.* **2001**, *219*, 17.
[2] F. Lebrun, M.Sc. Dissertation, Université de Sherbrooke, **2001**.
[3] D. Perreault, M. Drouin, A. Michel, P. D. Harvey, *Inorg. Chem.* **1992**, *31*, 3688.
[4] D. Fortin, M. Drouin, M. Turcotte, P. D. Harvey, *J. Am. Chem. Soc.* **1997**, *119*, 531.
[5] D. Fortin, M. Drouin and P.D. Harvey, *Inorg. Chem.* **2000**, *39*, 2758.
[6] D. Fortin, M. Drouin, P.D. Harvey, F.G. Herring, D.A. Summers, R.C. Thompson, *Inorg. Chem.* **1999**, *38*, 1253.
[7] P. D. Harvey, M. Drouin, A. Michel, D. Perreault, *J. Chem. Soc. Dalton Trans.* **1993**, 1365.
[8] D. Perreault, M. Drouin, A. Michel, P.D. Harvey, *Inorg. Chem.* **1993**, *32* , 1903.
[9] M. Turcotte, P.D. Harvey, *Inorg. Chem.* **2002**, *41*, 1739.
[10] D. Fortin, M. Drouin, P.D. Harvey, *J. Am. Chem. Soc.* **1998**, *120*, 5351.
[11] D. Perreault, M. Drouin, A. Michel, V. M. Miskowski, W. P. Schaefer, P. D. Harvey, *Inorg. Chem.* **1992**, *31*, 695.
[12] J. Gagnon, M. Drouin, P.D. Harvey, *Inorg.Chem.* **2001**, 40, 6052.
[13] P.D. Harvey, *Coord.Chem Reviews*, **2002**, *233-234*, 289.
[14] E. Elisabeth, L.J. Barbour, G.W. Orr, K.T. Holman, J.L. Atwood, *Supramol. Chem.* **2000**, *12*, 317.
[15] T. Zhang, M. Drouin, P.D. Harvey, *Inorg. Chem.* **1999**, *38*, 1305.
[16] T. Zhang, M. Drouin, P.D. Harvey, *Inorg. Chem.* **1999**, *38*, 957.
[17] T. Zhang, M. Drouin, P.D. Harvey, *Inorg. Chem.*, **1999**, *38*, 4928.
[18] D. Fortin and P.D. Harvey, *Coord. Chem. Rev.* **1998**, *171*, 351.

Phosphine-Based Coordination Cages and Nanoporous Coordination Polymers

Stuart L. James,[*1] *Xingling Xu,*[1] *Robert V. Law*[2]

[1]School of Chemistry, Queens University Belfast, David Keir Building, Stranmillis Road, Belfast, Northern Ireland, BT9 5AG, UK
E-mail: s.james@qub.ac.uk

[2]Department of Chemistry, Imperial College London, South Kensington Campus, London, SW7 2AZ, UK

Summary: Recent findings in the use of multidentate phosphines to synthesise porous coordination polymers (metal-organic frameworks) and their possible precursor cages are reviewed. Additional recent investigations into using large adamantoid Ag_6 cages as polymer vertices in giant diamandoid structures are also presented. The results are discussed in terms of possible strategies for the controled synthesis of porous coordination polymers.

Keywords: cage; coordination; phosphine; polymer; porous; silver

Introduction

The field of coordination polymers and metal-organic frameworks has advanced very rapidly over the past decade.[1] Examples of highly porous and intricate interpenetrated structures not found in other types of solid have been characterised and initial investigations made of sorption behaviours with regard to eventual applications as chemical separation and storage media. However, virtually all the work reported has been based on N- or O-donor ligands in particular oligopyridines and carboxylates such as those in Figure 1. The rigidity of these ligands tends to place the metals in well-defined positions. This is advantageous since it gives some control over the polymer structure. A related field which has grown greatly in recent years is that of discrete multimetallic ring or cage structures based on bridging organic ligands.[2] Again, almost all work here has been based on N- or O-donor bridging ligands.

© 2003 WILEY-VCH Verlag GmbH & KGaA, Weinheim CCC 1022-1360/00/$ 17,50+.50/0

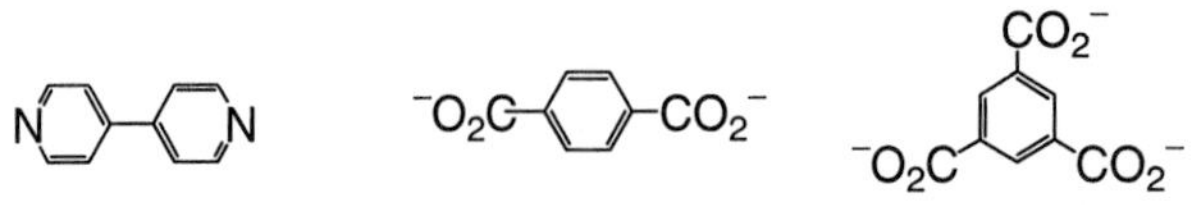

Figure 1. Some important examples of typical N- and O-donor bridging ligands used in coordination polymers and metal-organic frameworks. Note how the 180° mutual orientations of the lone pairs in 4,4'-bipyridine (left) are fixed by the rigidity of the ligand framework.

Against this background, our approach has been to investigate bridging multidentate phosphines, such as those shown in Figure 2, to generate unusual structures, behaviours and mechanistic insights in coordination rings, cages and polymers (at the outset very little work had been published on this[3]).

The key differences between these ligands and those in Figure 1 which have emerged are:

1. Soft donor character, leading to stable structures based on soft late d-block metals.
2. Rotational freedom of lone pairs due to pyramidal geometry of the phosphorus centres. Even where rigid-backboned diphosphines such as dppa or dppet, are used, rotation about P-C(backbone) bonds allows *syn* or *anti* disposition of the lone pairs, which can lead respectively to discrete or extended structures (Figure 2, bottom).
3. Presence of bulky aromatic substituents. Normally, for sufficient stability to air-oxidation (and possibly because they can increase the crystallinity of products), phenyl groups are the chosen substituents at phosphorus. The steric bulk of these groups can determine the structures (nulcearities) of coordination cages, and may also help to generate large pores within metal-organic polymers. The ability of multiple interaromatic contacts to stabilise structure may also be significant, but this has not yet been quantified.
4. Ability to monitor the donor atom directly by NMR spectroscopy. Especially in combination with spin-active metal nuclei, ^{31}P NMR spectroscopy gives valuable information on solution-state speciation and on the local structure of amorphous polymeric products.

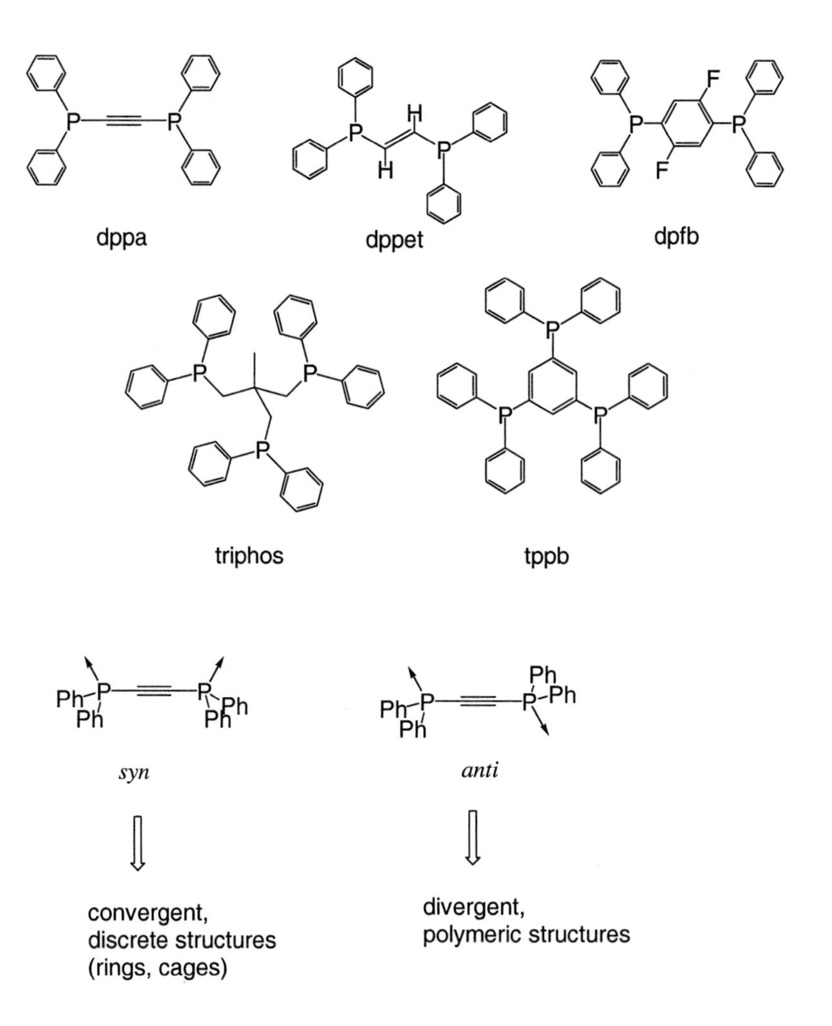

Figure 2. Phosphine P-donor bridging ligands which have been used in our studies, and the *syn* and *anti* conformations of 'rigid' connecting ligands such as dppa, which can give rise respectively to discrete or polymeric structures.

Taking tertiary phosphine ligands out of their familiar environments of organometallic chemistry and homogeneous catalysis, into the areas of coordination cages and polymers can therefore lead to new and interesting aspects in these fields. In this article are reviewed our findings in this research, as well as some previously unpublished results.

Review, Results and Discussion

The starting point for our work was the observation that flexible diphosphines $Ph_2P(CH_2)_nPPh_2$ (n = 3-6) give rise to large dinuclear ring structures on reaction with equivalent amounts of silver salts (Figure 3).[4]

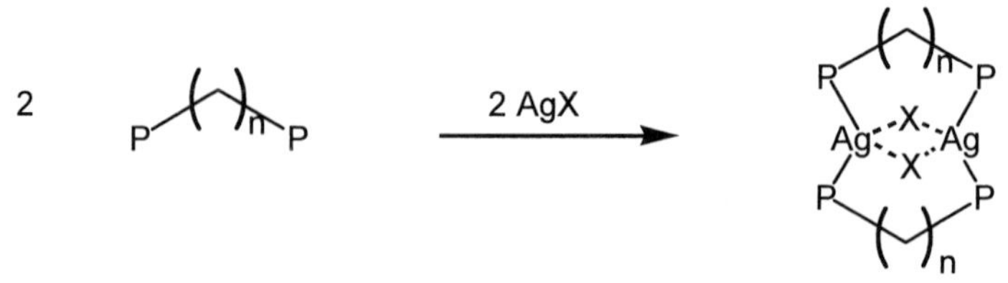

Figure 3. Disilver rings based on flexible bridging diphosphines (P = PPh_2).

Several such complexes have been characterised in the solid state by X-ray crystallography, and in some cases ^{31}P NMR spectroscopy was also consistent with the formation of such structures in solution. Possibly important to this behaviour is the ability of these flexible ligands to allow their lone pairs to point inward, and so set up two metal ions for anion bridging. In this way, anion bridges would be able to stabilise the large-ring dinuclear structure. We therefore examined the effect of changing to rigid-backboned ligands such as dppa and dppet (Figure 2), which disallowed the lone pairs from pointing inward. Accordingly, strikingly different solution-state speciation occurs for dppa in comparison to dppp (Figure 4).[5]

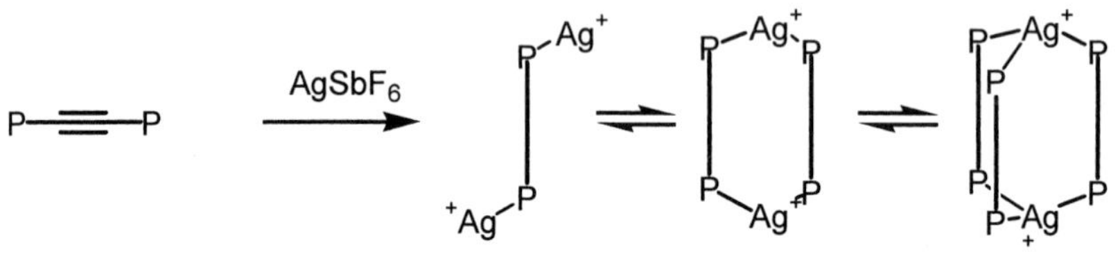

Figure 4. Reaction between the rigid bridging diphos dppa and $AgSbF_6$ at an equimolar ratio, to give a dynamic mixture of Ag_2L, Ag_2L_2 and Ag_2L_3 complexes (P = PPh_2).

In particular, one-two and three-coordinate complexes were observed consistent with the structures shown in Figure 4. At room temperature the complexes were dynamically interconverting on the NMR timescale, although at –60°C the interconversion was slow on the NMR timescale. However, altering the metal-to-ligand ratio to 2:3 revealed that for dppa, triply-bridged species would form then selectively (Figure 5).

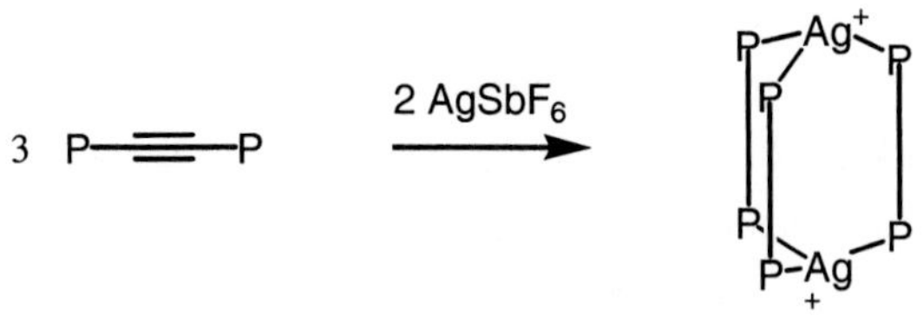

Figure 5. Selective formation of $[Ag_2(dppa)_3]^{2+}$ complexes at the appropriate metal to ligand ratio.

The structures of these simple cages were supported by a solution state molecular weight determination and by single crystal X-ray crystallography. This also revealed that the use of anions with different nucleophilicities lead to structures which varied in the bending strain present in the –C≡C– backbones of the bridging ligands (the less nucleophilic anions such as SbF_6 and BF_4 gave more strained structures than the more nucleophilic NO_3 or OTf anions). Interestingly, when crystals of the triflate complex were left in their supernatant for several weeks, they redissolved and a new compound crystallised, which was found to be the polymeric structure shown in Figure 6.[6]

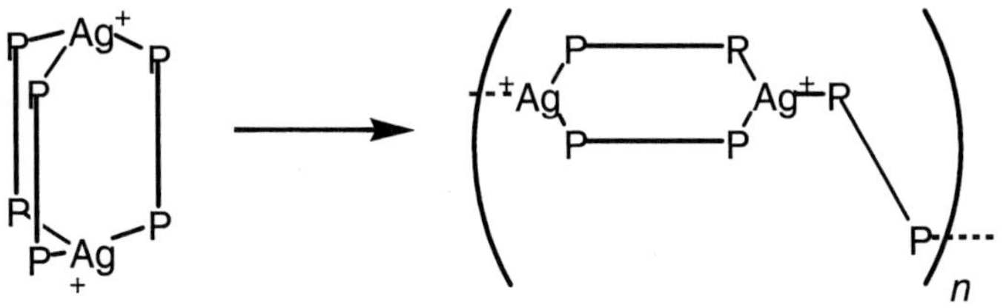

Figure 6. Triply bridged $[Ag_2(dppa)_3(OTf)_2]$cages and their formal ring-opening polymerisation to give $[Ag_2(dppa)_3(OTf)_2]_n$ polymers, P = PPh_2.

Of key interest is the fact that the precursor cage and polymer are formally related to each other by ring-opening polymerization (ROP). Despite the important role played by ROP methodology in main-group polymerizations, we were unaware of any instances where the ROP relationship had been pointed out for coordination cages and polymers (although other examples were pointed out simultaneously and later[7]). Given the difficulties in predicting and controlling the structures of metal-organic polymers, ROP methodology is potentially useful, particularly so if coordination cages can be formed in solution and rational ways then found for their controlled ring-opening. The question of whether coordination polymers represent thermodynamic or kinetic products is also raised by the observation of the ROP relationship. Coordination polymers have most often been assumed to be thermodynamic products because the metal-ligand bonds are normally highly labile, allowing, in principle, kinetic products to rearrange to more stable products. However, the relatively low solubility of coordination polymers could enable them to be trapped as kinetic products, which is in accord with the ROP relationship observed here.

A coordination cage or ring (a potential polymer precursor) might be deliberately destabilised (ring-opened) in a number of ways. Firstly, inter-ligand steric repulsion could be increased through use of bulky groups on the ligands, or potentially nucleophilic solvents, coligands or counterions might encourage P-M dissociation (the first step in ROP). We found that the *trans*-alkene-backboned ligand dppet, appeared to give less stable cages in solution than did dppa.[5] The qualitative evidence for this came from ^{31}P NMR spectroscopy, which at room temperature gave broad peaks most likely due to reversible P-Ag bond formation on the NMR timescale. At low temperatures, however, sharp spectra consistent with the triply-bridged cages were seen, suggesting the ligand-metal dissociation was less significant at lower temperatures. Consistent with the apparent lower stability of cages based on dppet was the fact that, under similar crystallisation conditions as those employed for the dppa cages, only the formal ring-opened polymers were obtained (Figure 7).

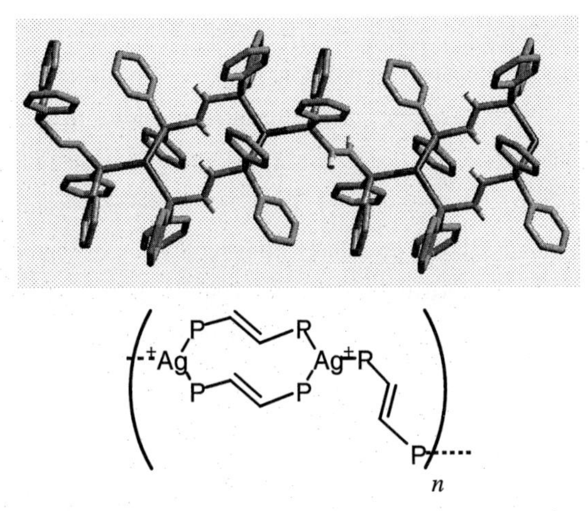

Figure 7. X-ray Crystal structure of the apparently ring-opened polymer $[Ag_2(dppet)_3(SbF_6)_2]_n$, P = PPh_2.

We have also investigated triphosphine ligands, such as those shown in Figure 2. The flexible tripodal neopentane-backboned triphosphine $CH_3C(CH_2PPh_2)_3$ (triphos, Figure 2) was expected to give Ag_3(triphos)$_2$-stoichiometry cages as shown in Figure 8. However, surprisingly, Ag_6(triphos)$_4$ structures with adamantanoid connectivity were obtained exclusively.[8] This initially surprising result seems most likely to be due to the bulky phenyl groups of the triphos ligands, since models suggest that interligand steric crowding would occur in the trinuclear complexes.

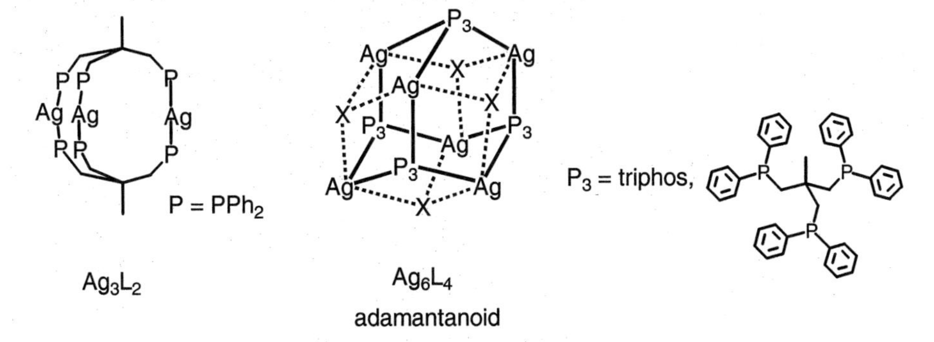

Figure 8. Postulated Ag_3L_2 cages based on tripodal triphosphines, and the actually observed adamantanoid Ag_6L_4 cages formed by $CH_3C(CH_2PPh_2)_3$ (X = $CF_3SO_3^-$, NO_3^-, ClO_4^-, but not BF_4^- or SbF_6^-).

These hexanuclear tetrahedral 2nm cages could potentially be connected into a diamandoid 3-D polymer by the use of bridging anions as shown in Figure 9. The large size of these cages made them of particular interest for generating porous structures, according to the strategy advanced by Yaghi *et al.*, which relies on the fact that large vertices make interpenetration of networks more difficult for steric reasons.[9] Interpenetration frequently mitigates against the generation of porosity coordination polymers. The potentially bridging dianion in Figure 9 has two tripodal sulfonate groups which were expected to template the adamantanoid cages and link them together (previous studies with monoanions showed that tripodal oxoanions such as nitrate, perchlorate and triflate are essential to template the hexanuclear cages $[Ag_6(triphos)_4(anion)_4]^{2+}$, and the use of fluoroanions such as SbF_6^- gave rise instead to dynamic mixtures, presumably due to their lower nucleophilicity and inability to stabilise the cages by bridging between metal ions[8]). Reaction between triphos, $AgSbF_6$ and the bridging dianion rapidly gave insoluble precipitates which XRPD showed to be amorphous. It was not possible to obtain single crystals by slowing down the mixing of the components. However, solid-state ^{31}P NMR spectroscopy (see Figure 10) did reveal that the phosphorus centres were, as hoped, all in very similar chemical environments with chemical shift and $^{109/107}Ag$-^{31}P coupling similar to those seen in the discrete cage in solution ($\delta = -9.32$ ppm, $J = 527$ Hz, assuming the latter to be J coupling rather than dipolar coupling, and that it would therefore be a weighted average of coupling due to ^{107}Ag (48%) and ^{109}Ag (52%), and given that $\gamma^{107}Ag/\gamma^{109}Ag = 1.149$, this corresponds to $^{1}J_{31P\text{-}109Ag} = 566$ Hz, which compares with 561 Hz for $[Ag_6(triphos)_4(O_3SCF_3)_4][O_3SCF_3]_2$[8]). Whilst it seems likely that the desired diamandoid structure was formed, to some degree, the lack of definitive structural determination hampered further progress in the study of these materials.

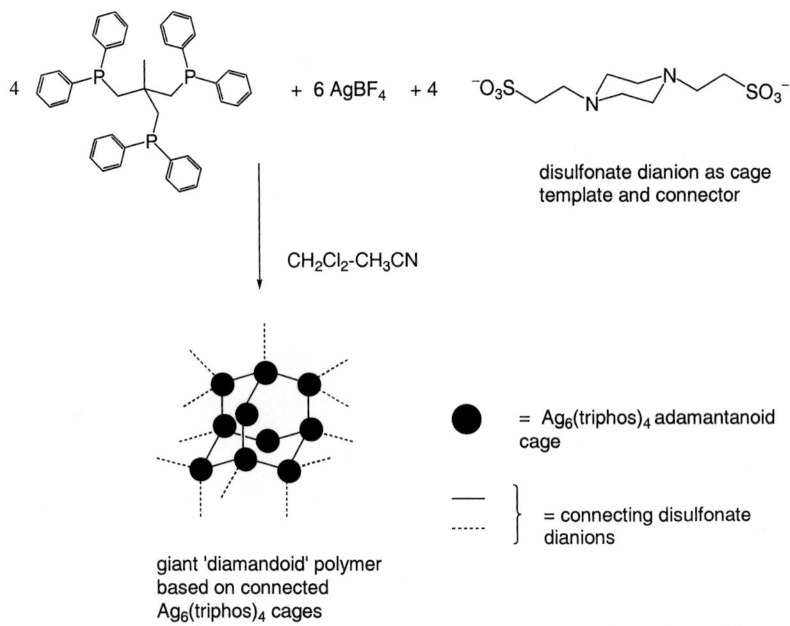

Figure 9. Attempted formation of a porous diamandoid material (**5**) by self-assembly of interconnected adamantanoid $Ag_6(triphos)_4$ cages.

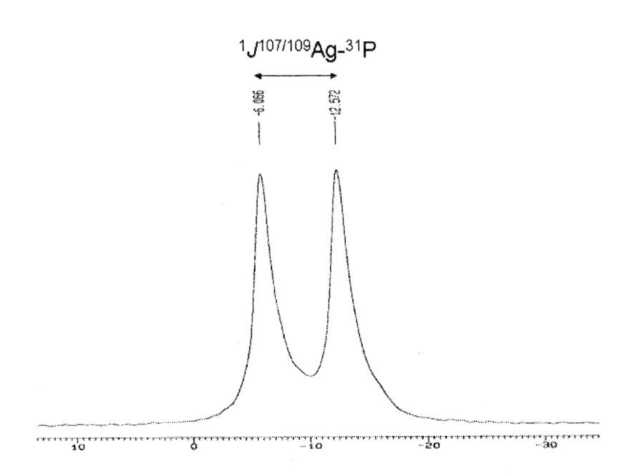

Figure 10. Solid state MAS ^{31}P NMR spectrum of the insoluble product from the reaction shown in Figure 9, showing the suspected J coupling between silver and phosphorus, similar to that seen in discrete cages in solution.

Use of the rigid triphosphine 1,3,5-*tris*(diphenylphosphino)benzene, tppb, did however, give rise to a crystalline large pore open polymer.[10] Reaction between this ligand and AgOTf, in dichloromethane-nitromethane followed by crystallisation by diffusion of diethyl ether vapour into this solution gave hexagonal crystals in *ca.* 10% yield, along with powdery material which was found to be amorphous by XRPD. The X-ray crystal structure of the hexagonal crystals was determined to be that shown in Figure 11.

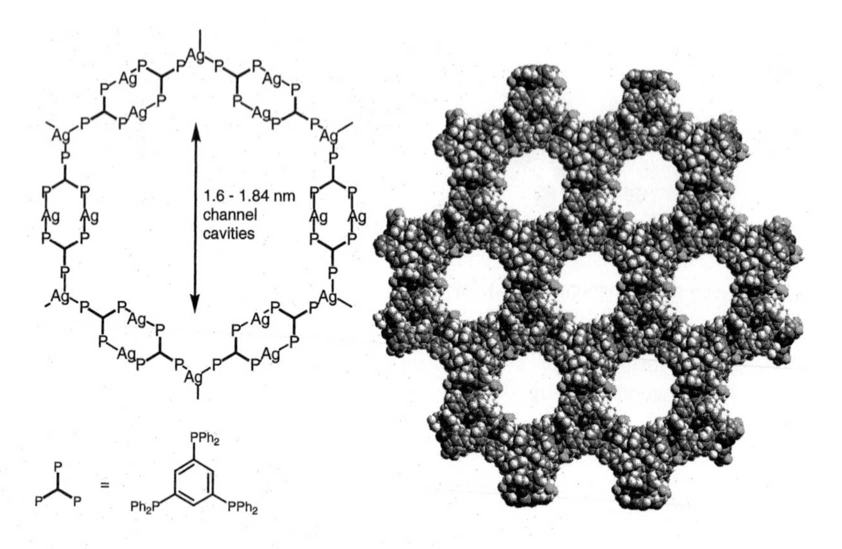

Figure 11. Diagramatic structure of the large fused hexagonal rings in the chickenwire polymer polymer $Ag_4(tppb)_3(OTf)_4$ (left), and the crystal structure of this material (right).

It consists of a fused hexagonal (chickenwire) layers, in which the nodes are trigonal planar AgP_3 centres, and the connectors are $Ag_2(tppb)_2$ units, which contain 12-membered rings. The large hexagonal rings are 72-membered. There are two types of triflate anions – those associated with the trigonal silver nodes, and those associated with the dinuclear rings in the $Ag_2(tppb)_2$ connectors. The latter triflates have their F atoms pointing towards the centres of the large rings. Despite this there remains an very large cavity with a transannular F..F distance of 1.901(3) nm, which corresponds to 1.6 nm when the F van der Waals radius is taken into account. The diameter of the rings is slightly larger between the triflates with an H…H distance of 2.08 nm, corresponding to 1.84 nm when the H van der Waals radius is taken into account. Crucially, the chickenwire layers pack in an eclipsed manner. Therefore the holes line up to form channels through the crystal structure (Figure 11, right). The 1.6-1.84 nm width of the channels is unusually large for a stable crystalline material. The compound was insoluble in chlorinated and aromtaic solvents, but was redissolved by the strongly metal-coordinating acetonitrile. The X-ray crystal structure showed no evidence of guests, however small residual electron density peaks were ascribed to disordered solvent molecules within the channels. The presence of solvent was confirmed with a sample which was washed with ethanol, then dissolved in d_3-acetonitrile for ^1H NMR spectroscopy. Signals for the previously included ethanol corresponding to *ca.* 75% of the available channel volume (using the molar volume of pure ethanol) were observed. It was also possible to exchange the included solvent for diethylether by immersion of the ethanol-loaded material in that solvent. A material which was apparently free of included solvent was obtained by immersion in dichloromethane for 24 hours, followed by heating to 170°C at 0.1 mmHg, although it did apparantly contain some residual water, however the ^1H NMR signal of the latter corresponded to only *ca.* 4% of the channel volume. Very importantly, the XRPD pattern of this material confirmed that the structure was intact, despite the lack of solvent.

In attempting to understand the formation of this particular structure the nature of the solution-state precursors was investigated. ^{31}P NMR spectroscopy revealed that several species were present in the solution, depending on the concentration, each being apparently highly symmetrical and having AgP_2 coordination. The structures were therefore most likely cage complexes of general formula $\{Ag_3L_2\}_n$ (n = 1,2 etc), *i.e.* corresponding to the Ag_3L_2 and Ag_6L_4 structures discussed above for the flexible triphos ligand together, possibly, with higher nuclearity structures. However, puzzlingly, none of these symmetrical cages could have given rise to the hexagonal polymer directly by ROP since the stoichiometries of the polymer $\{Ag_4(tppb)_3\}$ and these symmetrical precursors $\{Ag_3(tppb)_2\}_n$ are different. However, intriguingly, mass spectrometry did show, in addition to peaks corresponding to the $n = 1$ and $n = 2$ structures, a peak corresponding to the $Ag_4(tppb)_3(OTf)_2^{2+}$ cation, which is a potential precursor to the porous network polymer.

Conclusions

The use of multidentate phosphines to generate coordination cages and polymers has lead to some unusual cage and polymeric structures and pointed to a novel formation mechanism for coordination polymers (ROP of coordination cages). The isolation of a stable nanoporous polymer indicates that there is great potential in this field for novel structural chemistry, particularly if it proves possible to control the polymerisation reactions. Current and future aims in our group are to further clarify and exploit the relationship between coordination polymers and their solution-based precursors, as well as to investigate the characteristics of porous coordination polymers with regard to their inclusion behaviour and related chemistry. Funding from EPSRC, McClay Trust and The Leverhulme Trust for the work described here as well as the Questor Centre, Belfast, for funding other projects on porous coordination polymers for environmental remediation and gas storage applications is gratefully acknowledged.

Experimental

The MAS NMR study was conducted on ^{31}P nuclei at resonance frequency of 80 MHz (4.7 T), using a NMR spectrometer (Avance-200 Bruker, Germany). The powder was packed into a zirconia rotor with 4mm of outer diameter and 20mm long and sealed with a Kel-F endcap. Spinning rates of the sample at the magic angle (54.74°) were 5.0 kHz. The spectra were acquired with single pulse excitation and the recycle time between pulses was 5.0 s for ^{31}P. Reference material for ^{31}P chemical shift (in ppm) was relative to 85%H_3PO_4 at zero ppm.

Preparation of $[Ag_6(triphos)_4(O_3SCH_2CH_2NC_4H_8NCH_2CH_2SO_3)_2][BF_4]_2$

$AgBF_4$ (350mg, 1.80 mmol) and triphos (749 mg, 1.20 mmol) were dissolved in a mixture of acetonitrile (1 ml) and dichloromethane (5 ml). To this solution was added a solution of $[NBu_4]_2[O_3SCH_2CH_2NC_4H_8NCH_2CH_2SO_3]$ (468 mg, 0.60 mmol) in acetonitrile (5 ml) dropwise with stirring over 5 minutes. The mixture was allowed to stand for 24 hours, and a white precipitate was collected by filtration and dried in air for 24 hours. Yield 1.08g (86%, based on $[Ag_6(triphos)_4(O_3SCH_2CH_2NC_4H_8NCH_2CH_2SO_3)_2][BF_4]_2$). Elemental analysis % for $[Ag_6(triphos)_4(O_3SCH_2CH_2NC_4H_8NCH_2CH_2SO_3)_2][BF_4]_2.2CH_2Cl_2$ calc. C 53.98, H 4.74, N 1.38; found C 53.76, H 4.92, N 1.90.

[1] M. Eddaoudi, D. B. Moler, H. Li, B. Chen, T. M. Reineke, *Acc. Chem. Res.* **2001**, *34*, 319. O. M. Yaghi, H. Li, C. Davis, D. Richardson, T. L. Groy, *Acc. Chem. Res.* **1998**, *31*, 474. G. B. Gardner, D. Venkataraman, J. S. Moore, S. Lee, *Nature* **1995**, *374*, 792-795. S.-I. Noro, S. Kitagawa, M. Kondo, K. Seki, *Angew. Chem. Int. Ed.* **2000**, *39*, 2082. K. Biradha, Y. Hongo, M. Fujita, *Angew. Chem. Int. Ed.* **2000**, *39*, 3843-3845. S. S.-Y. Chui, S. M.-F. Lo, J. P. H. Charmant, A. G. Orpen, I. D. Williams, *Science* **1999**, *283*, 1148-1150. C. Banglin, M. Eddaoudi, S. T. Hyde, M. O'Keeffe, O. M. Yaghi, *Science* **2000**, *291*, 1021. T. M. Reineke, M. Eddaoudi, D. Moler, M. O'Keeffe, O.M. Yaghi, *J. Am. Chem. Soc.* **2000**, *122*, 4843. T. M. Reineke, M. Eddaoudi, M. O'Keeffe, O. M. Yaghi, *Angew. Chem. Int. Ed.* **1999**, *38*, 2590. T. M. Reineke, M. Eddaoudi, M. Fehr, D. Kelley, O. M. Yaghi, *J. Am. Chem. Soc.* **1999**, *121*, 1651. L. Pan, E. B. Woodlock, X. Wang, *Inorg. Chem.* **2000**, *39*, 4174. S. R. Batten, R. Robson, *Angew. Chem. Int. Ed.* **1998**, *37*, 1460.

[2] D. L. Caulder, K. N. Raymond, *Acc. Chem. Res.* **1999**, *32*, 975-982; P. J. Stang, *Chem. Eur. J.* **1998**, *4*, 19; M. Fujita, *Chem. Soc. Rev.* **1998**, *27*, 417; C. J. Jones, *Chem. Soc. Rev.* **1998**, *27*, 289; Transition Metals in Supramolecular Chemistry: R. W. Saalfrank, B. Demleitner in *Perspectives in Supramolecular Chemistry, Vol. 5* (Ed.: J. P. Sauvage), Wiley-VCH, Weinheim, 1999, pp. 1-51; S. Leininger, B. Olenyuk, P. J. Stang, *Chem. Rev.* **2000**, *100*, 853.

[3] P.M. Van Calcar, M.M. Olmstead, A.L. Balch, *J. Chem. Soc., Chem. Commun.*, **1996**, 2597. E. Lindner, C. Hermann, G. Baum, D. Fenske, *Eur. J. Inorg. Chem.*, **1999**, 679. M.J. Irwin, J.J. Vittal, R.J. Puddephatt, *Organometallics*, **1997**, *16*, 3541. G.C. Jia, R.J. Puddephat, J.D. Scott and J.J. Vittal, *Organometallics*, **1993**, *12*, 3565.

[4] S. Kitagawa, M. Kondo, S. Kawata, S. Wada, M. Maekawa, M. Munakata, *Inorg. Chem.*, **1995**, *34*, 1455; E.R.T. Tiekink, *Acta. Crystallogr.*, Sect. C, **1990**, *46*, 1933; D.M. Ho, R. Bau, *Inorg. Chem.*, **1983**, *22*, 4073. S.P. Neo, Z.-Y. Zhou, T.C.W. Mak, T.S.A. Hor, *Inorg. Chem*, **1995**, *34*, 520; A.F.M.J. van der Ploeg, G. van Koten G., A.L. Spek, *Inorg. Chem.*, **1979**, *18*, 1052; A.F.M.J. van der Ploeg, G. van Koten, *Inorg. Chim. Acta*, **1981**, *51*, 225; Y. Ruina, Y.M. Hou, B.Y. Xue, D.M. Wang, D.M. Jin, *Transition Met. Chem.*, **1996**, *21*, 28; A. Cassel, *Acta Crystallogr., Sect B*, **1976**, *32*, 2521; F. Caruso, M. Camalli, H. Rimml, L.M. Venanzi, *Inorg. Chem.*, **1995**, *34*, 673; A.A.M. Aly, D. Neugebauer, O. Orama, U. Schubert, H. Schmidbaur, *Angew. Chem. Int. Ed. Engl.* **1978**, *17*, 125-126. b) A.L. Airey, G.F. Swiegers, A.C. Willis, S.B. Wild, *Inorg. Chem*, **1997**, *36*, 1588-1597. C.M. Che, H.K. Yip, V.W.W. Yam, P.Y. Cheung, T.F. Lai, S.J. Shieh, S.M. Peng, *J. Chem. Soc., Dalton Trans.*, **1992**, 427-433.

[5] E. Lozano, M. Niewenhuyzen, S.L. James, *Chem. Commun.* **2000**, 617.

[6] E. Lozano, M. Niewenhuyzen, S.L. James, *Chemistry – A European Journal*, **2001**, *12*, 1644.

[7] M. C. Brandys, R.J. Puddephatt, *J. Am. Chem. Soc.* **2001**, *123*, 4839. Z.Q. Qin, M.C. Jennings, R.J. Puddephatt, *Chemistry - A European Journal,* **2002**, *8*, 735. M.C. Brandys, R.J. Puddephatt, *J. Am. Chem. Soc.*, **2002**, *124*, 3946.

[8] S. L. James, D. M. P. Mingos, A. J. P. White, D. J. Williams, *Chem. Commun.*, **1998**, 2323. X. Xu, E. J. MacLean, S. J. Teat, M. Nieuwenhuyzen, M. Chambers, S. L. James, *Chem. Commun.* **2002**, 78. P.W. Miller, M. Nieuwenhuyzen, S.L. James *Chem. Commun.* **2002**, 2008.

[9] T.M. Reineke, M. Eddaoudi, D. Moler, M. O'Keeffe, O.M. Yaghi, *J. Am. Chem. Soc.*, **2000**, *122*, 4843.

[10] X.Xu, M. Niewenhuyzen, S.L. James, *Angewandte Chemie, Int. Ed.*, **2002**, *41*, 764.

Macromol. Symp. ***196***, *201–211 (2003)*

Preparation of Synthons for Carborane Containing Macromolecules

*Matthew C. Parrott, Erin B. Marchington, John F. Valliant, Alex Adronov**

Department of Chemistry, McMaster University, Hamilton, Ontario, Canada
E-mail: adronov@mcmaster.ca, valliant@mcmaster.ca

Summary: The modification of *para*-carborane with appropriate functionalities for incorporation within a dendrimer framework was accomplished by functionalizing the carbon centers with protected alcohol and free acid groups. These compounds are excellent candidates for utilization as functional linkers between two generations of an aliphatic polyester dendrimer structure. Future assembly of these structures will result in dendritic macromolecules containing carboranes within their interior and will be enveloped by hydrophilic groups (hydroxyls) to maintain their water solubility and biocompatibility. These structures have potential applications in Boron Neutron Capture Therapy and Synovectomy. Additionally, carboranes were coupled to polymerizable acrylate structures, and it was shown that the resulting carborane monomers could be polymerized using living free-radical polymerization techniques.

Keywords: atom transfer radical polymerization; carboranes; dendrimers; drug delivery; polymers

Introduction

Carboranes, icosahedral clusters with molecular formula $C_2B_{10}H_{12}$, have attracted a great deal of attention for more than 40 years. Their extraordinary thermal stability, hydrophobicity, and three dimensional "aromaticity" impart interesting properties to the molecules and materials that they comprise.[1] Through functionalization at the carbon centers, carboranes have been incorporated in a wide array of coordination compounds, catalysts, polymers, and pharmaceuticals.[2] Perhaps the most interesting aspect of boron-rich compounds is the nuclear reaction that results upon irradiation of ^{10}B nuclei with thermal neutrons, namely the $^{10}B(n,\alpha)^{7}Li$ reaction.[3] Due to the high neutron capture cross section of ^{10}B, and its high relative natural abundance (20%), this α-particle emitting reaction can be induced with reasonably high yields. The low relative penetration of α-radiation, equal to approximately the diameter of a single human cell, allows this reaction to cause highly localized damage to boron-containing environments. In 1936, Locher proposed that neutron irradiation of ^{10}B delivered to malignant cells could be used to treat

© 2003 WILEY-VCH Verlag GmbH & KGaA, Weinheim CCC 1022-1360/00/$ 17,50+.50/0

diseases such as cancer, and this treatment was dubbed Boron Neutron Capture Therapy (BNCT).[4] More recently, a similar approach has been investigated for the treatment of refractory cases of rheumatoid arthritis (RA), in which the neutron capture reaction is used to destroy the inflamed synovial lining of a joint that causes debilitating pain to the patient.[5] This process is termed Boron Neutron Capture Synovectomy (BNCS). Despite the great potential of BNCT and BNCS in treating disease, clinical success with both of these therapies has been limited. The main challenge to these approaches has been the delivery of adequate amounts of boron (10-30 μg ^{10}B g^{-1} tumour) to the site of interest, and especially into the cells that are to be destroyed.[2] Carboranes, with their high B content have been the molecules of choice for B-delivery, but are themselves inadequate due to their hydrophobicity. Conjugation with various hydrophilic biological molecules, such as peptides, carbohydrates, nucleosides, and monoclonal antibodies has been investigated, but only limited success has been achieved due to rapid clearance of these structures from the circulation.[2] Conjugation with non-biological molecules, such as hydrophilic polyamines, has been shown to target specific intracellular sites (i.e., DNA) *in vitro*, but these studies have revealed substantial cytotoxicity and a general inability to deliver adequate quantities of boron to tumours *in vivo*.[6]

Dendrimers are a class of polymers that are precisely synthesized, monodisperse, and can exhibit a large number of functionalizable end-groups.[7] This multivalency of dendrimers makes them attractive scaffolds for drug delivery applications, as a high loading of pharmacophores can be achieved on each structure. In addition, if the pharmacophore is incorporated within the interior of the dendrimer, the periphery of the macromolecule can be tailored for solubility and specific targeting functions.[8]

The current goals of our research program are the synthesis, characterization, and application of well-defined carborane-functionalized macromolecules. This paper describes the preparation of synthons for aliphatic polyester dendrimers, as well as linear block copolymers, in which the location and loading of carboranes can be easily controlled. The application of living radical polymerization techniques to the synthesis of carborane containing block-copolymers has the potential to yield boron-rich macromolecules with tuneable properties. In addition, aliphatic polyester dendrimers are water soluble in the neutral state, non-toxic, bio-compatible, and easily derivatized with a wide range of functionality, making them ideal for drug-delivery applications.[9] We have explored the use of functionalized carborane cages as linkers between two generations on the interior of the dendrimer, so as to effectively encapsulate them within the hydrophilic scaffold. This strategy will allow us to maximize water solubility and biocompatibility, control the degree of carborane incorporation, and possibly attach active targeting agents to the periphery of the final macromolecule.

Results and Discussion

Dendrimer Synthesis. The synthetic methodology for the preparation of aliphatic polyester dendrimers, based on 2,2-bis(hydroxymethyl) propanoic acid (bisMPA) as the monomer unit, has been documented in recent literature and involves standard esterification chemistry.[10-13] This family of dendrimers poses a number of key advantages, including the low cost of the starting materials, the possibility of utilizing convergent or divergent growth strategies, and the high-yields of the synthetic steps.[14] We have chosen to utilize the divergent growth strategy involving a highly active anhydride form of the protected bisMPA monomer. The steps involved in the preparation of a first-generation (G-1) hydroxyl terminated dendrimer are outlined in Scheme 1. Higher generation dendrimers were obtained through repetition of the coupling and deprotection steps.

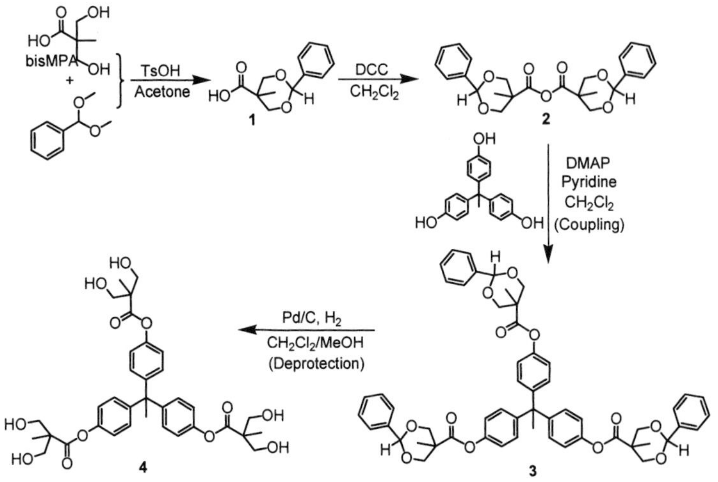

Scheme 1.

Carborane Functionalization. In order to incorporate the carborane cage structure into the polyester dendrimer synthesis, it was necessary to introduce an acid functionality for coupling to the terminal hydroxyls of the dendrimer, as well as a protected alcohol functionality that could be subsequently utilized to react with **2** for further dendrimer growth. In this way, the functionalized carborane acts as an inter-generation linker, allowing its incorporation at any desired stage of dendrimer synthesis. Preparation of functionalized *para*-carboranes was accomplished using literature procedures,[15] and is illustrated in Scheme 2. Deprotonation of *para*-carborane with n-butyllithium (n-BuLi) followed by addition of oxetane produced **5** in 50% yield. Protection of the alcohol with TBDPS was accomplished in quantitative yield under standard conditions and was followed by a second n-BuLi deprotonation and reaction with CO_2. The desired acid (**7**) was obtained in 42% yield over the three steps. The benzylated bisMPA (**8**) was prepared in a single step through the reaction of bisMPA with benzyl bromide under basic conditions,[13] and was used as a model compound to investigate the esterification chemistry between acid **7** and the eventual hydroxyl functionalized dendrimer periphery.

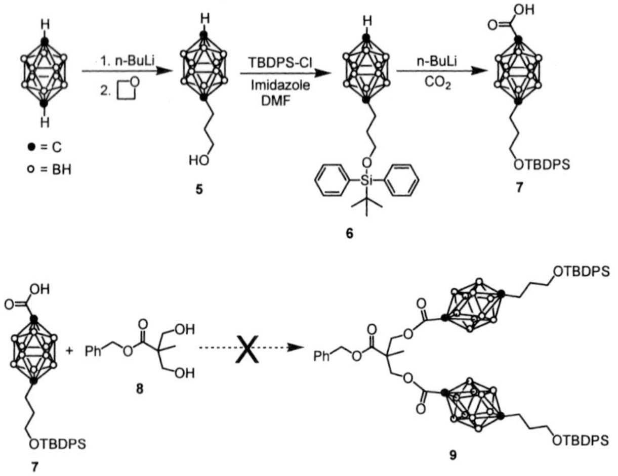

Scheme 2.

Unfortunately, all attempts to couple **7** and **8** using carbodiimide chemistry were unsuccessful, presumably due to deactivating electronic and steric effects of the proximal carborane cluster. It was hypothesized that increasing the distance between the acid group and the carborane of compound **7** should greatly improve the coupling efficiency. This was tested with model compound **10**,[16] the acid derivative of *ortho*-carborane having a two carbon spacer between the acid and the carborane. Coupling of **10** with **8** proceeded smoothly using 1-(3-dimethylaminopropyl)-3-ethylcarbodiimide hydrochloride (EDC) and the 1:1 salt of 4-dimethylaminopyridine and toluene sulfonic acid (DPTS), yielding **11** in quantitative yield.

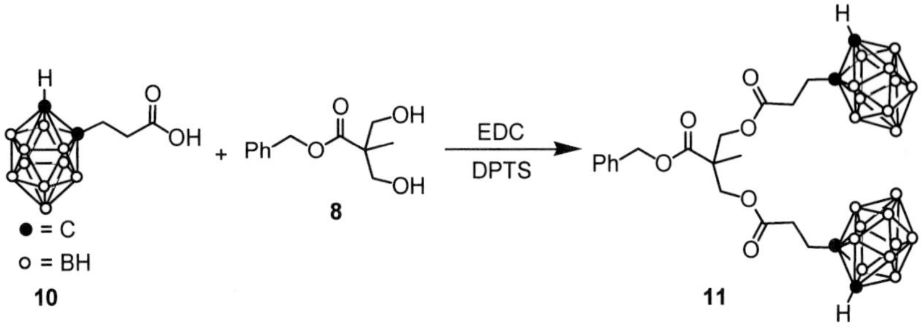

Scheme 3.

This successful model chemistry indicates that extending the distance between the acid group and the carborane in **7** by two methylene groups should restore its reactivity toward esterification. Preparation of the required carborane synthon, starting with protected alcohol **6**, involved a repetition of the n-BuLi/oxetane reaction, which was carried out in 61% yield, followed by oxidation to the acid (Scheme 4). The final step of this sequence has not yet been completed, but once accomplished, will provide the necessary structures for incorporation of *para*-carboranes into the aliphatic polyester dendrimer synthesis. The G-3 target structure is illustrated in Scheme 4.

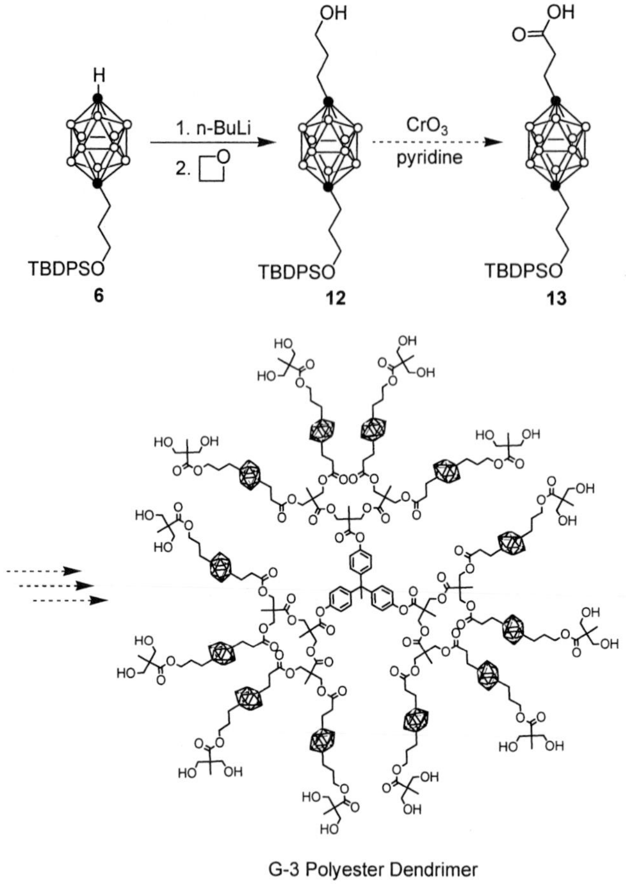

G-3 Polyester Dendrimer

Scheme 4.

Carborane-Containing Polymers. The carborane derivatization strategy described above was easily adapted to the preparation of a polymerizable carborane-containing acrylate monomer. This was accomplished by esterifying alcohol **5** with acryloyl chloride in triethyl amine to produce monomer **14** (Scheme 5). It was found that this monomer readily polymerizes by atom transfer radical polymerization, using CuBr/N,N,N',N'',N''-pentamethyldiethyltriamine (PMDETA) as the catalyst/ligand and ethyl-2-bromopropionate (2-EBP) as the initiator, resulting in narrow-polydispersity living polymers (Scheme 5). The polymerization product was precipitated in MeOH to yield a white powder with Mw = 11,600 and PDI = 1.18 (Figure 1). The isolated polymer retained its living character and could be chain extended with various monomers, such as t-butyl acrylate. The formation of block copolymers, where one block contains carborane side chains, opens up the possibility for the preparation of a wide array of carborane-containing macromolecules, including amphiphilic structures capable of self-assembly in aqueous and/or organic media to produce micelles and vescicles. Investigation of the self assembly properties of block copolymers incorporating carboranes is currently underway in our laboratory.

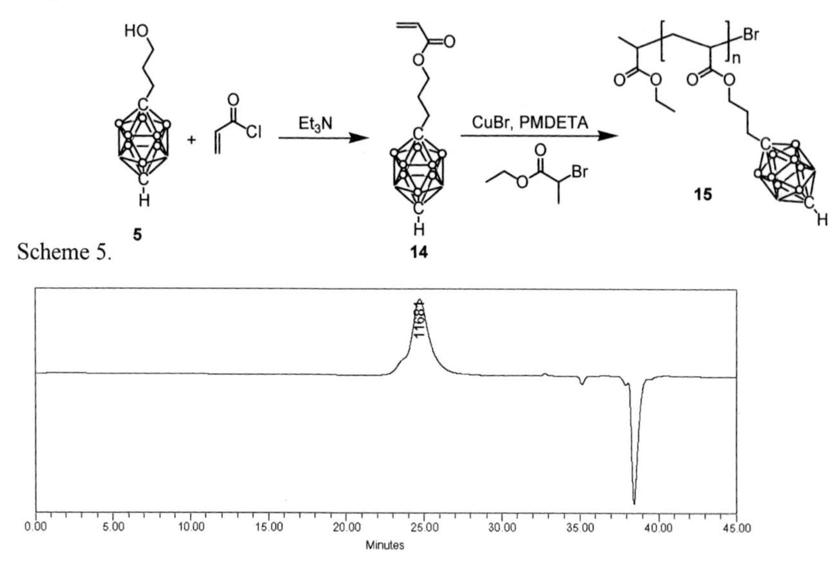

Scheme 5.

Figure 1. Size exclusion chromatogram for polymer 15.

Conclusions and Outlook

The preparation of appropriate synthons for the incorporation of carborane cages within a dendritic structure was accomplished. It was found that an acid group attached directly to the carborane was not active toward esterification, but one separated by a two carbon linker could be esterified in good yield. Once the final carborane synthon is prepared, it will be possible to produce a series of dendrimers in which the carborane loading can be controlled by varying the generation at which it is incorporated. The solubility and biocompatibility properties of the dendrimers will be controlled by the hydroxyl groups present at the dendrimer periphery after deprotection, which can be further derivatized with biomolecules to impart active targeting characteristics to the structures. Once made, the effectiveness of these molecules in BNCT and BNCS will be evaluated. In addition, it was found that functionalization of carboranes with acrylate groups resulted in carborane-containing monomers that could undergo atom transfer radical polymerization. The resulting polymer had low polydispersity and could be chain extended with t-butyl acrylate, indicating that the polymerization of the initial block was living. Therefore, it will be possible to utilize this monomer to prepare a variety of block copolymers that will differ in their carborane loading, solubility, and mechanical properties, depending on the identity of the comonomers. Due to the unusual properties of carboranes, a number of new and interesting materials will likely result from this initial investigation.

Experimental

Preparation of 1-Hydroxypropyl-p-carborane (5)

Para-carborane (300 mg, 2.08 mmol) was dissolved with 60 mL of dry tetrahydrofuran and placed in a flame dried round bottom flask with an argon atmosphere. The reaction mixture was cooled to 0°C, and butyllithium (945 μL, 2.08 mmol) was added slowly to the reaction mixture. After 1 hour, trimethylene oxide (135 μL, 2.08 mmol) was added and allowed to react for an additional 2 hours at room temperature. The excess butyllithium was quenched with 20 mL of HCl (0.2 M) for 30 minutes. After quenching, 50 mL of dichloromethane and 50 mL of water were used to extract the product. The organic layer was separated, dried over $MgSO_4$, and evaporated to yield a white glass: 0.211 g, 1.04 mmol (50%). IR (cm^{-1}): 3620, 2610. ^{1}H NMR (500 MHz, $CDCl_3$): δ = 1.29 (s) 1.43 (m), 1.72 (m), 2.63 (s), 3.47 (t , 2, J = 6). ^{13}C NMR (125 MHz, $CDCl_3$): δ = 32.25, 35.32, 58.07, 61.85, 84.20. ^{11}B NMR (160 MHz, $CDCl_3$, proton-decoupled): δ = -14.29, -11.78.

Silyl Protection of 1-Hydroxypropyl-p-carborane (6)

Compound 5 (1.00 g, 4.94 mmol) and imidazole (0.673 g, 9.89 mmol) were dissolved in 60 mL of DMF. The mixture was allowed to cool to 0°C, and tert-butylchlorodiphenylsilane (2.04 g, 7.41 mmol) was added to the mixture and allowed to react for 3 hours. The reaction mixture was quenched with 20 mL of water and was washed ether. The ether layers were collected and additionally washed with 2 M HCl, 10% NaOH, and brine. The product was isolated by column chromatography (9:1 hexane : ethyl acetate), yielding a white glass: 2.12 g, 4.82 mmol (98%). ^{1}H NMR (500 MHz, $CDCl_3$): δ = 1.04 (s, 9), 1.43 (m), 1.72 (m), 2.63 (s), 3.50 (t , 2, J = 6), 7.44(m, 6), 7.62 (d, 4, J = 8) . ^{13}C NMR (125 MHz, $CDCl_3$): δ = 19.15, 26.81, 32.25, 35.51, 57.95, 62.76, 84.59, 127.63, 129.62, 133.72, 135.51. ^{11}B NMR (160 MHz, $CDCl_3$, proton-decoupled): δ = -14.29, -11.72.

Carboxylation of the Silyl Protected 1-Hydroxypropyl-p-carborane (7)

Compound 6 (0.199 g, 0.452 mmol) was dissolved in 9 mL of dry THF. The reaction vessel was cooled to 0°C and n-butyl lithium (0.224 mL, 0.477 mmol) was added and allowed to react for 1 hour. The mixture was then cooled to –78°C and CO_2 was bubbled into the mixture. The formation of a precipitate was observed after ca. 15- 30 min. After 3 hrs of bubbling, the reaction was quenched using 30 ml of HCl (0.2 M). This was followed by the addition of 50 ml of H_2O and extraction of the resulting mixture using CH_2Cl_2 (2 x 50 mL). The organic phases were collected, dried (Na_2SO_4), filtered, and evaporated to dryness in vacuo to yield a white crystalline solid (184 mg, 84%). ^{1}H NMR (200 MHz, $CDC1_3$): δ = 1.05 (s, 9 H), 1.43 (m, 2 H), 1.76 (t, 2 H), 3.50 (t, 2 H), 7.42 (m, 5), 7.61 (m, 5), 9.41 (s, 1).

Hydroxypropylation of the Silyl Protected 1-Hydroxypropyl-p-carborane (12)

Compound 6 (0.219 g, 0.498 mmol) was dissolved in 100 mL of dry THF. The reaction vessel was cooled to 0°C and n-butyl lithium (0.553 mL, 0.996 mmol) was slowly added and allowed to react for 4 hours. The reaction mixture was cooled to 0°C once more and the trimethylene oxide (0.0486 mL, 0.747 mmol) was added and allowed to react overnight. After the reaction was completed the THF was removed by roto-evaporation and the product was extracted with 2 M HCL and ether. The product was isolated by column chromatography using 100% DMC (to remove staring material) and then 8:2 (DCM : Ethyl Acetate). The product was isolated as clear oil: 0.1503 g, 0.301 mmol (61%).

^{1}H NMR (500 MHz, $CDCl_3$): δ = 1.02 (s, 9), 1.41 (m, 4), 1.73 (m, 4), 3.47 (t , 2, J = 6), 7.41 (m,

6), 7.59 (d, 4, $J = 8$) . ^{13}C NMR (125 MHz, $CDCl_3$): $\delta = 19.07$, 26.60, 32.36, 34.10, 61.86, 62.72, 78.46, 78.98, 127.55, 129.53, 133.65, 135.43. ^{11}B NMR (160 MHz, $CDCl_3$, proton-decoupled): $\delta = -11.97$.

Coupling of O-Carborane-1-propionic acid with Benzyl-2-2-bis(oxy methyl) Propionate (11)

O-carborane-1-propionic acid (0.150 g, 0.694 mmol), benzyl-2-2-bis(oxy methyl) propionate (0.071 g, 0.315 mmol) and 4-(dimethylamino)-pyridine/p-toluene sulphonic salt (0.037 g, 0.126 mmol) were dissolved in 10 mL of dichloromethane. EDC (0.151 g, 0.788 mmol) was added to the reaction and stirred for 18 hours at room temperature. The reaction mixture was diluted with 40 mL of DCM and washed with distilled water. The organic layers were collected and removed by roto-evaporation. The product was isolated through column chromatography using 100% DCM as the solvent system. The coupling product was isolated as a glass: 0.1795 g, 0.290 mmol (92%). ^{1}H NMR (500 MHz, $CDCl_3$): $\delta = 1.28$ (s, 3), 2.38 (m), 2.48 (m), 3.66 (s, 2), 4.24 (q, 4, $J = 11$), 5.17 (s, 2) 7.34 (m, 5). ^{13}C NMR (125 MHz, $CDCl_3$): $\delta = 17.81$, 32.45, 33.05, 46.27, 61.55, 65.89, 67.03, 73.72, 128.27, 128.59, 128.66, 135.40, 170.81, 172.05. ^{11}B NMR (160 MHz, $CDCl_3$, proton-decoupled): $\delta = -11.98$, -11.32, -10.80, -8.48, -4.57, -1.17.

1-p-Carborane Propyl Acrylate (14)

Compound 5 (1.00 g, 4.943 mmol) and triethylamine (1.00 g, 9.886 mmol) were dissolved in 100 mL of dry DCM. After 10 minutes, acryloyl chloride (0.671 g, 7.414 mmol) was added and allowed to react for two hours. The reaction mixture was diluted with 50 mL of ether and washed with distilled water. The organic layers were collected and removed by roto-evaporation. The product was isolated through column chromatography using 100% DCM as the solvent system. The product was isolated as white crystals: 1.22 g, 4.764 mmol (96%). IR (cm^{-1}): 2610, 1732. ^{1}H NMR (500 MHz, $CDCl_3$): $\delta = 1.54$ (m, 2), 1.70 (m, 2), 2.64 (s, 1), 3.97 (t, 2, $J = 6$), 5.80 (d, 1, $J = 1.3$), 6.06 (q, 1, $J = 10.5$, $J = 17$) 6.36 (d, 1, $J = 1.3$). ^{13}C NMR (125 MHz, $CDCl_3$): $\delta = 28.36$, 35.33, 58.22, 63.26, 83.65, 128.21, 130.76, 165.89. ^{11}B NMR (160 MHz, $CDCl_3$, proton-decoupled): $\delta = -14.25$, -11.82.

Poly(1-p-Carborane Propyl Acrylate) via ATRP (15)

Compound 14 (1.00 g, 3.900 mmol), pentamethyldiethylenetriamine (14 mg, 78 μmol), and ethyl 2-bromopropionate (14 mg, 78 μmol) were dissolved in 1 mL of DMF. Copper (I) bromide (11 mg, 78 μmol) was placed in a separate reaction vessel. After 30 minutes of evacuation with argon on both reaction vessels, the monomer/ligand/initiator mixture was added to the copper bromide catalyst. The reaction mixture was heated to 60°C and allowed to react for 18 hours. Polymeric products were separated by precipitation in methanol and water (1:1). The isolated product was a light blue glass: 0.91 g (91% conversion). GPC: Mw = 11,681g/mol, Mn = 10,285 g/mol. ^{1}H NMR (500 MHz, $CDCl_3$): δ = 1.51 (s, 145), 1.70 (s), 2.69(s, 141), 2.88 (s, 3), 2.96 (s, 4), 3.84 (s, 100). ^{13}C NMR (125 MHz, $CDCl_3$): δ = 28.33, 35.20, 41.27, 58.27, 63.50, 83.70, 173.88. ^{11}B NMR (160 MHz, $CDCl_3$, proton-decoupled): δ = -14.20, -11.77.

[1] J. Plesek, *Chemical Reviews* **1992**, *92*, 269.
[2] J. F. Valliant, K. J. Guenther, A. S. King, P. Morel, P. Schaffer, O. O. Sogbein, K. A. Stephenson, *Coordination Chemistry Reviews* **2002**, *232*, 173.
[3] M. F. Hawthorne, *Angewandte Chemie-International Edition* **1993**, *32*, 950.
[4] G. L. Locher, *Am. J. Roentgenol. Radium Ther.* **1936**, *36*, 1.
[5] J. C. Yanch, S. Shortkroff, R. E. Shefer, S. Johnson, E. Binello, D. Gierga, A. G. Jones, G. Young, C. Vivieros, A. Davison, C. Sledge, *Medical Physics* **1999**, *26*, 364.
[6] R. F. Barth, D. M. Adams, A. H. Soloway, F. Alam, M. V. Darby, *Bioconjugate Chemistry* **1994**, *5*, 58.
[7] D. A. Tomalia, J. M. J. Frechet, *Journal of Polymer Science Part a-Polymer Chemistry* **2002**, *40*, 2719.
[8] H. R. Ihre, O. L. Padilla De Jesus, F. C. J. Szoka, J. M. J. Frechet, *Bioconjugate Chemistry* **2002**, *13*, 443.
[9] O. L. P. De Jesus, H. R. Ihre, L. Gagne, J. M. J. Frechet, F. C. Szoka, *Bioconjugate Chemistry* **2002**, *13*, 453.
[10] H. Ihre, E. Soderlind, A. Hult, *Abstracts of Papers of the American Chemical Society* **1995**, *210*, 187.
[11] H. Ihre, A. Hult, E. Soderlind, *Journal of the American Chemical Society* **1996**, *118*, 6388.
[12] H. Ihre, A. Hult, *Abstracts of Papers of the American Chemical Society* **1997**, *214*, 12.
[13] H. Ihre, A. Hult, J. M. J. Frechet, I. Gitsov, *Macromolecules* **1998**, *31*, 4061.
[14] H. Ihre, O. L. Padilla De Jesus, J. M. J. Fréchet, *JACS* **2001**, *123*, 5908.
[15] A. Herzog, C. B. Knobler, M. F. Hawthorne, A. Maderna, W. Siebert, *Journal of Organic Chemistry* **1999**, *64*, 1045.
[16] J. F. Valliant, P. Morel, P. Schaffer, J. H. Kaldis, *Inorganic Chemistry* **2002**, *41*, 628.

Design, Synthesis and Structural Diversity in Coordination Polymers

*Heba Abourahma, Gregory J. McManus, Brian Moulton, Rosa D. Bailey Walsh Michael J. Zaworotko**

University of South Florida, Department of Chemistry, 4202 E Fowler Ave. (SCA 400), Tampa, FL 33620, USA
E-mail: xtal@usf.edu

Summary: The fundamental precept of crystal engineering is that all the information necessary for the design of a network is already stored in the molecular building blocks used. Coordination polymers represent an example of how crystal engineering has become a paradigm for the synthesis of new architectures and compositions. We report herein ten structures that are supramolecular isomers of one another. The structures are prepared from the same building blocks under mild reaction conditions. The modularity of coordination polymers imparts structural diversity that otherwise would not be possible.

Keywords: coordination polymers; dimetal tetracarboxylate; isomerism; SBU; supramolecular

Introduction

Supramolecular chemistry was defined by Lehn as the chemistry beyond the molecule.[1,2] The concepts of supramolecular chemistry are derived from nature and rely on the phenomena of molecular recognition and self-assembly. Molecules recognize complementary sites (functionality, geometry, size, *etc.*) on other molecules and associate into larger entities, supermolecules, via weaker, non-covalent interactions such as hydrogen bonding and π-π interactions. It should be stressed that supramolecular approaches to synthesis provide an alternate paradigm to sequential multi-step synthesis that is particularly attractive since it can be accomplished in a single step, often in high yield. In short, one can design a supermolecule with predictable architecture by selecting the appropriate set of building blocks that have complementary sites.

Crystal engineering[3-8] is inherently linked to supramolecular chemistry. Schmidt demonstrated in his early work that crystals can be thought of as the sum of a series of molecular recognition events and self-assembly, rather than the need to "avoid a vacuum".[3] The field of crystal

© 2003 WILEY-VCH Verlag GmbH & KGaA, Weinheim CCC 1022-1360/00/$ 17,50+.50/0

engineering developed further in the 1980s with the work of Etter[9-11] and Desiraju,[8,12] who concentrated upon applying the Cambridge Structural Database[13] (CSD) to the analysis, interpretation and design of non-covalent bonding patterns in organic solids.

In this contribution we focus upon how coordination polymers exemplify how crystal engineering has become a paradigm for the design of new supramolecular architectures. The work of Wells[14,15] serves as an excellent starting point concerning how one can define crystal structures in terms of their topology by reducing them to a series of points (nodes) of a certain geometry (tetrahedral, trigonal planar, *etc.*) that are connected to a fixed number of other points. The resulting structures, which can also be calculated mathematically, can either be discrete (0D) polyhedra or infinite (1-, 2-, or 3-D) periodic nets. Robson and co-workers[16] delineated the "Wellsian" work on inorganic network structures into the realm of metal-organic coordination polymers, in turn facilitating the rapid development of the field. In this context, the "node and spacer" approach has been remarkably successful for producing predictable network architectures. The metal typically acts as the "node" and an organic ligand acts as the "spacer" that propagates this node. Scheme 1 illustrates some of the simplest architectures that can be generated using commonly available metal moieties and linear organic "spacer" ligands. It is obvious from the scheme that each network consists of at least two components; i.e. the structure

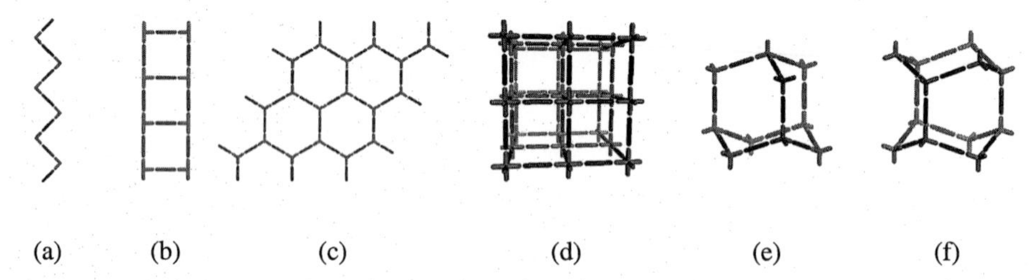

Scheme 1. A schematic representation of some of the simple 1D, 2D and 3D network architectures that have been structurally characterized for metal-organic polymers: (a) zigzag chain; (b) ladder; (c) honeycomb; (d) octahedral; (e) cubic diamondoid; (f) hexagonal diamondoid.

is modular.[17] In principle, these network structures can be regarded as blueprints for the construction of networks from a diverse range of chemical components, provided the components are predisposed to self-assembly. The modular approach is powerful because it facilitates fine-tuning of the structural and functional features. In addition, since, in principle, it is possible for a given set of building blocks to arrange into more than one possible superstructure the phenomenon of supramolecular isomerism may arise.[18] This concept is illustrated by structures (e) and (f) in scheme 1, which represent cubic and hexagonal diamondoid structures, respectively. It is important to note that, although supramolecular isomerism affords structural diversity, it is limited by the number of possible architectures that can be generated *rationally* from the molecular components present in a network.

The architecture alone often affords information that allows the chemist to predict some of the bulk properties. For example, most of the structures in scheme 1 inherently generate cavities that are based upon the size and length of the spacer ligand. The 3D architectures (d)-(f) represent in some ways the ultimate challenge in terms of crystal engineering since they lead directly to crystal structure prediction. It should therefore be unsurprising that diamondoid[17] and octahedral[19-21] frameworks have attracted considerable attention. In general, for 3D architectures one would expect rigidity to be coupled with porosity (zeolite-like properties).[17,21-23] For 2D architectures, one would expect inherent ability to intercalate guest (clay-like properties).[5,24-27] In the case of 1D structures one would normally expect close-packing variability in the context of how adjacent networks pack with respect to one another.[28-33] It should be noted that the construction of networks from single component systems is also an active area of research, but that the two approaches are conceptually different.

In this contribution we explore further the field of coordination polymers by highlighting examples of 2D and 3D coordination polymers that exhibit supramolecular isomerism. The molecular components that we targeted to build our coordination polymers are the metal cluster dizinc tetracarboxylate and the angular ligand benzene-1,3-dicarboxylate (bdc), both illustrated in

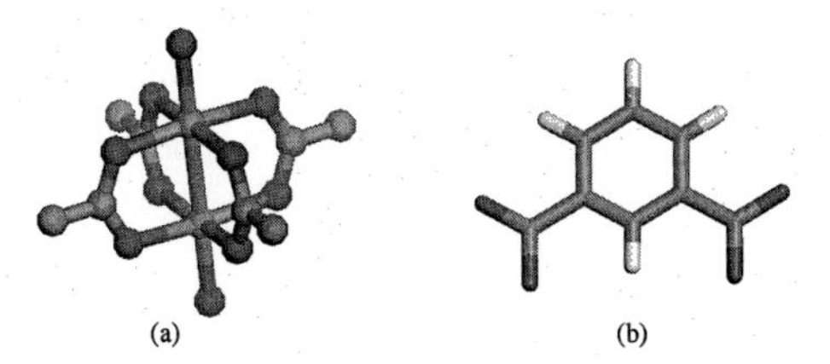

Figure 1. (a) The dimetal tetracarboxylate cluster can be considered a molecular square when viewed down the four-fold axis, (b) The angular bdc ligand.

Figure 1. All structures reported herein have the same empirical chemical formula, $Zn_2(bdc)_2(L)_2$ (L= coordinated base), yet they have dramatically different architectures. The dimetal tetracarboxylate[34] secondary building unit (SBU)[35] is a ubiquitous metal cluster that is present in the CSD[13] for over 1000 structures involving 21 transition metals. However, SBUs based on Zn(II) remain rare as evident by the presence only 10 structures in the CSD.[36-42] Furthermore, only one of these structures utilizes the Zn(II) SBU as a building block in a coordination polymer.[42] The use of such SBUs as building blocks for the generation of discrete and infinite networks is a relatively recent phenomenon. Nevertheless, it is already clear that its structural and functional versatility makes it a particularly attractive molecular building block: the metal can be varied with resulting property changes; the carboxylate moiety has many permutations; the axial ligand can be varied almost at will. Two approaches to the use of this SBU have thus far been delineated: as a linear spacer when coordination occurs at the axial positions,[43-47] or as a molecular square linked at the vertices (as suggested when viewing the SBU down the four-fold axis in Fig. 1).[42,48] To most effectively utilize the SBU as a molecular square building block simple polycarboxylate ligands can be exploited. Benzene-1,4-dicarboxylate[49] (a linear dicarboxylate), benzene-1,3-dicarboxylate[42,48,50-52] (an angular dicarboxylate) and benzene-1,3,5-tricarboxylate[53] (a three-directional carboxylate) have all afforded self-assembled structures with rational, if not predictable, topologies.

We have focused our efforts on the angular ligand bdc, which has the two carboxylate moieties that are predisposed rigidly at 120°, and we have demonstrated that it can facilitate the assembly of square SBUs into square (a cluster of four SBUs) or triangular (a cluster of three SBUs) nanoscale SBUs, nSBUs.[42] These nSBUs can further assemble into discrete faceted polyhedra[48,50] or infinite networks. The former contains both types of nSBUs (square and triangular), whereas two types of the latter result: one that contain square nSBUs only,[42] and one that contain triangular nSBUs only.[52] We have reported these findings for structures based on the SBU derived from Cu(II). In this contribution we demonstrate that structures previously reported for Cu(II) are also accessible when using Zn(II).

Experimental

General methods: All materials were used as received; solvents were purified and dried according to standard methods. TGA data were obtained on a TA instruments 2950 TGA at high resolution with N_2 as purge gas. XRPD data were obtained on a Rigaku MiniFlex$^+$ Diffractometer. Formulations of the coordination polymers are based upon nSBUs rather than embirical units.

Synthesis of $\{[Zn_2(bdc)_2(pyridine)_2]_4$•Benzene $\}_n$ (1)

Colorless crystals of **1** were obtained from slow diffusion via layering an ethanolic solution (10.0mL) of 1,3-H_2bdc (82mg, 0.49mmol) and zinc nitrate hexahydrate (150mg, 0.50mmol) through an ethanolic layer (10.0mL) onto an ethanolic solution (10.0mL) of pyridine (0.12mL, 1.5mmol) and benzene (2.0mL). Crystals formed within days in 3.5% yield. The most intense peaks observed in the X-ray powder diffraction (XPD) patterns from the bulk sample are consistent with those calculated from single crystal diffraction data.

Synthesis of $\{[Zn_2(bdc)_2(2\text{-picoline})_2]_4$•2Nitrobenzene $\}_n$ (2)

Colorless crystals of **2** were obtained from slow diffusion via layering a methanolic solution (10.0mL) of 1,3-H_2bdc (84mg, 0.50mmol) and zinc nitrate hexahydrate (150mg, 0.50mmol) through a methanolic layer (10.0mL) onto a methanolic solution (10.0mL) of 2-picoline (0.15mL, 1.5mmol) and nitrobenzene (2.0mL). Crystals formed within days in 14% yield. The crystals are thermally stable upto 120 °C after which the TG curve shows a mass loss of about 39% between 140 and 240 °C, which is consistent with and corresponds to the loss of guest and axial ligands. Further heating leads to decomposition of the structure above 300 °C. The most intense peaks observed in the X-ray powder diffraction (XPD) patterns from the bulk sample are consistent with those calculated from single crystal diffraction data.

Synthesis of $\{[Zn_2(bdc)_2(4\text{-picoline})_2]_4$•*o*-Dichlorobenzene $\}_n$ (3a)

Colorless crystals of **3a** were obtained from slow diffusion via layering a methanolic solution (20.0mL) of 1,3-H_2bdc (166mg, 0.999mmol) and zinc nitrate hexahydrate (297mg, 0.998mmol) through a methanolic layer (10.0mL) onto a methanolic solution (10.0mL) of 4-picoline (0.30mL, 3.1mmol) and *o*-dichlorobenzene (2.0mL). Crystals formed within days in 6.1% yield. The crystals are thermally stable upto 190 °C after which the TG curve shows a mass loss of about 40% between 220 and 320 °C, which is consistent with and corresponds to the loss of guest and axial ligands. Further heating leads to decomposition of the structure above 380 °C. The most intense peaks observed in the X-ray powder diffraction (XPD) patterns from the bulk sample are consistent with those calculated from single crystal diffraction data.

Synthesis of {$[Zn_2(bdc)_2(4\text{-picoline})_2]_4$•Nitrobenzene}$_n$ (3b)

Colorless crystals of **3b** were obtained from slow diffusion via layering a methanolic solution (20.0mL) of 1,3-H_2bdc (166mg, 0.999mmol) and zinc nitrate hexahydrate (297mg, 0.998mmol) through a methanolic layer (10.0mL) onto a methanolic solution (10.0mL) of 4-picoline (0.30mL, 3.1mmol) and nitrobenzene (2.0mL). Crystals formed within days in 28% yield. The most intense peaks observed in the X-ray powder diffraction (XPD) patterns from the bulk sample are consistent with those calculated from single crystal diffraction data.

Synthesis of {$[Zn_2(bdc)_2(3,5\text{-lutidine})_2]_4$•*o*-Dichlorobenzene }$_n$ (3c)

Colorless crystals of **3c** were obtained from slow diffusion via layering a methanolic solution (10.0mL) of 1,3-H_2bdc (84mg, 0.50mmol) and zinc nitrate hexahydrate (150mg, 0.50mmol) through a methanolic layer (10.0mL) onto a methanolic solution (10.0mL) of 3,5-lutidine (0.17mL, 1.5mmol) and *o*-dichlorobenzene (2.0mL). Crystals formed within days in 32% yield. The most intense peaks observed in the X-ray powder diffraction (XPD) patterns from the bulk sample are consistent with those calculated from single crystal diffraction data.

Synthesis of {$[Zn_2(bdc)_2(3,5\text{-lutidine})_2]_4$•Benzene }$_n$ (3d)

Colorless crystals of **3d** were obtained from slow diffusion via layering a methanolic solution (10.0mL) of 1,3-H_2bdc (84mg, 0.50mmol) and zinc nitrate hexahydrate (150mg, 0.50mmol) through a methanolic layer (10.0mL) onto a methanolic solution (10.0mL) of 3,5-lutidine (0.17mL, 1.5mmol) and benzene (2.0mL). Crystals formed within days in 47% yield. The most intense peaks observed in the X-ray powder diffraction (XPD) patterns from the bulk sample are consistent with those calculated from single crystal diffraction data.

Synthesis of {[Zn_2(bdc)$_2$(4-methoxypyridine)$_2$]$_4$•Benzene }$_n$ (3e)

Colorless crystals of **3e** were obtained from slow diffusion via layering an ethanolic solution (10.0mL) of 1,3-H_2bdc (82mg, 0.49mmol) and zinc nitrate hexahydrate (150mg, 0.50mmol) through an ethanolic layer (10.0mL) onto an ethanolic solution (10.0mL) of 4-methoxypyridine (0.15mL, 1.5mmol) and benzene (2.0mL). Crystals formed within days in 19% yield. The crystals are thermally stable upto 180 °C after which the TG curve shows multiple mass losses, which are consistent with and corresponds to the loss of guest and axial ligands. Further heating leads to decomposition of the structure above 300 °C. The most intense peaks observed in the X-ray powder diffraction (XPD) patterns from the bulk sample are consistent with those calculated from single crystal diffraction data.

Synthesis of {[Zn_2(bdc)$_2$(isoquinoline)$_2$]$_4$•*o*-Dichlorobenzene }$_n$ (3f)

Colorless crystals of **3f** were obtained from slow diffusion via layering a methanolic solution (10.0mL) of 1,3-H_2bdc (84mg, 0.50mmol) and zinc nitrate hexahydrate (150mg, 0.50mmol) through a methanolic layer (10.0mL) onto a methanolic solution (10.0mL) of isoquinoline (0.18mL, 1.5mmol) and *o*-dichlorobenzene (2.0mL). Crystals formed within days in 38% yield. The most intense peaks observed in the X-ray powder diffraction (XPD) patterns from the bulk sample are consistent with those calculated from single crystal diffraction data.

Synthesis of {[Zn_2(bdc)$_2$(4-methoxypyridine)$_2$]$_4$•guest}$_n$ (4)

Colorless crystals of **4** were obtained from slow diffusion via layering a methanolic solution (10.0mL) of 1,3-H_2bdc (82mg, 0.49mmol) and zinc nitrate hexahydrate (150mg, 0.50mmol) through a methanolic layer (10.0mL) onto a nitrobenzene solution (10.0mL) of 4-methoxypyridine (0.15mL, 1.5mmol). Crystals formed within days in 1.4% yield. The most intense peaks observed in the X-ray powder diffraction (XPD) patterns from the bulk sample are consistent with those calculated from single crystal diffraction data.

Synthesis of {[Zn_2(bdc)$_2$(quinoline)$_2$]$_4$•Nitrobenzene }$_n$ (5)

Colorless crystals of **5** were obtained from slow diffusion via layering a methanolic solution (20.0mL) of 1,3-H_2bdc (166mg, 0.999mmol) and zinc nitrate hexahydrate (297mg, 0.998mmol) through a methanolic layer (10.0mL) onto a methanolic solution (10.0mL) of quinoline (0.24mL,

2.0mmol) and nitrobenzene (2.0mL). Crystals formed within days in 13% yield. The crystals are thermally stable upto 190 °C after which the TG curve shows multiple mass losses, which are consistent with and corresponds to the loss of guest and axial ligands. Further heating leads to decomposition of the structure above 300 °C. The most intense peaks observed in the X-ray powder diffraction (XPD) patterns from the bulk sample are consistent with those calculated from single crystal diffraction data.

Crystal Structure Determination

Single crystals suitable for X-ray crystallographic analysis were selected following examination under a microscope. Intensity data were collected on a Bruker-AXS SMART APEX/CCD diffractometer using Mo_{ka} radiation (λ = 0.7107 Å). The data were corrected for Lorentz and polarization effects and for absorption using the SADABS program. The structures were solved using direct methods and refined by full-matrix least-squares on $|F|^2$. All non-hydrogen atoms were refined anisotropically and hydrogen atoms were placed in geometrically calculated positions and refined with temperature factors 1.2 times those of their bonded atoms. All crystallographic calculations were conducted with the SHELXTL 5.1 program package. Table 1 provides crystallographic data for compounds **1-5**. CCDC-# 206112 - 206120 and 202849 contain the supplementary crystallographic data for this paper. These data can be obtained free of charge via www.ccdc.cam.ac.uk/conts/retrieving.html (or from the Cambridge Crystallographic Data Centre, 12 Union Road, Cambridge CB2 1EZ, UK; fax: (+44)1223-336-033; or deposit@ccdc.cam.ac.uk).

Table 1. Crystallographic data for structures **1-5**.

Compound	**1**	**2**	**3a**	**3b**	**3c**
Formula	$C_{32}H_{20}Zn_2N_2O_8$	$C_{40}H_{32}Zn_2N_4O_{12}$	$C_{34}H_{26}Zn_2N_2O_8Cl_2$	$C_{34}H_{27}Zn_2N_3O_{10}$	$C_{36}H_{30}Zn_2N_2O_8Cl_2$
MW	691.24	891.44	792.21	768.33	820.26
Color, habit	Colorless, column	Colorless, column	Colorless, column	Colorless, column	Colorless, column
Crystal system	Tetragonal	Orthorhombic	Monoclinic	Monoclinic	Monoclinic
Space group, Z	P4/ncc, 8	Pbcm, 4	C2/c, 4	C2/c, 4	C2/c, 4
a (Å)	19.0707 (7)	10.5778 (10)	19.2652 (15)	19.796 (2)	18.8097 (17)
b (Å)	19.0707 (7)	19.6493 (18)	12.9285 (10)	12.2755 (15)	13.1619 (12)
c (Å)	16.2329 (11)	19.1872 (18)	14.4703 (11)	14.8457 (18)	14.3915 (13)
α (°)	90°	90°	90°	90°	90°
β (°)	90°	90°	115.3510 (10)°	116.064 (2)°	109.109 (2)°

γ (°)	90°	90°	90°	90°	90°
V	5903.8 (5)	3988.0 (6)	3257.0 (4)	3240.7 (7)	3366.6 (5)
Temp (K)	100 (2)	100 (2)	100 (2)	100 (2)	100 (2)
M (mm^{-1})	1.68	1.271	1.693	1.544	1.641
Theta range	1.51 – 28.28°	1.93 – 28.26°	1.96 - 28.26°	2.02 – 28.25°	1.93 – 28.24°
Reflections collected	33447	24385	13748	9716	6391
Independent refelections (R_{int})	3612 (0.0765)	4931 (0.0388)	7080 (0.0237)	5974 (0.0257)	3302 (0.0226)
R (F), R_w(F)	0.0534, 0.1311	0.0557, 0.1475	0.0338, 0.0849	0.0359, 0.0931	0.0322, 0.0817

Compound	**3d**	**3e**	**3f**	**4**	**5**
Formula	$C_{36}H_{32}Zn_2N_2O_8$	$C_{34}H_{28}Zn_2N_2O_{10}$	$C_{40}H_{26}Zn_2N_2O_8Cl_2$	$C_{29}H_{26}Zn_2N_2O_{11}$	$C_{34}H_{22}Zn_2N_2O_8$
MW	751.38	755.32	864.27	709.26	717.28
Color, habit	Colorless, column	Colorless, column	Colorless, column	Colorless, hexagonal	Colorless, column
Crystal system	Monoclinic	Monoclinic	Monoclinic	Trigonal	Rhombohedral
Space group, Z	C2/c, 4	C2/c, 4	C2/c, 4	P-3m1	R-3c
a (Å)	19.206 (2)	19.291 (7)	20.416 (3)	18.9565(8)	30.2394 (11)
b (Å)	13.0543 (16)	13.439 (5)	13.2052 (18)	18.9565(8)	30.2394 (11)
c (Å)	14.339 (18)	13.864 (5)	13.8901 (19)	11.1246(9)	18.2791 (13)
α (°)	90°	90°	90°	90	90°
β (°)	109.542 (2)°	114.797 (6)°	109.447 (2)°	90	90°
γ (°)	90°	90°	90°	120	120°
V	3386.8 (7)	3263 (2)	3530.4 (8)	3462.0(3)	14475.4 (13)
Temp (K)	100 (2)	100 (2)	100 (2)	100(2)	100 (2)
M (mm^{-1})	1.471	1.532	1.569	1.080	1.545
Theta range	1.92 – 28.34°	1.91 – 23.27°	1.87 – 28.25°	1.83 – 28.33 °	2.33 – 28.28°
Reflections collected	10274	6724	10475	21996	28023
Independent refelections (R_{int})	3963 (0.0538)	2354 (0.0567)	4104 (0.0534)	3028 (0.0396)	3916 (0.0489)
R (F), R_w(F)	0.0442, 0.0954	0.0867, 0.1727	0.0859, 0.1993	0.0796, 0.2729	0.0437, 0.1149

Results and Discussion

Three supramolecular isomers based on dizinc tetracarboxylate SBU and bdc have been obtained: (a) 2D sheet structures based on square nSBUs only, $\{[Zn_2(bdc)_2(L)_2]_4\}_n$, (b) 2D sheet structure based on triangular nSBUs only, $\{[Zn_2(bdc)_2(L)_2]_3\}_n$ (Kagomé lattice), (c) 3D structure with a $6^5.8$ topology, $\{[Zn_2(bdc)_2(L)_2]\}_n$. Figure 2 illustrates the schematic for the three polymeric networks. The geometry around the zinc ion in all structures is that of a square pyramid. The basal positions are occupied by carboxylate moieties and the apical positions are occupied by the base used. Bond distances and angles in the Zn SBU are consistent with those reported in the literature.[36-42] Structures **1-5** are discussed in detail below.

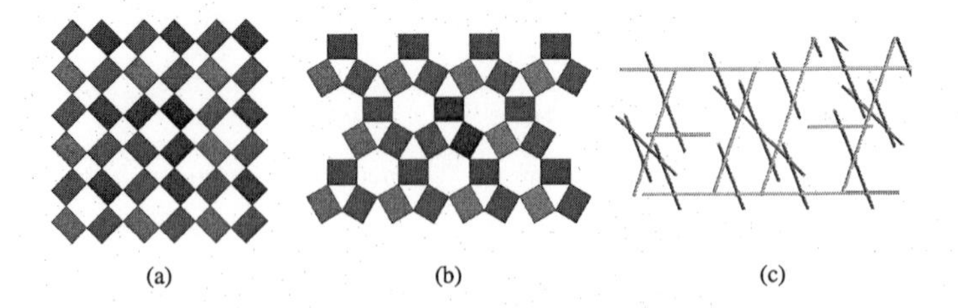

Figure 2. Schematic representaion of the three network topologies of structures 1-5. (a) 2D tetragonal network (1-3), (b) 2D trigonal network (4), (c) 3D 6^5.8 network (5).

(a) 2D sheet structures based on square nSBUs only

Compounds **1-3** are supramolecular isomeric 2D sheet structures that result from the self-assembly of square nSBUs only. In this series of compounds four SBUs are linked in an angular fashion by bdc ligands and the resulting nSBU therefore possess curvature and, in principle, torsional flexibility. Close examination of the square nSBU reveals that it resembles a calix[4]arene in its shape and ability to adopt different conformations. Calixarenes have been isolated in different conformations (atropisomers)[54-58] that have been designated by Gutsche and coworkers[59] as the cone, partial cone, 1,2-alternate and 1,3-alternate. The nomenclature refers to the orientation of the arene rings with respect to one another. We have previously observed the square nSBU in the cone and 1,3-alternate conformations.[42] Structures **1-3** are polymeric structures based on the metal-organic calix-like monomeric nSBUs shown in Figure 3.

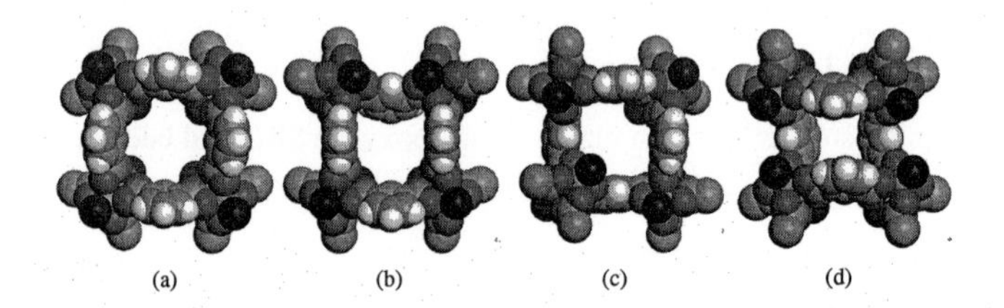

Figure 3. Monomeric nSBU constituents of polymeric structures 1-3. (a) Cone, (b) partial cone, (c) 1,2-alterante, (d) 1,3-alternate.

Compound **1**, $\{[Zn_2(bdc)_2(py)_2]_4\}_n$, contains two atropisomeric nSBUs, namely the cone and the 1,3-alternate (Figure 4). The nSBUs self-assemble, alternating between the cone and the 1,3-alternate conformations and thereby yield an undulating sheet structure. Each cone has an outer diameter of 0.94 nm and a depth of 0.84 nm (measured from the center of the base to the midpoint of a line joining the top hydrogen atoms on opposite bdc moieties). When the sheets stack, they eclipse one another so that the cones sit inside one another and the 1,3-alternate nSBUs define hour-glass-shaped channels. Disordered guest molecules occupy both types of cavities. The potential solvent accessible-volume in one unit cell is 5.5%, but would be increased to 18.1% upon removal of guest molecules.[60] We have previously reported an isostructural Cu(II) version[42] where Cu(II) nSBUs self-assemble into undulating sheets containing both conformers of the nSBU.

Compound **2**, $\{[Zn_2(bdc)_2(2\text{-pic})_2]_4\}_n$, results from the self-assembly of partial cone nSBUs into an undulating sheet structure (Figure 4). The partial cone is the result of a bdc ligand lying in a plane almost perpendicular to the plane defined by the three others. The outer diameter of the partial cone (measured from opposite bdc ligands that lie in one plane) is 0.96 nm and the depth is 0.47 nm (measured from the centers of the lines joining the top and bottom hydrogen atoms of opposite bdc moieties that are in the same plane). Each partial cone in **2** hosts two disordered nitrobenzene molecules. The sheets stack eclipsed so that the partial cones sit inside one another with interlayer separation of *ca.* 1.06 nm. The unit cell contains no residual solvent accessible area; however, the potential solvent area, upon removal of guest, is 35.2%.[60]

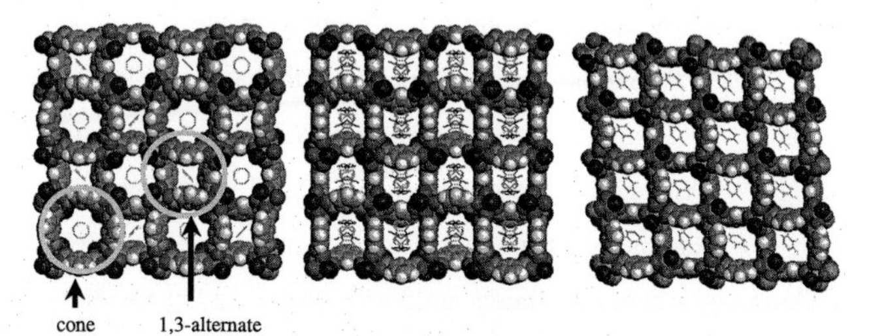

Figure 4. Crsytal structures of 1 (left), 2 (middle) and 3a (right), shown in space filling mode. The view illustrates the resulting 2D sheets hosting guest molecules (shown in line mode). The cavities contain disordered guest molecules in 1 and 2 (nitrobenzene) and ordered dichlorobenzene in 3a. Note that the aromatic rings of the axial ligands have been removed for clarity.

In compounds **3a-f**, $\{[Zn_2(bdc)_2(L)_2]_4\}_n$, (L = 4-picoline for **3a** and **3b**; 3,5-lutidine for **3c** and **3d**; 4-methoxypyridine for **3e** and isoquinoline for **3f**), the nSBU adopts the 1,2-alternate conformation and hosts guest molecules (*o*-dichlorobenzene for **3a**, **3c** and **3f** (solvent molecules are disordered for **3c** and **3f**); nitrobenzene for **3b**; and benzene for **3d** and **3e**) in the resulting cavity. The nSBUs self-assemble into 2D corrugated square grids that propagate in the YZ-plane (Figure 4) and stack along the X-axis in an ABCD fashion with an interlayer separations ranging between 0.87 and 0.96 nm. The effective dimensions of the grids range between 1.1 x 1.1 and 1.2 x 0.95 nm (distance from Zn-Zn mid point of opposite SBU units in the nSBU taking into account the van der Waals radii of zinc). The unit cell of all structures contains no residual solvent accessible area except for **3e** (2.1%); however, the average potential solvent accessible area for all structures upon removal of guest is 25.1%.[60]

1, **2** and **3a-f** therefore, collectively, exhibit all four atropisomers of the nSBU likely to be generated due to the angularity of the bridging bdc ligand. The curvature of the cone nSBU affects the "trans" carboxylate moieties in the basal position of the SBUs, which twist slightly from planarity (dihedral angle between planes of trans carboxylates θ = 7.74°). On the other hand, since in the partial cone and the 1,2-alternate nSBUs the bdc ligand is rotated, all "trans" carboxylates are planar (θ = 0°), except for **3a** and **3b**, where the dihedral angle is 23.3° and 23.7°, respectively.

(b) 2D structure based on triangular nSBUs only (Kagomé lattice)

Compound **4**, $\{[Zn_2(bdc)_2(4\text{-methoxypyridine})_2]_3\}_n$, results from the self-assembly of triangular nSBUs only into a 2D sheet structure. The resulting structure has a topology that is an example of a Kagomé lattice.[61] A Kagomé lattice is a spin-frustrated lattice where spins are located on the vertices of triangles. There are few examples of molecular Kagome lattices[62-64] and we have recently reported the first example of a nanoscale Kagome lattice based on Cu(II) SBU.[52] Zn(II) produces the same topology where Zn_2 dimers are positioned at the lattice points and are bridged by the bdc ligands, thereby generating large hexagonal cavities within the layer. The bowl-shaped nSBUs pack eclipsing one another thereby generating hexagonal channels with effective diameter of 0.93 nm. The potential solvent accessible area in the unit cell is 31.6%, which can be increased further to 46.3 % upon removal of guest molecules.[60] The resulting 2D sheet is undulated due to the curvature of the nSBU imparted by the angularity of the bdc ligand. It is important to note that this lattice would not have magnetic properties since Zn(II) is a d^{10} ion.

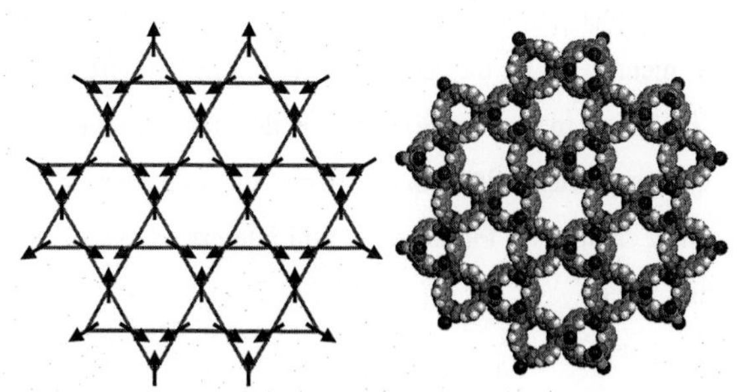

Figure 5. Schematic representaion (left) and crystal structure (right) of the Kagome lattice 4 .

(c) 3D structure with $6^5.8$ topology $\{[Zn_2(bdc)_2(quinoline)_2]_n\}$

We have analyzed the above two supramolecular isomers as molecular squares connected at the vertices. Although we can do the same for this isomer, little will be gained for our understanding of it in terms of topology. To derive the schematic illustrated in Fig 2c, we treated the dimetal cluster as a single node located between the two metal ions, and connecting the nodes according to the connectivity defined by the bdc bridging ligands. This yields a $6^5.8$ 3D network that is sustained by distorted square planar nodes. The distortion is the result of treating the angular bdc ligand as a linear connection. Although the schematic can be modified to illustrate this it provides no significant insight into the structure or properties, and the simplified illustration therefore seems appropriate. **5** does not possess notable pores or cavities (effective diameter is ca. 0.25 nm) and contains no residual solvent accessible area; however, the potential solvent-accessible area upon removal of guest is 17.3%.[60]

Conclusions

In this contributiuon, we have presented examples of coordination polymers that illustrate the type of control and structural diversity that is possible in such compounds. It should be noted that the three different architectures reported herein are dramatically different event though they are prepared from the same set of building blocks. Furthermore, the reaction conditions used are mild and utilize commercially available starting materials. It is simply fine-tuning of the reaction conditions that effects changes in the architecture of the resulting supramolecular structure and in turn the chemical and physical properties such as porosity, guest-encapsulation, *etc*. We have also demonstrated that metals can be interchanged without affecting the structural possibilities. This is particularly important in the context of the development of functional coordination polymers in which the properties arise directly from the presence of transtion metals (e.g. magnetism, luminescence, catalysis). Future work will attempt to exapnd the range of structures and compositions even further.

Acknowledgment

We gratefully acknowledge the financial support of the National Science Foundation (DMR-0101641).

[1] J. M. Lehn, Supramolecular Chemistry: Concepts and Perspectives, VCH, Weinheim 1995.
[2] J. M. Lehn, *Pure Appl. Chem.* **1978,** *50*, 871.
[3] G. M. J. Schmidt, Pure Appl. Chem. **1971,** *27*, 647.
[4] B. Moulton, M. J. Zaworotko, *Chem. Rev.* **2001,** *101*, 1629.
[5] M. J. Zaworotko, *Chem. Commun.* **2001,** 1.
[6] G. R. Desiraju, A. Gavezzotti, *Chem. Commun.* **1989,** 621.
[7] G. R. Desiraju, *Acc. Chem. Res.* **1991,** *24*, 290.
[8] G. R. Desiraju, *Angew. Chem., Int. Ed. Engl.* **1995,** *34*, 2311.
[9] M. C. Etter, *Acc. Chem. Res.* **1990,** *23*, 120.
[10] M. C. Etter, J. C. MacDonald, J. Bernstein, *Acta Crystallogr.* **1990,** *B46*, 256.
[11] M. C. Etter, *J. Phys. Chem.* **1991,** *95*, 4601.
[12] G. R. Desiraju, Crystal Engineering: the Design of Organic Solids, Elsevier, Amsterdam 1989.
[13] F. H. Allen, O. Kennard, *Chem. Des. Autom. News* **1993,** *8*, 31.
[14] A. F. Wells, Three-dimensional Nets and Polyhedra, Wiley, New York, 1977.
[15] A. F. Wells, Structural Inorganic Chemistry, Oxford University Press, Oxford, 1975
[16] B. F. Abrahams, B. F. Hoskins, R. Robson, *J. Am. Chem. Soc.* **1991,** *113*, 3606.
[17] M. J. Zaworotko, *Chem. Soc. Rev.* **1994,** *23*, 283.
[18] T. L. Hennigar, D. C. MacQuarrie, P. Losier, R. D. Rogers, M. J. Zaworotko, *Angew. Chem. Int. Ed. Engl.* **1997,** *36*, 972.
[19] S. Subramanian, M. J. Zaworotko, Angew. Chem., *Int. Ed. Engl.* **1995,** *34*, 2127.
[20] H. Li, M. Eddaoudi, M. O'keeffe, O. M. Yaghi, *Nature* **1999,** *402*, 276.
[21] S. Noro, S. Kitagawa, M. Kondo, K. Seki, *Angew. Chem., Int. Ed. Engl.* **2000,** *39*, 2082.
[22] Y. Aoyama, *Top. Curr. Chem.* **1998,** *198*, 131.
[23] O. R. Evans, Z. Y. Wang, R. G. Xiong, B. M. Foxman, W. B. Lin, *Inorg. Chem.* **1999,** *38*, 2969.
[24] K. Biradha, Y. Hongo, M. Fujita, *Angew. Chem., Int. Ed. Engl.* **2000,** *39*, 3843.
[25] S. R. Batten, *Curr. Opin. Solid State Mater. Sci.* **2001,** *5*, 107.
[26] R. H. Groeneman, L. R. MacGillivray, J. L. Atwood, *Chem. Commun.* **1998,** 2735.
[27] R. W. Gable, B. F. Hoskins, R. Robson, *Chem. Commun.* **1990,** 1677.

[28] J. Cernak, K. A. Abboud, *Acta Crystallogr.* **2000,** *C56*, 783.
[29] E. Colacio, M. Ghazi, R. Kivekas, M. Klinga, F. Lloret, J. M. Moreno, *Inorg. Chem.* **2000,** *39*, 2770.
[30] B. F. Abrahams, K. D. Lu, B. Moubaraki, K. S. Murray, R. Robson, *Dalton Trans.* **2000,** 1793.
[31] M. Maekawa, K. Sugimoto, T. Kuroda-Sowa, Y. Suenaga, M. Munakata, *Dalton Trans.* **1999,** 4357.
[32] H. Abourahma, B. Moulton, V. Kravtsov, M. J. Zaworotko, *J. Am. Chem. Soc.* **2002,** *124*, 9990.
[33] Z. M. Sun, P. K. Gantzel, D. N. Hendrickson, *Polyhedron* **1998,** *17*, 1511.
[34] F. A. Cotton, R. A. Walton, Multiple Bonds Between Metal Atoms, Oxford University Press, Oxford, 1993.
[35] D. W. Breck, Zeolite Molecular Sieves: Structure, Chemistry and Use, Wiley Interscience, New York, 1974.
[36] F. Demirhan, J. Gun, O. Lev, A. Modestov, R. Poli, P. Richard, *Dalton Trans.* **2002,** 2109.
[37] H. Necefoglu, W. Clegg, A. J. Scott, *Acta Crystallographica Section E-Structure Reports Online* **2002,** *58*, m121.
[38] Q. D. Zhou, T. W. Hambley, B. J. Kennedy, P. A. Lay, P. Turner, B. Warwick, J. R. Biffin, H. L. Regtop, *Inorg. Chem.* **2000,** *39*, 3742.
[39] B. Singh, J. R. Long, F. F. Debiani, D. Gatteschi, P. Stavropoulos, *J. Am. Chem. Soc.* **1997,** *119*, 7030.
[40] W. Clegg, P. A. Hunt, B. P. Straughan, *Acta Crystallogr.* **1995,** 51, 613-617.
[41] W. Clegg, I. R. Little, B. P. Straughan, *Dalton Trans.* **1986,** 1283-1288.
[42] S. A. Bourne, J. Lu, A. Mondal, B. Moulton, M. J. Zaworotko, *Angew. Chem., Int. Ed. Engl.* **2001,** *40*, 2111.
[43] F. A. Cotton, C. Lin, C. A. Murillo, *Acc. Chem. Res.* **2001,** *34*, 759.
[44]G. S. Papaefstathiou, L. R. Macgillivray, *Angew. Chem., Int. Ed. Engl.* **2002,** *41*, 2070.
[45] B. Moulton, J. J. Lu, M. J. Zaworotko, *J. Am. Chem. Soc.* **2001,** *123*, 9224.
[46] F. A. Cotton, C. Lin, C. A. Murillo, *Chem. Commun.* **2001,** 11.
[47] S. R. Batten, B. F. Hoskins, B. Moubaraki, K. S. Murray, R. Robson, *Chem. Commun.* **2000,** 1095.
[48] B. Moulton, J. Lu, A. Mondal, M. J. Zaworotko, *Chem. Commun.* **2001,** 863.
[49] M. Eddaoudi, H. L. Li, O. M. Yaghi, *J. Am. Chem. Soc.* **2000,** *122*, 1391.
[50] H. Abourahma, A. W. Coleman, B. Moulton, B. Rather, P. Shahgaldian, M. J. Zaworotko, *Chem. Commun.* **2001,** 2380.
[51] J. Lu, A. Mondal, B. Moulton, M. J. Zaworotko, *Angew. Chem., Int. Ed. Engl.* **2001,** *40*, 2113.
[52] B. Moulton , J. Lu, R. Hajndl, S. Hariharan, M. J. Zaworotko, *Angew. Chem., Int. Ed. Engl.* **2002,** *41*, 2821.
[53] S. S. Y. Chui, S. M. F. Lo, J. P. H. Charmant, A. G. Orpen, I. D. Williams, *Science* **1999,** *283*, 1148.
[54] B. Zimmer, V. Bulach, C. Drexler, S. Erhardt, M. W. Hosseini, A. De Cian, *New J. Chem.* **2002,** *26*, 43.
[55] A. Ikeda, S. Shinkai, *Chem. Rev.* **1997,** *97*, 1713.
[56] V. Bohmer, *Angew. Chem., Int. Ed. Engl.* **1995,** *34*, 713-745.
[57] C. D. Gutsche, B. Dhawan, J. A. Levine, K. H. No, L. J. Bauer, *Tetrahedron* **1983,** *39*, 409.
[58] G. M. L. Consoli, F. Cunsolo, C. Geraci, E. Gavuzzo, P. Neri, *Organic Letters* **2002,** *4*, 2649.
[59] C. D. Gutsche, *Acc. Chem. Res.* **1983,** *16*, 161.
[60] A. L. Spek, PLATON, Ultrecht University, The Netherlands, **2000**.
[61] I. Syozi, *Prog. Theor. Phys.* **1951,** *6*, 306.
[62] K. Awaga, T. Okuno, A. Yamaguchi, M. Hasegawa, T. Inabe, Y. Maruyama, N. Wada, *Phys. Rev. B* **1994,** *49*, 3975.
[63] A. P. Ramirez, J. *Appl. Phys.* **1991,** 70, 5952.
[64] M. G. Townsend, G. Longworth, E. Roudaut, *Phys. Rev. B* **1986,** *33*, 4919.

Conjugated Metallopolymers

*Alfred C. W. Leung, Jonathan H. Chong, Mark J. MacLachlan**

Department of Chemistry, University of British Columbia, 2036 Main Mall, Vancouver, BC, Canada V6T 1Z1
E-mail: mmaclach@chem.ubc.ca

Summary: Conjugated polymers incorporating transition metals have enormous potential for chemically-tuneable conducting and semiconducting polymers. Unfortunately, progress in this field has been limited by the challenge of preparing suitable metal-containing monomers, the difficulty of polymerizing them in a controlled manner, and the limited solubility of the resulting polymers. We are developing new, soluble conjugated metal-containing polymers that contain transition metals in a conjugated organic backbone.

Keywords: conjugated polymers; fluorescent; metal-containing polymers; porphyrins

Introduction

Recently, there has been a great deal of interest in the development of specialty functional polymers for applications as diverse as catalysts, chemical sensors, and biomedical implants. Conducting polymers containing a conjugated organic backbone are fascinating materials that combine properties associated with both organic and inorganic elements. Although they can be processed like other organic polymers (e.g., polyesters, polyethylene), these polymers may have semiconducting or metallic properties as a virtue of their conjugated backbones. It is for this unusual combination of properties that conjugated polymer researchers received the 2000 Nobel Prize in Chemistry.[1]

Although conducting polymers have been investigated for many years, it is much more recently that researchers have begun to investigate the incorporation of inorganic elements into conjugated polymer backbones.[2,3] The addition of metals to conjugated organic polymers is anticipated to deliver new properties to the material. For example, the polymers may exhibit enhanced thermal stability necessary for increasing lifetimes in polymeric LEDs. Alternatively, the interaction of the metal's d-orbitals may modify the electronic properties of the conjugated polymers. Three key areas that conjugated metallopolymers appear promising are for the development of new catalysts, sensors, and electroluminescent materials.

Unfortunately, the preparation of conjugated metallopolymers has been hampered by the

© 2003 WILEY-VCH Verlag GmbH & KGaA, Weinheim CCC 1022-1360/00/$ 17,50+.50/0

difficulty of preparing suitable monomers, of finding suitable polymerization methods, and of making the resulting polymer soluble. Solubility is often the most difficult problem to overcome as rigid metallopolymers are generally very insoluble. For this reason, most conjugated polymers have been synthesized on an electrode via electropolymerization.

In this paper, some prominent recent investigations of conjugated metallopolymers are described.

Catalysis

One clear application of conjugated metallopolymers is in new catalytic materials. Planar, conjugated molecules (e.g., porphyrins) that are known catalysts may be incorporated into conjugated polymers to modify their characteristics, or to combine the favourable electronic and optical properties of the polymer with the catalyst. Cobalt-containing polymers have excellent electrocatalytic properties and good stability. Electropolymerizable cobalt porphyrin and salen derivatives have been polymerized on an electrode (e.g., polymer **1**) and demonstrated to undergo highly efficient 4 electron reduction of oxygen to water without significant formation of H_2O_2.[4,5] Polymer **2**, formed by Ni(0) catalyzed coupling of 5,5'-dibromo-2,2'-bipyridine, is an active catalyst for the photoevolution of H_2 from water.[6] These reactions are relevant to energy conversion and storage.

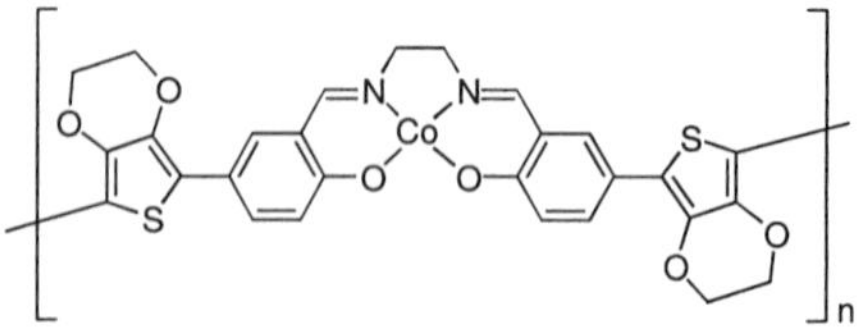

1 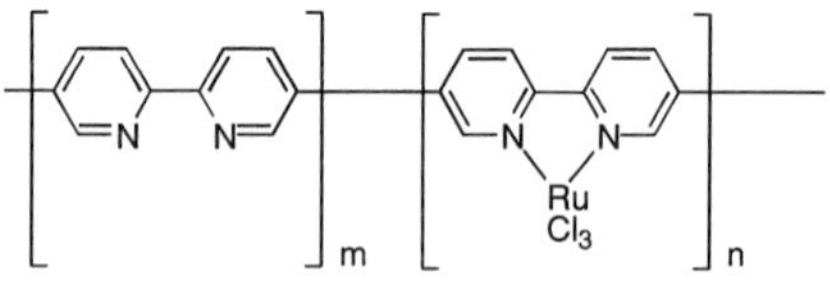 **2**

Sensing

In addition to catalysis, the availability of empty coordination sites on a conjugated metallopolymer offers the opportunity to affect the polymers' properties by coordination of a ligand to the metal site. In this way, a response may be measured and the polymer may serve as a sensor for Lewis bases. Polymer **1**, formed by electropolymerization of a cobalt salen with 3,4-ethylenedioxythiophene groups, behaves as a sensor for Lewis bases.[7] When exposed to pyridine or 2,6-lutidine, this polymer showed a significant reduction in conductivity, attributed to coordination of the Lewis base to the metal sites.

Another recent example that highlights the ability of conjugated metallopolymers to function as sensors, as well as supramolecular building blocks, can be found in the conjugated butadiyne-linked porphyrin polymers recently reported by Screen *et al.*[8] Polymer **3**, which is soluble in chloroform, undergoes changes in its optical spectrum when a Lewis base, such as 1,4-diazabicyclo[2.2.2]octane (DABCO), binds to the zinc atoms. Anderson and coworkers explored the use of rigid bidentate ligands (e.g., 4,4'-bipyridine) to link the linear polymers **3** into ladder polymers **4**. When the ratio of Zn:bpy is 1:0.5, the polymer forms a double stranded ladder structure which shows a considerable redshift (75 nm) relative to the coordinated linear polymer, indicating enhanced conjugation. The enhanced conjugation present in the ladder polymer, presumably due to restricted rotation of the porphyrins along the polymer chains, results in a 7-fold increase in the third-order nonlinear optical properties ($\chi^{(3)}$, measured by degenerate four-wave mixing, DFWM). With the addition of more 4,4'-bipyridine, the polymers may link theoretically into a 2-D grid. Such grids may show size- and shape-selective adsorption of analytes, making them useful for chemical sensors.

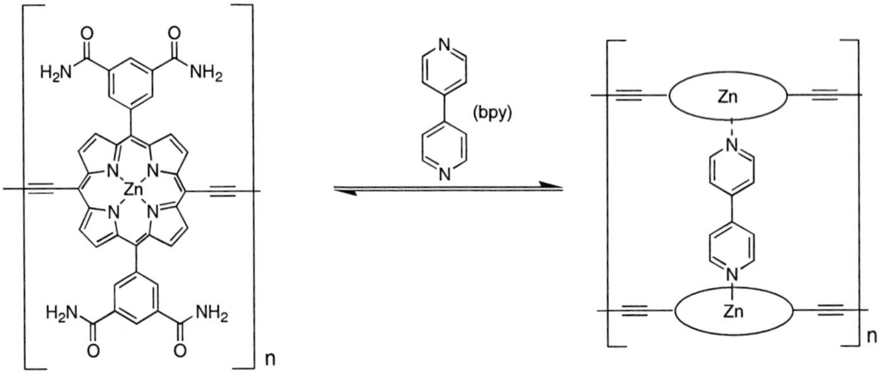

H_2N NH_2 O O N N Zn N N H_2N NH_2 O O n

(bpy)

Zn N N Zn n

3 **4**

Electroluminescence

Conjugated polymers have received enormous attention for application in organic LED technology, and are now the basis of numerous start-up companies producing flexible displays. It is surprising that metal-containing conjugated polymers have received very little attention for their electroluminescent properties, particularly with growing attention to phosphorescent LED materials.[9] Such polymers may offer the ability to engineer the bandgap of the polymer on the basis of the metal choice and ligand variation. Schanze and coworkers have been investigating poly(phenyleneethynylene)s incorporating Re and Ru complexes on the backbone (e.g., **5**).[10] These polymers exhibit fluorescence from the conjugated polymer, as well as phosphorescence from the metal-to-ligand charge transfer (MLCT) band of the complex. The efficiency of energy transfer from the polymer to the metal complex depends on the concentration of metal centres along the polymer backbone. In this case, incorporation of transition metals into the conjugated polymer backbone dramatically affects the photophysical properties of the polymer.

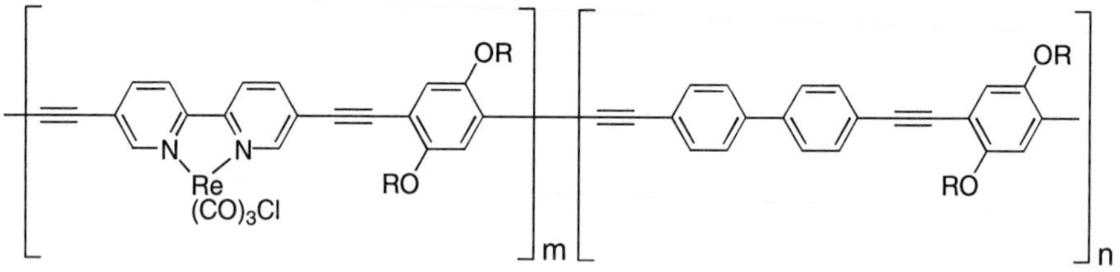

5

Lavastre *et al.* recently synthesized a large number of poly(phenyleneethynylene) polymers using a combinatorial approach.[11] Their results indicated that conjugated poly(phenyleneethynylene)s **6** incorporating Ni(salphen) or Zn(salphen) complexes, are highly luminescent and may be excellent candidates for metal-organic LED materials. These polymers are worthy of further study.

6 (M = Zn, Ni)

Our Research

It is clear that conjugated metallopolymers have interesting properties as a virtue of their conjugated organic backbone and transition metals. These materials may be important for developing new catalysts, nonlinear optical (NLO) devices, electroluminescent displays, or supramolecular architectures. We are investigating new soluble conjugated polymers that can be prepared on a bulk scale. Figure 1 shows a photograph of a new conjugated metallopolymer we have recently synthesized. This red polymer, which has a fully conjugated organic backbone with metals interspersed, is soluble in THF and readily forms free-standing films. We are investigating its electroluminescence.

Figure 1. Photograph of thin films of a conjugated metallopolymer prepared in our laboratory. The coin is provided as a size reference.

[1] A. J. Heeger, *Angew. Chem. Int. Ed. Engl.* **2001**, *40*, 2591.
[2] R. P. Kingsborough, T. M. Swager, *Prog. Inorg. Chem.* **1999**, *48*, 123.
[3] P. G. Pickup, *J. Mater. Chem.* **1999**, *9*, 1641.
[4] J. E. Bennett, A. Burewicz, D. E. Wheeler, I. Eliezer, L. Czuchajowski, T. Malinski, *Inorg. Chim. Acta* **1998**, *271*, 167.
[5] R. P. Kingsborough, T. M. Swager, *Chem. Mater.* **2000**, *12*, 872.
[6] T. Yamamoto, T. Maruyama, Z. Zhou, T. Ito, T. Fukuda, Y. Yoneda, F. Begum, T. Ikeda, S. Sasaki, H. Takezoe, A. Fukuda, K. Kubota, *J. Am. Chem. Soc.* **1994**, *116*, 4832.
[7] R. P. Kingsborough, T. M. Swager, *Adv. Mater.* **1998**, *10*, 1003.
[8] T. E. O. Screen, J. R. G. Thorne, R. G. Denning, D. G. Bucknall, H. L. Anderson, *J. Am. Chem. Soc.* **2002**, *124*, 9712.
[9] B. W. D'Andrade, J. Brooks, V. Adamovich, M. E. Thompson, S. R. Forrest, *Adv. Mater.* **2002**, *14*, 1032.
[10] K. D. Ley, C. E. Whittle, M. D. Bartberger, K. S. Schanze, *J. Am. Chem. Soc.* **1997**, *119*, 3423.
[11] O. Lavastre, I. Illitchev, G. Jegou, P. H. Dixneuf, *J. Am. Chem. Soc.* **2002,** *124,* 5278.

Iridium(III) Complexes as Polymer Bound Oxygen Sensors

Maria C. DeRosa, Peter J. Mosher, Christopher E. B. Evans, Robert J. Crutchley**

Chemistry Department, Carleton University, ON, K1S 5B6, Canada and the National Research Council of Canada, Ottawa, ON, Canada
E-mail: robert_crutchley@carleton.ca

Summary: New luminescent oxygen sensors have been prepared by covalent attachment of iridium complex luminophores to a silicone polymer. The oxygen sensor properties of these novel materials were compared to related sensors in which the luminophore is dispersed within the polymer matrix. Covalently bound luminophore materials showed increased sensitivity to oxygen over dispersions in pure silicone polymer as well as in blends with polystyrene, which was added to improve the mechanical properties of the material.

Keywords: luminescence; metal-polymer complexes; oxygen sensors; photochemistry; silicones

Introduction

It has been known for many years that the luminescent excited state of many molecules is quenched by oxygen, forming highly reactive singlet oxygen.[1] This is of great concern to polymer chemists as singlet oxygen formation is thought to be the main cause of polymer photodegradation. On the other hand, researchers in the field of sensor technology have taken advantage of this behavior to create oxygen sensors for *in vivo*, groundwater and barometric measurements.[2,3] The latter is of particular interest to aerospace and automotive industries as an understanding of aerodynamics is crucial to stability and fuel efficiency concerns. Traditional methods for measuring surface pressure require pressure taps or transducers. These are time consuming and expensive to employ and impractical for moving parts (i.e. propeller blades). A far more efficient method is to employ pressure sensitive paints (PSP) in which a luminescent molecule (a luminophore) is dispersed in the polymer matrix/solvent that composes the bulk of the PSP. The luminescence derived from a model painted with a PSP will ideally be in direct proportion to the air pressure over the model's surface and this is illustrated schematically in Figure 1.

© 2003 WILEY-VCH Verlag GmbH & KGaA, Weinheim CCC 1022-1360/00/$ 17,50+.50/0

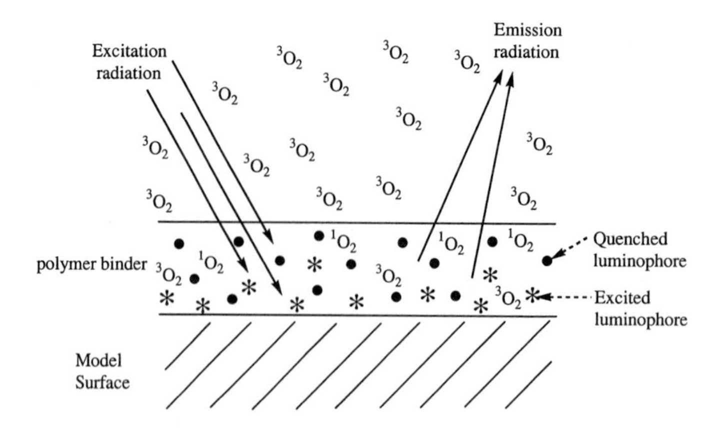

Figure 1. Schematics of a pressure sensitive paint coating a model surface. Filled circles are ground state luminophores while asterisks are excited state luminophores.

Bimolecular collisions between excited luminophores and ground state (triplet) oxygen quench the luminophore non-radiatively, producing singlet oxygen and ground state luminophore. The relationship between the luminophore's emission intensity, I, or excited state lifetime, τ, and oxygen concentration is given by the Stern-Volmer equation (1),

$$\frac{I^0}{I} = \frac{\tau^0}{\tau} = 1 + K_{SV}[O_2] \qquad (1)$$

in which I^0 and τ^0 are, respectively, the luminophore's intensity and lifetime in the absence of oxygen, and K_{SV} is the Stern-Volmer constant. Equation 1 assumes a single luminophore environment, and deviations from the expected linear behavior occur when this is not the case, or when quenching occurs by mechanisms other than bimolecular collisions.

In luminescence barometry applications, changes in ambient pressure alter the concentration of dissolved oxygen within the sensor film. Measurements at zero oxygen concentration are often impractical, and it is therefore convenient to use an alternate point of reference, such as $P = 1$ atmosphere. In this case the relationship between luminescence intensity and air pressure becomes

$$\frac{I_{ref}}{I} = A + Q_S \frac{P}{P_{ref}} \qquad (2)$$

in which P_{ref} is the reference pressure at which the luminophore's intensity is I_{ref}. The slope of the Stern-Volmer plot, Q_S, is nominally constrained to values between 0 and 1 and is a measure of the sensitivity of the PSP to changes in oxygen pressure.[4]

Pressure sensitive paints must meet a number of criteria before they are of practical value. These are:

1. Luminophore excitation wavelengths should be at long wavelengths
2. Emission properties should be largely temperature independent
3. The luminophores Stokes shift should be large
4. Good stability
5. Short response time
6. Appropriate sensitivity to oxygen concentration

Excitation at short wavelengths is likely to increase photodegredation of the polymer and the cost and safety concerns associated with the light source are also disadvantageous. Temperature independence of luminophore emission is a very desirable property because without it small changes in intensity of emission due to oxygen quenching could be equivalent to that created by changes in temperature. Deconvoluting this temperature effect from that produced by varying oxygen pressure is not straightforward and can lead to only qualitative results. In practice, all luminophores show some temperature dependence to their emission. The goal is to find a luminophore with a low temperature dependence so that the correction is reasonable and leads to quantitative results. If the overlap between the absorption and emission band envelopes is significant, the luminophore in the PSP will absorb emission radiation. A large Stokes shift, a measure of the separation energy between absorption and emission energies, will minimize this effect. The PSP must be stable during the entire course of the experiment in order to demonstrate reproducibility of the measurements. It is especially important that the luminophore be resistant to reaction with the singlet oxygen that is produced during the experiment. The establishment of equilibrium conditions within the PSP for a given pressure of oxygen over its surface determines the response time (time required afterwhich measurement is constant). This is directly related to the permeability of the PSP to oxygen or more specifically the diffusion of oxygen molecules within the PSP layer. The required response time and oxygen sensitivity depend upon the application. For example, the short air bursts in blow-down wind tunnels require PSP response times of ≤ 1 second and, for the changes in oxygen pressures created, Q_s values of 0.4 to 0.7 are the norm.[3] For ground water measurements, where changes in concentration of oxygen are small, greater sensitivity is required but response times can be much longer.

Low gas-permeability polymers such as polystyrene (PS) and poly(methylmethacrylate) (PMMA) have good mechanical stability but PSPs based on these polymers would have very long response times. Fluorination is known to increase gas permeability, and PSPs using polymers such as poly(hexafluoroisopropyl-*co*-heptafluoro-*n*-butyl methacrylate) and poly(2,2,2-trifluoroethyl-*co*-isobutyl methacrylate)[5] have significantly shorter response times. Silicones (polysiloxanes, *e.g.* polydimethylsiloxane, PDMS) were investigated from very early on due to their extremely large gas permeabilities. Unlike high glass transition temperature (T_g) organic polymers such as PS (90 °C) or PMMA (110 °C), many silicones have extremely low T_g (*e.g.* PDMS, -127 °C) and thus form films with rather poor mechanical properties unless cured (*i.e.* crosslinked) or mixed with hardening additives.

Commonly used luminophores include polycyclic aromatic hydrocarbons,[6] metalloporphyrins,[7] or polypyridyl transition metal complexes.[8] Sensors based *e.g.* on pyrene, PtOEP (OEP = octaethylporphyrin), and $[Ru(dpp)_3]Cl_2$ (dpp = 4,7-diphenylphenanthroline) have all been studied. The sensors studied to date have predominantly been ones in which the luminophore is dispersed within the host matrix as a solid solution. This can lead to the following behaviors that can lead to undesirable properties:

1. Aggregation
2. Leaching
3. Solvation by blend polymers

Luminophore aggregation can lead to self-quenching and loss of emission. Further emission losses can occur by luminophore leaching or sublimation from the PSP layer.[9] If additives or polymer blends are desired to improve PSP properties, the luminophore may partition between the various components, leading to complex or unexpected behavior.[4,10] These problems could be eliminated or at least moderated by covalently attaching the luminophore to the polymer matrix.[11]

Recent studies by Manners and Winnik et al[12] have demonstrated the efficacy of this approach. These authors covalently attached ruthenium phenanthroline complexes to polythionylphosphazenes and found improved Stern-Volmer behavior and significantly increased sensitivity compared to analogous two-component materials where luminophores are dispersed in a polythionylphosphazene matrix.

In this study, we present our results for PSPs composed of luminophores covalently attached to silicones. Silicones have very high gas permeabilities relative to most organic polymers.[13] The main drawback of silicones is that, without curing, they form films with poor mechanical properties. To provide mechanical stability, silicones can be blended with high T_g polymers and/or rigidifying additives such as silica or alumina. The multitude of microenvironments created in these heterogenous materials can create unexpected behaviors in PSP applications. For example, it has been noted that luminophores such as $[Ru(dpp)_3]Cl_2$ have a strong tendency to aggregate within these blended matrices,[14] or to adsorb to the surface of silica.[10] Covalent attachment of the luminophore to silicones should alleviate aggregation and lead to improved PSP properties.

As our luminophore we have selected cyclometallated complexes of iridium(III). The complex $[Ir(ppy)_3]$ (ppy = 2-phenylpyridine, τ^0 = 2 μs, ϕ = 0.4)[15] has previously been examined as a luminescent oxygen sensor,[15,16] and provides the advantage over analogous ruthenium complexes such as $[Ru(bpy)_3]Cl_2$ (bpy = 2,2'-bipyridine, τ^0 = 0.6 μs, ϕ = 0.042)[17] or $[Ru(phen)_3]Cl_2$ (phen = 1,10-phenanthroline, τ^0 = 0.74 μs, ϕ = 0.08)[14] of a longer lifetime and much higher quantum yield. In addition, the non-ionic iridium complexes should be soluable in a greater range of organic solvents compared to charged luminophores, increasing its synthetic utility, and avoiding any deleterious electrostatic interactions.

Synthesis

Cyclometallated complexes of Ir(III) are well known and the synthetic route to the family of luminophores used in this study are shown in Figure 2 and 3.

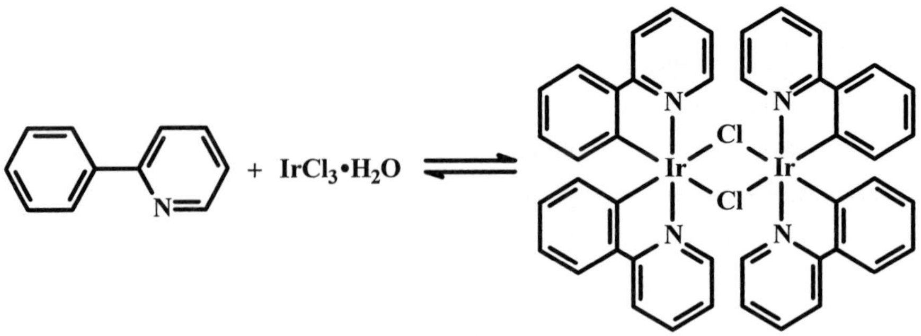

Figure 2. Preparation of **1.**

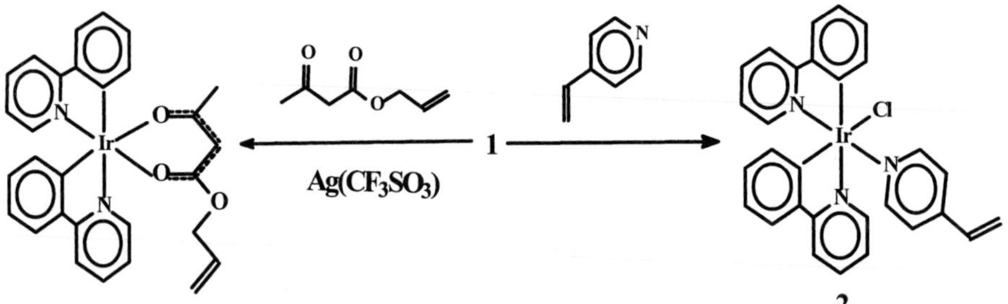

Figure 3. Preparation of the Iridium(III) luminophores.

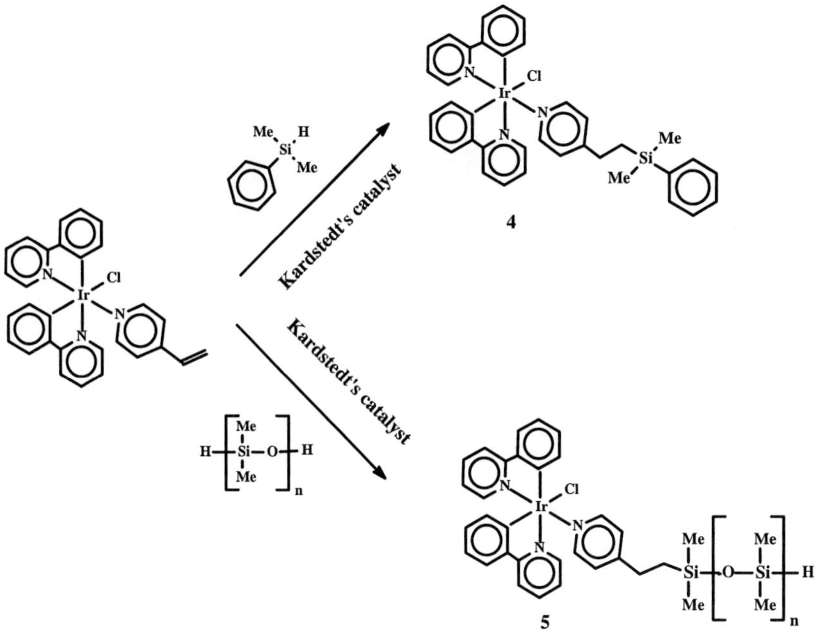

Figure 4. Hydrosilation reaction for the attachment of luminophores to silicones.

Transition metal catalyzed hydrosilation is a facile route to the covalent attachment of luminophores to a silicone backbone as there are a number of commercially available hydride-containing and hydride-terminated silcones. Of these silcones, the hydride-terminated polydimethylsiloxanes PDMS were the easiest to work with because cross-linking reactions, which had a tendency to generate insoluable polymers, were avoided. Figure 4 illustrates the general reaction scheme for luminophore attachment by hydrosilation. In addition to **4**, hydrosilation reactions of complexes **2** and **3**, with hydride-terminated silicones yielded the novel PSP materials **5** and **6,** respectively.

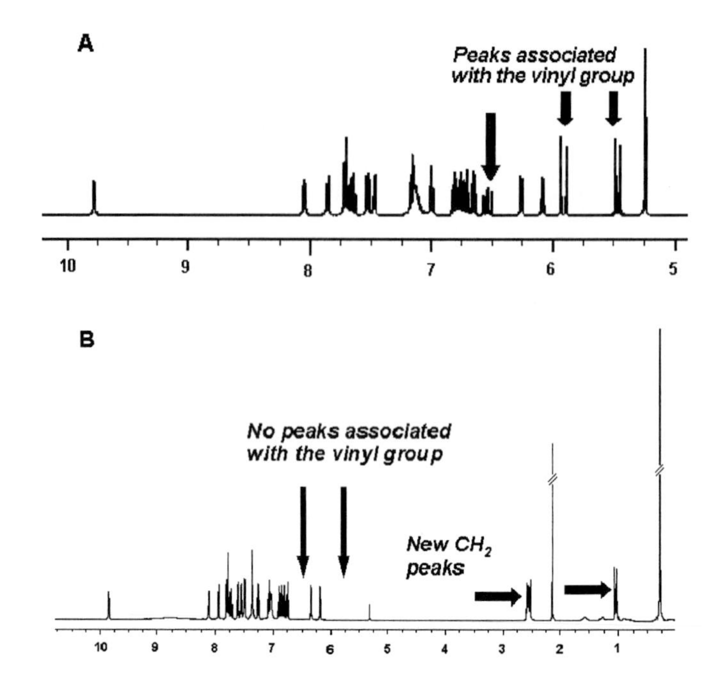

Figure 5. ^{1}H NMR spectra of complexes **2** and **4** in deuterated methylene chloride.

Characterization

^{1}H NMR of **2** and the model complex **4** (see Figure 5) provide evidence that the hydrosilation reaction between the luminophore and silicones is indeed possible. In Figure 5a, the ^{1}H NMR spectrum of **2** shows clearly the presence of vinyl proton chemical shifts. Hydrosilation to form **4** results in the lost of these chemical shifts (Figure 5b) and the appearance of methylene chemical shifts. Unfortunately, the low concentration of the luminophore in **5** and **6**, meant that NMR could not be used to probe the attached luminophore directly. Luminophore attachment was, however, evident by the different behavior of **5** and **6** relative to a dispersion of **2** and **3** in PDMS. For example, while PDMS and **5** are both soluble in hexanes, **2** is not. A sample of **5** can thus be wholly dissolved in hexanes, whereas with a dispersion of **2** in PDMS (prepared by solvent stripping a dichloromethane solution of the two components to dryness) only the polymer dissolves, leaving behind a suspension of **2**. Figure 6 shows the absorption and emission spectra of **5** in acetonitrile solution. The concentration of the luminophore in **5** and **6** were estimated from the π-π* absorption band (a combination of phenyl and pyridine π-π* transitions) that is seen in the ultraviolet region by using the quantitative electronic absorption spectra of **2** and **3** and assuming equivalency of spectral properties and no absorption from PDMS. In this way, the concentrations of luminophore complex in **5** and **6** were estimated to be 3.6 and 3.5 mM, respectively. For **5**, 100% attachment of the luminophore would have resulted in a luminophore concentration of 18.9 mM. The hydrosilation reaction yield is therefore approximately 20%.

Method of Analyses

Each sample was treated as follows: a solution of the sample was applied to an aluminum plate previously covered with a layer of Tristar Starpoxy fluid resistant white epoxy primer (DHMS C4.01 Ty3) using a conventional airbrush. The painted plate was

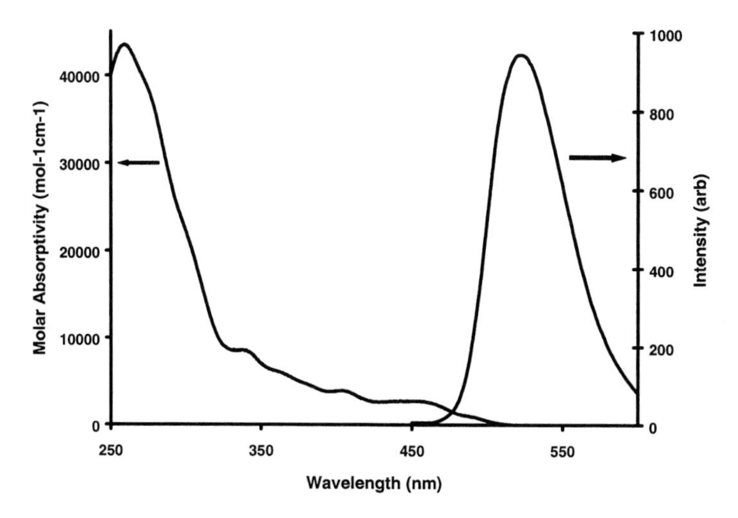

Figure 6. Absorption and emission spectra of **6** in acetonitrile solution. The molar absorptivity has been estimated (see text) and the emission spectrum is uncorrected.

then mounted in an in-house designed pressure chamber. Pressure was controlled via a Scanivalve Corp model PCC100 Pressure Calibrator/Controller in manual mode, and the temperature of the mounting plate was controlled via a thermoelectric cooler coupled to a model LFI-3551 Temperature Controller using a model TCS650 thermistor in the 10 μA range, both from Wavelength Electronics. Excitation was provided by a Hamamatsu Lightningcure LC5 200W model L8333 Hg/Xe source via a 10 m × 8 mm Oriel UV-Vis Liquid Light Guide (transmission window 300-650 nm). The source was equipped with 011FG09 and 300FS40 filters from Melles-Griot, transmitting approximately in the range 280-320 nm. Emission was measured by a 512×512 Photometrics CH350 12-bit CCD camera equipped with a 500 ± 40 nm bandpass filter (500FS80-50, Melles Griot). Measurements were taken at 10 °C, scanning from low to high pressure. At each pressure, multiple measurements were taken to ensure the oxygen concentration in the film had come to equilibrium. The time required for this to occur provided a qualitative measure of response time. At the end of each calibration a second reading at 15 psi was taken to ensure no significant photodegradation had occurred on the timescale of the experiment. Backgrounds were measured on a section of primed plate. Spectra were also collected for various calibrations using an Acton Research Corporation SpectruMM CCD Detection System. A fiber optic lightguide (LG-455-020) equipped with a Kodak #3 gelatin filter passed the emission through a SpectraPro-150 Imaging Dual Grating Monochromator/Spectrograph onto a 16-bit Hamamatsu 1024×256 CCD.

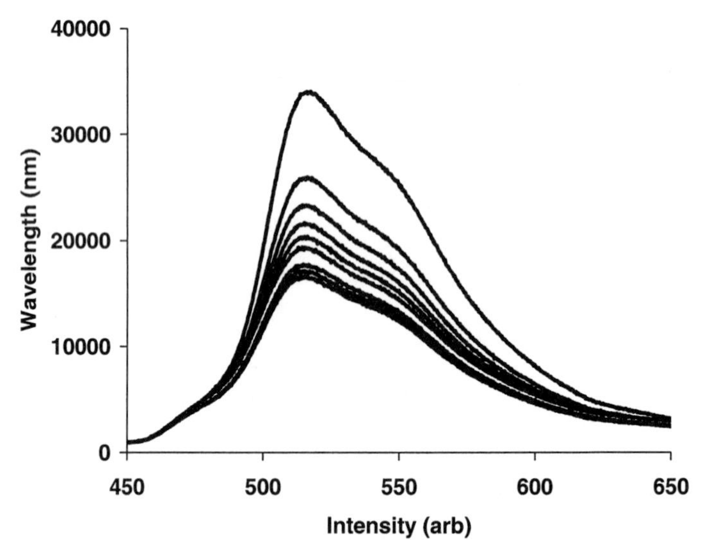

Figure 7. Emission spectra of a thin film of 1:9 **6**/PS showing the loss of intensity with increasing total air pressure. Spectra were taken at 5 psi increments over the range 0.05 to 45 psi and have been corrected for background.

Oxygen Sensitivity

Figure 7 shows the effect of changing air pressure (0 to 45 psi) on the emission spectrum of a thin film of a 1:9 blend of **6** with PS. Similar behavior was seen for **5**. It is clear that the sensitivity of the emission spectrum to oxygen is greatest at low pressures. This is more quantitatively represented by the Stern-Volmer plots (Eqn. 2) in Figure 8.

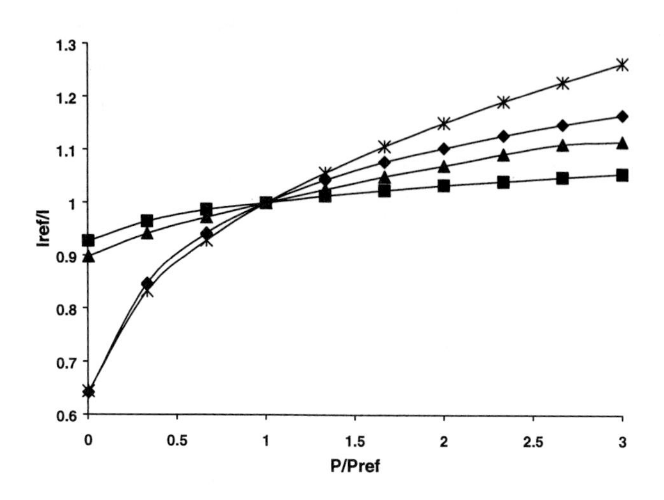

Figure 8. Stern-Volmer Calibration Curves for films of **5** (◆), **2** dispersed in PDMS (■), **2** dispersed in PS (▲) and **5** blended with PS 1:9 (✳)

The marked downward curvature of the Stern-Volmer plots of **5** and its blend with PS has been seen for transition metal complexes in PDMS by other researchers, and has been ascribed to the luminophore being solvated in different microenvironments within the polymer matrix, having different accessibilities to oxygen.[14,18,19] Thus at low pressures only the most accessible luminophore molecules interact with oxygen, leading to high sensitivity, whereas at higher pressures the easily accessible luminophore molecules are largely quenched, and the quenching sensitivity is increasingly dominated by the luminophore molecules in the less accessible domains. Often the luminescence emission decay profile in these cases can be modelled with two exponentials (luminophore lifetimes), and the downward curvature of the Stern-Volmer plots is then attributed to the 'two-site model'. This is, of course, a great oversimplification, and it is generally accepted that a range of different solvation microenvironments is likely to occur in any given sample.[18]

As shown by Figure 8, the oxygen sensitivity of a thin film of **5** is greater than the sensitivity of films of dispersions of **2** in PDMS or PS. Importantly, the blending of PS with **5** to increase the mechanical properties of the film has resulted in a further increase in oxygen sensitivity compared to the film made up of **5** alone. This unexpected albeit gratifying result is suggested to arise from an increase in lifetime of the luminophore because of a reduction in its freedom of movement.

The oxygen sensitivity of various films is given in Table 1 and, where curvature in the Stern-Volmer plots exists, Q_s values are given for an approximately linear range of data. The data clearly illustrate the advantage of polymer bound luminophores over dispersions of luminophores in a given polymer matrix. However, the overall Q_s values of **5** and **6** at pressures greater than 10 psi are too low to be suitable for wind tunnel experiments (Q_s values ranging from 0.4 to 0.7 are preferred).[3] It will be necessary to research other polymer bound luminophores to yield PSPs with greater sensitivity.

Sensor response times were not measured quantitatively. Nevertheless, the 'dispersed-in-PS' luminophores exhibited markedly longer response times to changes in total pressure (on the order of tens of seconds) relative to the 'dispersed-in-PDMS' luminophores (which responded as quickly as the pressure could be changed). The response times of PS blends of **5** and **6** remained very short, however, being at most a few seconds. Maintaining a short response time in the presence of rigidifying additives is a distinct advantage brought about by attachment of the luminophore.

Table 1. Q_s Values from CCD luminescence quenching data over the pressure range 0.05 to 45 psi

	Matrix	
	PDMS	PS
2	0.11,[b] 0.04,[c] 0.02[d]	0.13,[b] 0.08,[c] 0.05[d]
3	0.18,[b] 0.07,[c] 0.05[d]	0.20[bcd]
5[a]	0.62,[b] 0.06,[c] 0.02[d]	0.57,[b] 0.19,[c] 0.12[d]
6[a]	1.00,[b] 0.17,[c] 0.07[d]	0.63,[b] 0.24,[c] 0.17[d]

a) For **5** and **6** the matrices are **5** and **6** alone and **5** and **6** blended with PS at a ratio of 1:9.
b) 0.05 to 5 psi
c) 10 to 45 psi
d) 25 to 45 psi

Conclusion

Luminescent oxygen sensors **5** and **6** based on cyclometallated iridium complexes have been synthesized via hydrosilation of luminophores **2** and **3** onto hydride-terminated PDMS. These novel PSP materials compared favorably with related luminophore-dispersed sensors both in sensitivity and versatility. Not only is the oxygen sensitivity of these luminophore-bound films greater than dispersions of the luminophore in PDMS, but the physical properties of the sensor films can be improved by blending without compromising sensor behavior (higher sensitivity and short response times).

The strategy of polymer bound luminophores offers exciting opportunities in PSP research that we expect to address shortly in future publications.

[1] M. C. DeRosa, R. J. Crutchley, *Coord. Chem. Rev.* **2002**, *233-234*, 351.
[2] D. T. McQuade, A. E. Pullen, T. M. Swager, *Chem. Rev.* **2000**, *100*, 2537.
[3] J. H. Bell, E. T. Schairer, L. A. Hand, R. D. Mehta, *Annu. Rev. Fluid Mech.* **2001**, *33*, 155.
[4] R. Ruffolo, C. E. B. Evans, X.-H Liu, Y. Ni, Z. Pang, P. Park, A. R. McWilliams, X. Gu, X. Lu, A. Yekta, M. A. Winnik, I. Manners, *Anal. Chem.*, **2000**, *72*, 1894.
[5] E. Puklin, B. Carlson, S. Gouin, C. Costin, E. Green, S. Ponomarev, H. Tanji, M. Gouterman, *J. Appl. Polym. Sci.*, **2000**, *77*, 2795; C. S. Subramanian, R. T. Amer, D. M. Oblesby, C. G. Burkett Jr., *AIAA J.*, **2002**, *40*, 582.
[6] K. A. Kneas, W. Xu, J. N. Demas, B. A. DeGraff, *Applied Spectr.*, **1997**, *51*, 1346; M. E. Cox, B. Dunn, *Applied Optics*, **1985**, *24*, 2114.
[7] P. Douglas, K. Eaton, *Sens. Actuators B*, **2002**, *82*, 200; V. V. Vasil'ev, S. M. Borisov, S. M. *Sens. Actuators B*, **2002**, *82*, 272; G. DiMarco, M. Lanza, *Sens. Actuators B*, **2000**, *63*, 42.

[8] A. Mills A. *Sens. Actuators B*, **1998**, *51*, 60; A. Mills, M. D. Thomas, *Analyst*, **1998**, *123*, 1135; W. Xu, K. A. Kneas, J. N. Demas, B. A. DeGraff, *Anal. Chem.*, **1996**, *68*, 2605.
[9] H. Wang, G. Xu, S. Dong, *Analyst*, **2001**, *126*, 1095; J. P. Hubner, B. F. Carroll, K. S. Schanze, H. F. Ji, *Exp. Fluids*, **2000**, *28*, 21; I. Klimant, O. S. Wolfbeis, *Anal. Chem.*, **1995**, *67*, 3160.
[10] X. Lu, I. Manners, M. A. Winnik, *Macromolecules*, **2001**, *34*, 1917.
[11] M. Yafuso, P. F. Korkowski, R. A. Mader, C. Yan, J. T. Carlock, *US Patent 5,296,381*, **1994**; R. R. Holloway, T. T. Kiang, *US Patent 5,070,158*, **1991**; L. Hsu, H. Heitzmann, *US Patent 4,712,865*, **1987**.
[12] Z. Wang, A. R. McWilliams, C. E. B. Evans, X. Lu, S. Chung, M. A. Winnik, I. Manners, *Adv. Func. Mat.* **2002**, *12*, 415.
[13] *The Polymer Handbook*, 3rd Ed., J. Brandrup, E. H. Immergut, Eds., Wiley, New York, **1989**; *Polymer Data Handbook*, J. E. Mark, Ed., Oxford University Press, New York, **1999**
[14] I. Klimant, O. S. Wolfbeis, *Anal. Chem.*, **1995**, *67*, 3160.
[15] E. Vander Donckt, B. Camerman, F. Hendrick, R. Herne, R. Vandeloise, *Bull. Soc. Chim. Belg.*, **1994**, *103*, 207.
[16] Y. Amao, Y. Ishikawa, I. Okura, *Anal. Chim. Acta*, **2001**, *445*, 177.
[17] A. Mills, A. Lepre, *Anal. Chem.*, **1997**, *69*, 4653.
[18] J. N. Carraway, J. N. Demas, B. A. DeGraff, J. R. Bacon, *Anal. Chem.*, **1991**, *63*, 337.
[19] J. N. Demas, B. A. DeGraff, *J. Chem. Educ.*, **1997**, *74*, 690.

Macromol. Symp. **196**, 249–270 (2003)

Cyclophosphazenes as Polymer Modifiers

Mario Gleria,[1] Riccardo Po,[2] Giorgio Giannotta,[2] Luisa Fiocca,[2] Roberta Bertani,[3] Luca Fambri,[4] Francesco Paolo La Mantia,[5] Roberto Scaffaro[5]*

[1]Istituto di Scienze e Tecnologie Molecolari del C.N.R., c/o Dipartimento CIMA dell'Università, via F.Marzolo 1, 35131 Padova, Italy
[2]Istituto G. Donegani, Polimeri Europa S.p.A., Via Fauser 4, 28100 Novara, Italy
[3]Dipartimento di Processi Chimici dell'Ingegneria, Via F. Marzolo 9, 35131 Padova, Italy
[4]Dipartimento di Ingegneria dei Materiali e Tecnologie Industriali, Università di Trento,
Via Mesiano 77, 38050 Trento, Italy
[5]Dipartimento di Ingegneria Chimica dei Processi e dei Materiali, Università di Palermo, Viale delle Scienze, 90128, Palermo, Italy

Summary: The utilization of cyclophosphazenes as polymer modifiers is reviewed, with particular concern to their exploitation as versatile chain extenders, possibly for recycle problems, crosslinkers, to enhance mechanical properties of polymeric materials, branchers, to selectively introduce ramifications in linear polymers, and compatibilizers, to favor the formation of blends between originally incompatible organic macromolecules. The great versatility of the synthetic methods put forward for these substrates, together with the ease of controlling their modification, functionalization and reactivity are important parameters for the evaluation of which type of use is more feasible for these trimers. The importance of cyclophosphazenes bearing organic polymeric chains, azide groups, 2-oxazoline derivatives and oxirane rings in connection with organic conventional macromolecules is critically highlighted.

Keywords: chain-extenders; compatibilizers; cyclophosphazenes; epoxide; 2-oxazoline

© 2003 WILEY-VCH Verlag GmbH & KGaA, Weinheim CCC 1022-1360/00/$ 17,50+.50/0

Introduction

Cyclophosphazenes (CPs) are relevant phosphazene compounds[1-3] having the structure reported below:

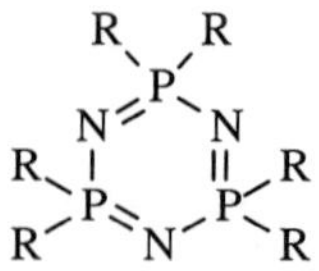

FORMULA 1

in which the phosphorus atoms in the cycle bear two substituent groups.

The large majority of cyclophosphazenes are formed by three -P=N- units[4], although less frequent superior cyclic homologues, formed by four[5-7], five[8,9], six [8-11], seven[12], etc., phosphazene units are known. The scientific importance of these products was restricted for long time to act as model compounds for the high molecular weight poly(organophosphazenes) (POPs)[13,14]

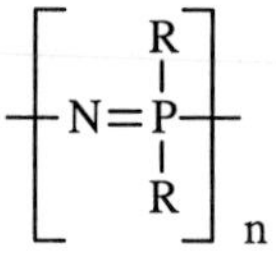

$$\left[-N=PR_2- \right]_n$$

FORMULA 2

in the sense that research on the chemical reactivity, substitutional kinetics, spectroscopic, photochemical, thermal, biological, etc. properties of these polymers was preliminary carried out at trimeric level, and only subsequently the results obtained were extended to the corresponding macromolecules.

However, it was realized very soon that cyclophosphazenes are very versatile materials, able to cover themselves a large variety of different applications[15]. Thus, they are excellent materials for the preparation of cyclomatrix[4] and cyclolinear[16,17] polymers; biologically important substrates (as antitumors[18-20], pesticides[21], insect chemosterilant[1,2], fertilizers[1,2,22-24]); products that show photochemical (as photoinitiators[25,26], photostabilizers[27-34]), antioxidant[35,36], and flame retardant[2,37] activity, with possibility of forming clathrates[38-40], and high temperature resistant fluids[41,42]. They are also important in supramolecular chemistry[43-48], electric conductivity[46,49], dendrimers[50-52], star polymers[50,53], and ion receptors[54-56].

Among the applications of cyclophosphazenes, the exploitation of these compounds as modifiers for conventional organic macromolecules looks very attractive. For their practical utilization, in fact, organic polymers very often need to be crosslinked, blended, extended, branched, compatibilized, in a controlled way, and cyclophosphazenes seem to be particularly useful to fulfill these goals. In this paper we would like to report on the advances of cyclophosphazene research in these domains, focusing on those carried out in our labs.

Discussion

Polymer modifiers are low-molecular weight mono- or multi-functional compounds which are reacted with conventional macromolecules both to modify the pristine structures and properties of the polymers and to introduce novel chemical and/or physical features in these materials. For this type of research, cyclophosphazenes lend themselves nicely. In fact they are multifunctional compounds containing, in the case of cyclotriphosphazenes, from 1 to 6 reactive groups "R", as reported in the formulae below:

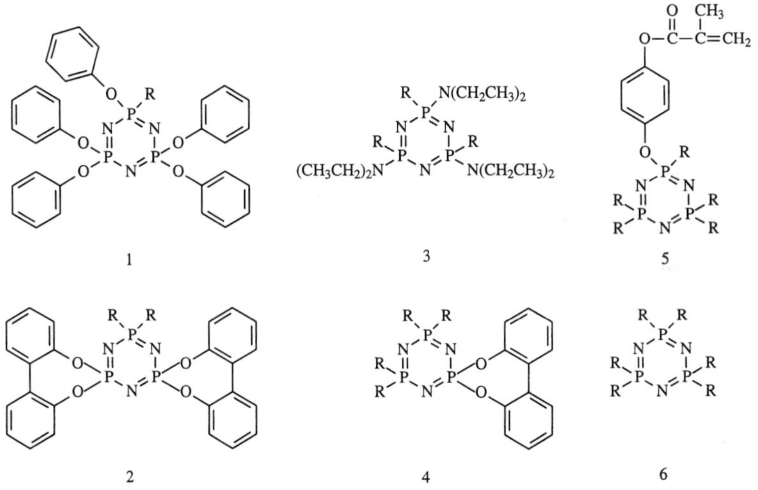

FORMULA 3

The number of reactive groups "R" in the cyclophosphazenes is able to determine the possible practical utilization of these substrates. In fact, cyclophosphazenes containing just one functional groups[57] can be used as monomers or as ***polymer modifiers***, as reported below:

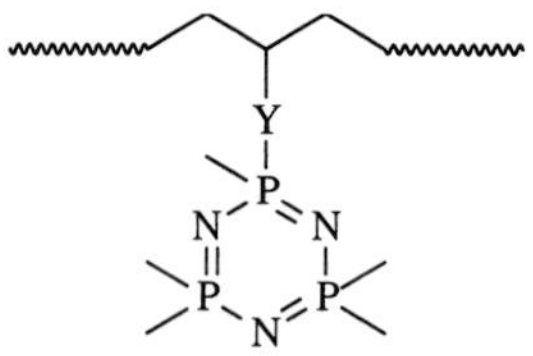

FORMULA 4

in which the trimeric substrate, attached to the polymer chain as a side group, may be able to impart to the macromolecule properties initially belonging to the trimer exclusively.

Cyclophosphazenes with two reactive substituents[58], may act as ***chain-extenders***, according to the following Formula:

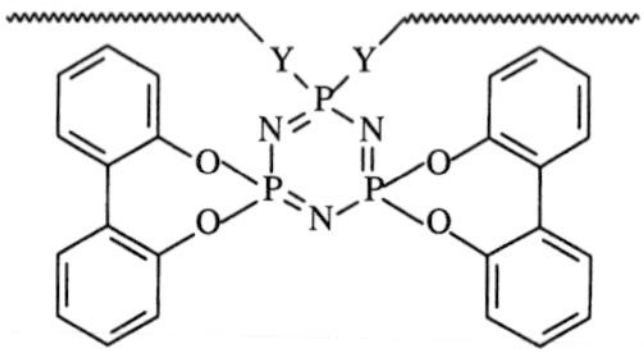

FORMULA 5

in which two polymeric chains are linked together through the cyclophosphazene thus inducing the increase in the molecular weight of the material and possibly recovering of mechanical properties.

In the case of cyclophosphazenes bearing from 3 to 6[57,59,60] reactive substituents, they can be used as ***branchers*** to introduce ramification of polymer chains in a controlled way, as illustrated in Formula 6 in which three polymer chains are bonded to the trimer;

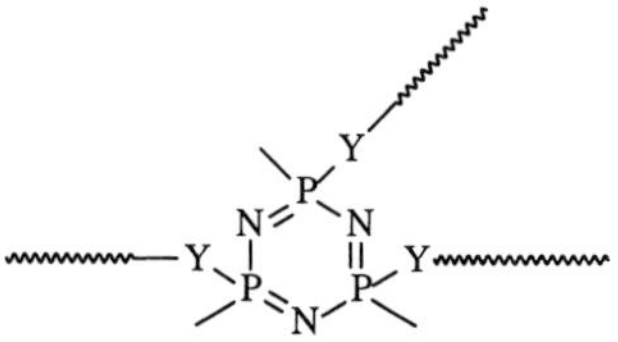

FORMULA 6

the same materials, moreover, may act ***crosslinkers*** inducing the complete reticulation of polymeric substrates through the formation of tridimensional networks having the general structure below:

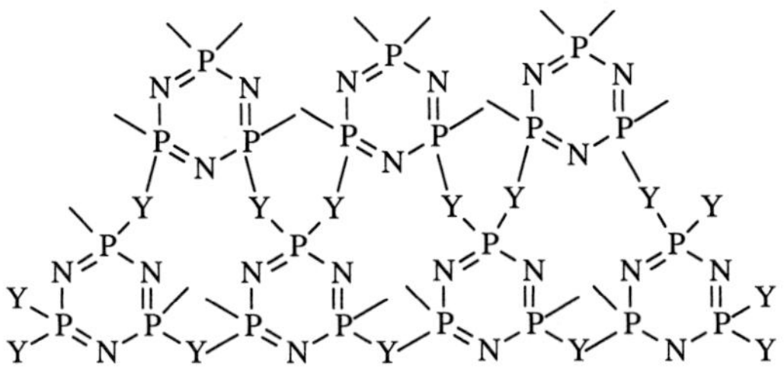

FORMULA 7

where "Y" represent the polymer chains included between two cyclophosphazene rings.

Finally, when on the same cyclophosphazene substrate are simultaneously present two different reactive groups[58], *e.g.* "R" and "R' ", these molecules can be exploited as ***polymer compatibilizers***, for instance, in reactive blending processes, linking chains of different polymers to the same cyclophosphazene ring and forming very efficient compatibilizing agents with the following general structure:

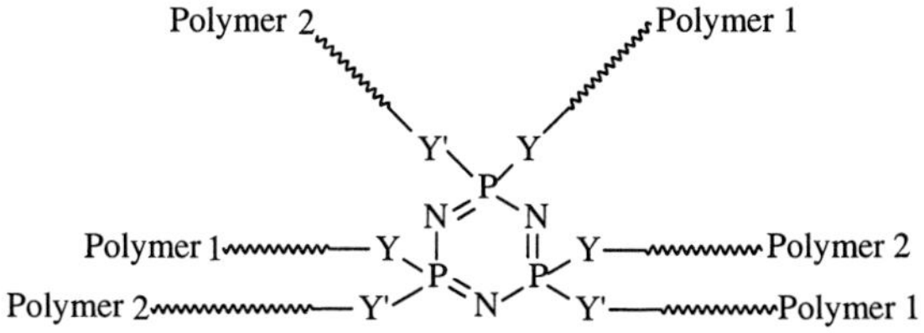

Polymer 2 Polymer 1 Y' Y P N N Polymer 1 Y P P Y Polymer 2 Polymer 2 Y' N Y' Polymer 1

FORMULA 8

Having in mind the guidelines exposed above, research on cyclophosphazenes as polymer modifiers started in 1996 when K.Inoue[61] proposed the use of cyclophosphazenes as compatibilizers for immiscible poly(2,6-dimethylphenylene-oxide) (PPO) and Nylon-6 (Polyamide-6, PA6) by means of the use of specially designed trimers containing polystyrene (PS) and polyamide chains in their chemical structure.

The cyclophosphazenes used as compatibilizers were prepared starting from hexakis(4-carboxyphenoxy)cyclophosphazene, $[NP(OC_6H_4\text{-}COOH)_2]_3$, chlorination of the compound to $[NP(OC_6H_4\text{-}COCl)_2]_3$ by treatment with thionyl chloride, and reaction with amino-terminated polystyrene, $PS\text{-}NH_2$, or with polystyrene anion, PS^-Li^+, to obtain cyclophosphazenes partially substituted at the phosphorus with PS chains. The materials obtained (still containing -COCl functions) was hydrolyzed with alkaline water to reform the original carboxylic groups, and eventually reacted with ε-caprolactame, to graft polyamide chains onto the cyclophosphazene platform. The chemical structures of the cyclophosphazenes obtained are reported below.

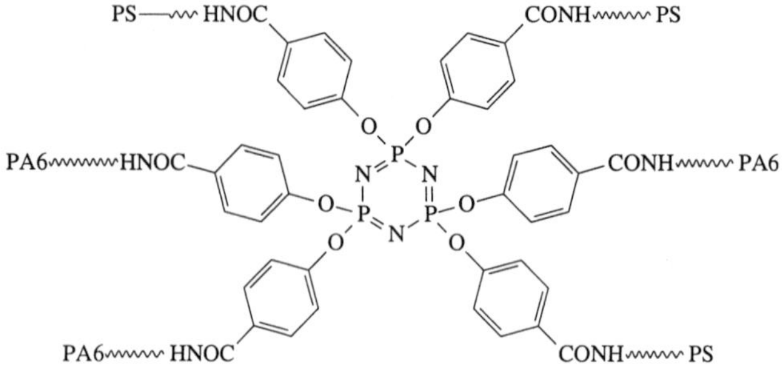

FORMULA 9

and

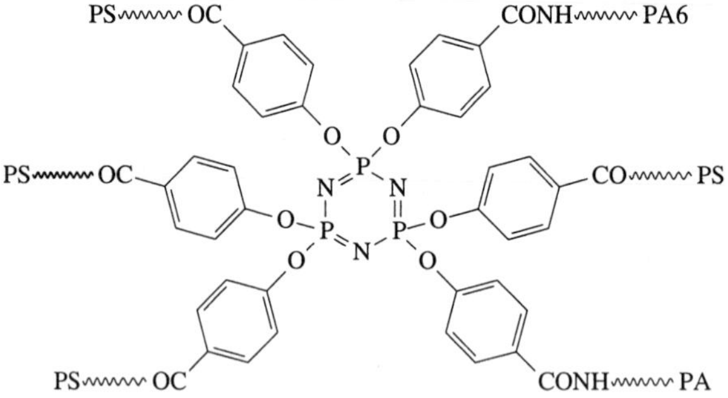

FORMULA 10

showing both polystyrene and polyamide chains attached to the phosphorus atoms of the cyclophosphazene ring.

The successive blending of PPO and Polyamide-6 polymers with variable percentages of these CPs induced the strong reduction of the surface tension between the two polymers due to the onset of interactions between the polystyrene and polyamide side arms of the cyclophosphazene with PPO and Polyamide-6 polymers, respectively. These facts were reflected by the presence of an unique T_g in the thermograms of blends prepared with a different

PPO/PA6/CP composition, by the decrease of the thermal stability of the final blends with the increase in the polystyrene content, and by the enhancement of tension modulus, strength and elongation of these products. After this initial step, research on the practical exploitation of CPs as modifiers for conventional polymers evolved towards the use of cyclophosphazenes substituted with azide groups.

Although in the past several articles have been published dealing with phosphazene substrates functionalized with azides[62-68], the use of these compounds as polymer crosslinkers was addressed a few years ago by H.R.Allcock[69-71] who succeeded in synthesizing the following series of compounds:

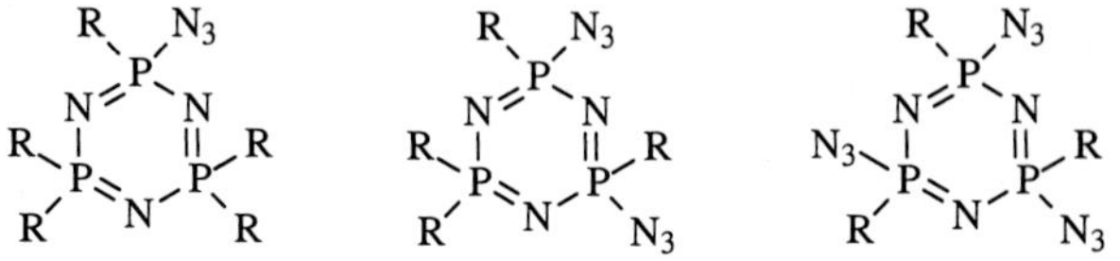

where: R may be C_6H_5O-; CF_3CH_2O- or $(CH_3)_2N$-

FORMULA 11

The synthesis has been carried out by reacting partially substituted chlorinated cyclophosphazenes with sodium azide in 2-butanone and in the presence catalytic amounts of tetrabutylammonium bromide[69-71].

The reactivity of these CPs was tested first by using monoazide derivatives in nitrene insertion reactions[69-71] induced both thermally[69-71] and photochemically[70,71]. It could be proved that thermal nitrene insertion takes place with $N_3P_3(OC_6H_5)_5(N_3)$ in the presence of 1-phenylnonane, while photolytic nitrene insertion could occur by irradiating $N_3P_3(OCH_2CF_3)_5(N_3)$ in cyclohexane. These types of reaction have been claimed to serve as a powerful tool for crosslinking polyolefins[69,71] and to chain extend syndiotactic polystyrene[72].

An alternative way for azide-functionalized CPs to induce modifications in conventional macromolecules is based on the utilization of the Staudinger reaction between triphenylphosphine-containing polystyrene[73-75] and/or silicones[74,75] and azido-functionalized cyclophosphazenes, according to the following Scheme:

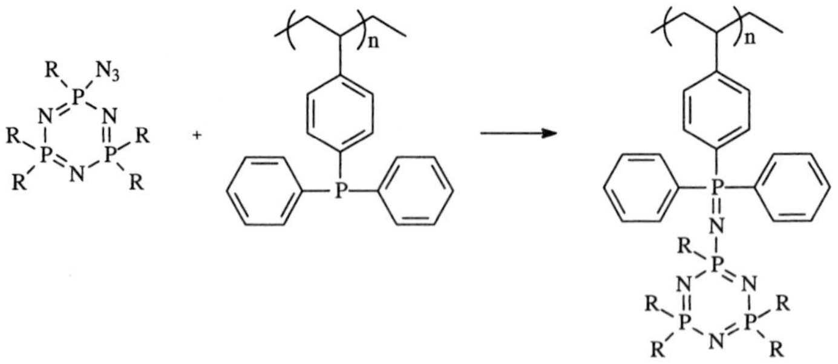

SCHEME 1

which might be extremely useful in the preparation of polymers containing high quantities of nitrogen and phosphorus atoms, particularly suitable to enhance the flame resistance and the thermal stability of these materials[73].

In our laboratory the utilization of cyclophosphazenes as polymer modifiers was faced using rather different point of view, *i.e.* by considering the synthesis of new CP derivatives functionalized with 2-oxazoline[76-81] and with epoxy groups[82].

2-Oxazolines are very interesting compounds used for many important practical applications, as monomers for polymerization processes[83], protective groups for carboxylic functions[84], core molecules for dendrimers[85,86], chain extenders[87,88] and blend compatibilizers[89-91]. A number of review articles covering the synthesis, modification, and utilization of 2-oxazolines have been published over the time[83,86,92-95].

The first combination between phosphazene materials and 2-oxazolines was reported on 1994 in the pioneering work by Chang[53,96,97], who used first an aryloxy-substituted cyclophosphazene containing six -CH_2Br functions in combination with 2-methyl-2-oxazoline to prepare star polymers[53] with a cyclophosphazene core, and successively a bromomethylated polyphosphazene to obtain POP-*g*-poly(2-methyl-2-oxazoline) grafted copolymers[96,97].

Expanding upon this research, we could succeed in inserting 2-oxazoline groups into polyphosphazene substrates by grafting a 2-oxazoline-containing maleate onto a poly[bis(4-methylphenoxyphosphazene], to obtain a grafted copolymer very reactive towards acidic compounds, that was successively blended with polyacrylic and/or poly methacrylic acids[76].

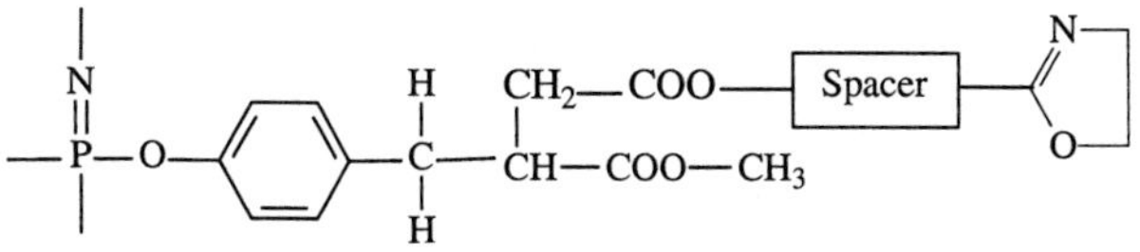

where "Spacer" is a long aliphatic chain

FORMULA 12

After these initial investigations, we developed a different strategy to combine 2-oxazoline derivatives and phosphazene materials, based on the use of 4-hydroxyphenyl-2-oxazoline[98]

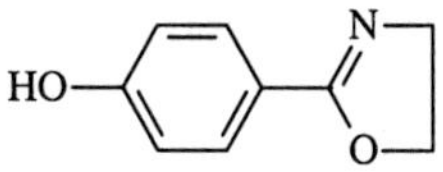

FORMULA 13

in combination with cyclophosphazenes containing a variable number of chlorine atoms[77-81,99].

Thus the reaction of 4-hydroxyphenyl-2-oxazoline with hexachlorocyclophosphazene led to the synthesis of hexakis(4-oxazolinophenoxy)cyclophosphazene (***C-6-OXA***)

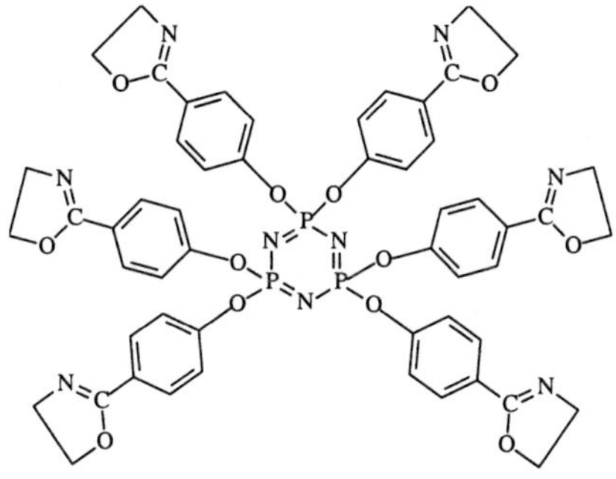

FORMULA 14

while the utilization of 2',2"-dichloro-4,4,6,6-bis[spiro(2',2''-dioxy-1',1''-biphenyl)]cyclotriphosphazene allowed the preparation of 2'2"-bis(4-oxazolinophenoxy)-4,4,6,6-bis[spiro(2',2"-dioxy-1',1"-biphenyl)]cyclophosphazene (***C-2-OXA***).

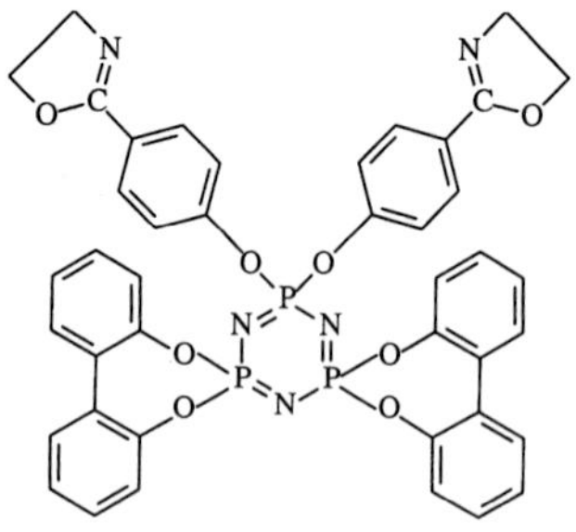

FORMULA 15

The reaction of ***C-2-OXA*** with low-molecular weight poly(ethylene terephthalate) (PET) in the internal chamber of a Haake Rheomix 600 Mixer at 270°C in the presence of about 5% of this trimer leads to the formation of a new material. The expected reaction is reported below.

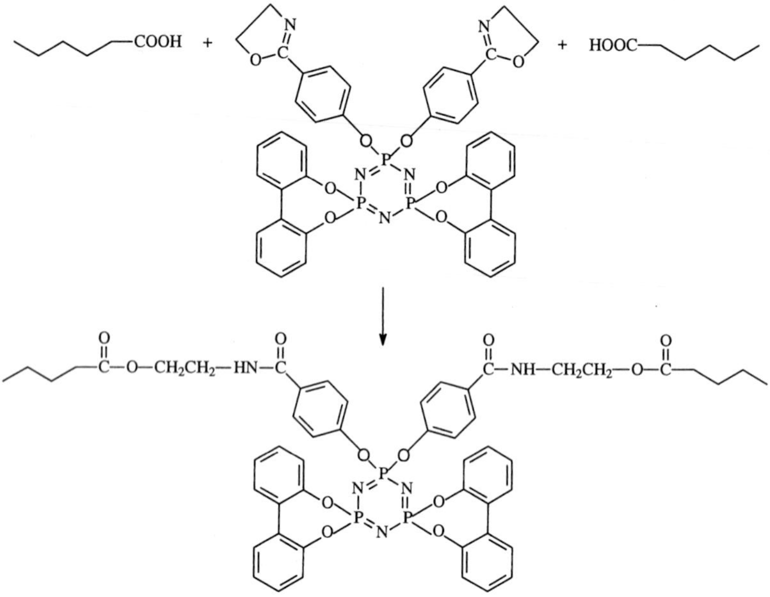

SCHEME 2

The polymers obtained showed different thermal and mechanical properties with respect to those of the starting material[81]. In fact, thermal analyses carried out by DSC on both virgin and treated PET showed that the T_g of the original polymer is always lower than that of the modified

PET, while the melt and crystallization enthalpies, and the cooling crystallization temperature of the starting polymer are always higher. At the same time, according to the data obtained by **D**ynamical **M**echanical **T**hermal **A**nalysis (DMTA), PET modified with ***C-2-OXA*** exhibited not only a higher modulus, but also a lower percentage of deformation above T_g, as reported in Figure 1:

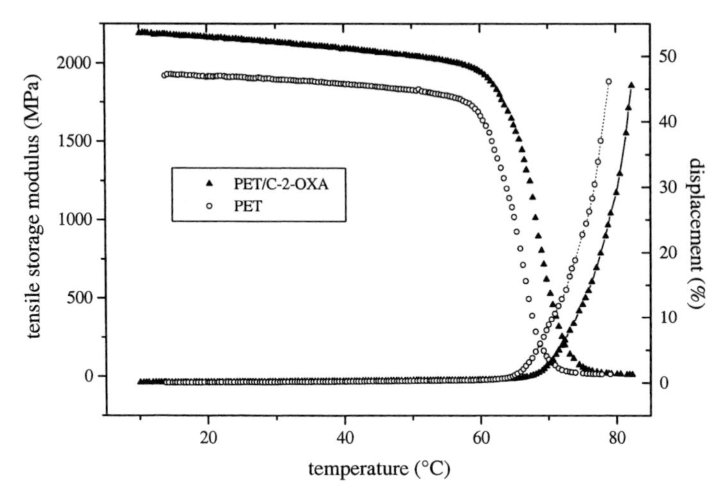

FIGURE 1. DMTA analysis for PET and PET treated with ***C-2-OXA*** (Tensile DMTA Mk II Polymers Laboratories: static stress 0.6 MPa, deformation amplitude 64 micron, frequency 1 Hz, heating rate 3 °C/min).

These experimental evidences of lower crystallinity, higher T_g, higher storage modulus, and reduction of mobility appear to be consistent with a remarkable increase of the molecular weight of the ***C-2-OXA*** modified polymers.

Even more striking are the results obtained by treating PET with ***C-6-OXA***, because of the presence of six 2-oxazoline residues in the trimer that are able to induce deep modifications in PET by simultaneously bonding several PET chains on the same cyclophosphazene ring. The general reaction is reported in the Scheme 3 below:

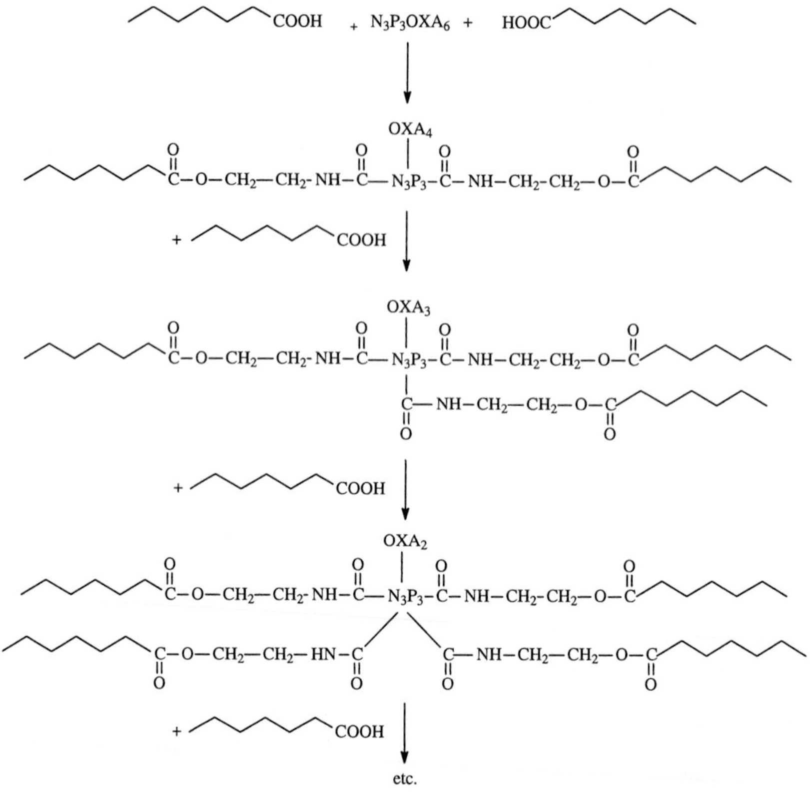

COOH + $N_3P_3OXA_6$ + HOOC

OXA_4

$C(=O)-O-CH_2-CH_2-NH-C(=O)-N_3P_3-C(=O)-NH-CH_2-CH_2-O-C(=O)$

+ COOH

OXA_3

$C(=O)-O-CH_2-CH_2-NH-C(=O)-N_3P_3-C(=O)-NH-CH_2-CH_2-O-C(=O)$

$C(=O)-NH-CH_2-CH_2-O-C(=O)$

+ COOH

OXA_2

$C(=O)-O-CH_2-CH_2-NH-C(=O)-N_3P_3-C(=O)-NH-CH_2-CH_2-O-C(=O)$

$C(=O)-O-CH_2-CH_2-HN-C(=O)$ $C(=O)-NH-CH_2-CH_2-O-C(=O)$

+ COOH

etc.

SCHEME 3

The rheological measurements reported in Figure 2, showed that the complex viscosity of ***C-6-OXA***-modified PET (upper curve) undergoes a dramatic enhancement with respect to that of the virgin polymer (lower curve) and that of a PET sample modified with a commercial bis-oxazoline derivative, *i.e.* (1-4 phenylene)-bis-2-oxazoline (***Ph-2-OXA***)[100] (middle curve), to suggest the onset of strong coupling processes between the cyclophosphazene and the polymer.

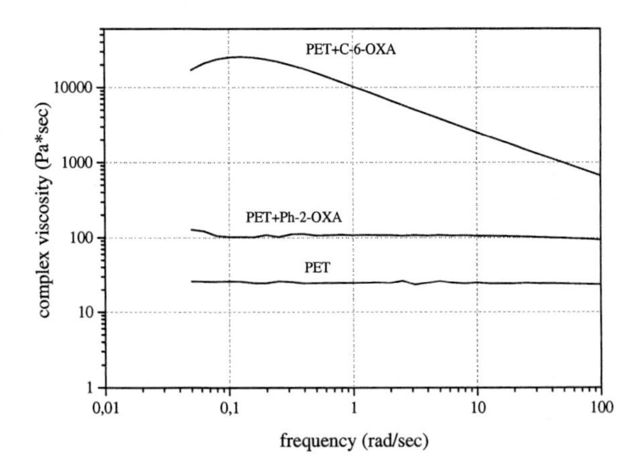

FIGURE 2. Complex viscosity variations at 270°C versus oscillation frequency of molten PET/***C-6-OXA***; PET/commercial bis-oxazoline (***Ph-2-OXA***); Virgin PET, after treatment in a mixer.

Similar results are obtained by analyzing the shear storage and loss shear moduli of molten polymers, as reported in Figures 3A and 3B where it can be easily seen that both shear storage and loss moduli of the virgin PET are much lower than those of same polymer treated with the bis-oxazoline and with ***C-6-OXA***, respectively. Moreover, the PET sample modified by reaction with ***C-6-OXA*** behave predominantly as a viscous polymer at low oscillation frequency, but it start to behave as an elastic materials when the applied frequencies increased.

The great usefulness of ***C-6-OXA*** as a polymer modifier could be confirmed also by using this trimer as a blend compatibilizer for immiscible polycarbonate (PC) and Polyamide-6 materials.

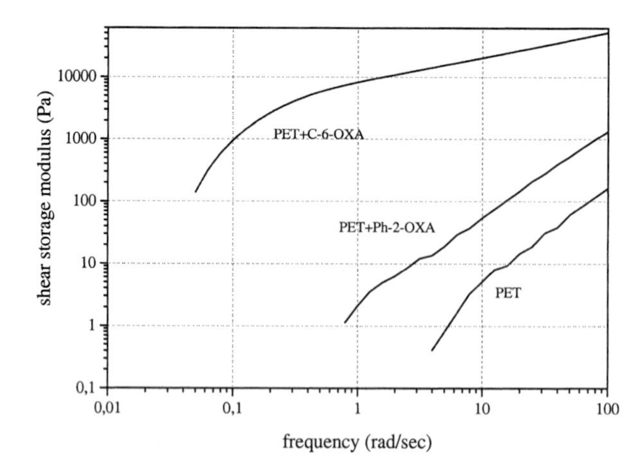

Figure 3A. Shear Storage Modulus G' variations at 270°C versus the oscillation frequency for molten PET/***C-6-OXA***; PET/commercial bis-oxazoline (***Ph-2-OXA***); virgin PET after treatment in a mixer.

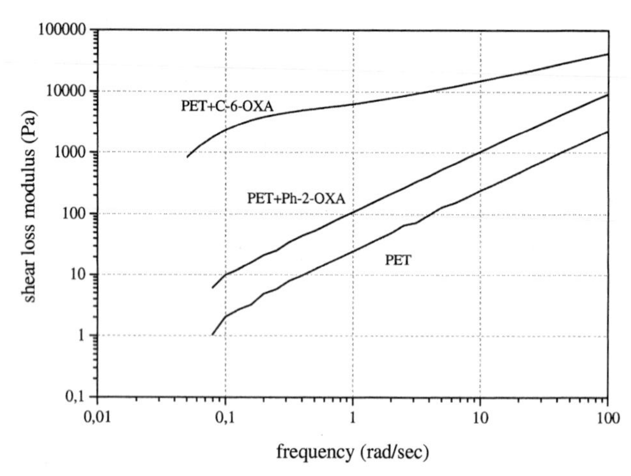

Figure 3B. Shear Loss Modulus G" variations at 270°C versus the oscillation frequency for molten PET/***C-6-OXA***; PET/commercial bis-oxazoline (***Ph-2-OXA***); virgin PET after treatment in a mixer

In general polyamides possess terminal -COOH groups that could react with 2-oxazolines, while PC does not contain any useful end groups on the skeleton because of the utilization of monophenols to control the molecular weight of the material during the synthesis. This polymer, however, is believed to thermally degrade during blending to produce free hydroxylic functions according to the following reaction[101,102]. These groups are reported to be reactive towards 2-oxazolines[94].

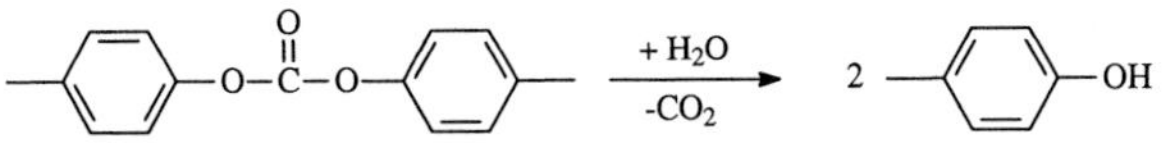

$$-C_6H_4-O-C(=O)-O-C_6H_4- \xrightarrow[-CO_2]{+H_2O} 2\ -C_6H_4-OH$$

SCHEME 4

As a matter of fact, three 50:50 blends have been prepared between PC and polyamide-6 containing no additives, (1,4-phenylene)-bis-2-oxazoline (***Ph-2-OXA***) and ***C-6-OXA*** (in similar molar 2-oxazoline content), respectively, using a Haake internal mixer at 270°C, which were analyzed from a morphological and mechanical point of view[99].

The results obtained showed that the mechanical mixture of the two polymers has a biphasic structure in which polyamide-6 acts as the matrix and PC as the dispersed phase. The structure becomes co-continuous when the commercial bis-oxazoline is added to the blend, while by treatment with ***C-6-OXA*** the situation is inverted, PC becoming the matrix and polyamide-6 the dispersed phase. At the same time, mechanical characterization revealed that the formation of PC/PA6/***C-6-OXA*** blend leads to an enhancement of the elastic modulus (possibly due to phase inversion and decreased effect of water as a plasticizing agent), while the toughness of the blends produced is substantially kept at a good level. These preliminary findings seem to indicate that the addition of ***C-6-OXA*** to the PC/PA6 blends alters in a significant way both the morphology and the physical properties of the mechanical mixture of the two polymers, possibly due to the formation in the reaction chamber of the internal mixer of a compatibilizing agent, according to the following Scheme:

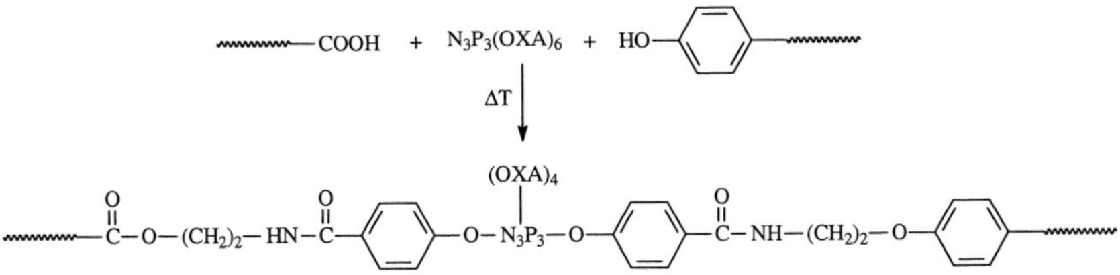

$$\sim\!\sim COOH + N_3P_3(OXA)_6 + HO-C_6H_4-\sim\!\sim \xrightarrow{\Delta T}$$

$$\sim\!\sim C(=O)-O-(CH_2)_2-HN-C(=O)-C_6H_4-O-N_3P_3(OXA)_4-O-C_6H_4-C(=O)-NH-(CH_2)_2-O-C_6H_4-\sim\!\sim$$

SCHEME 5

This hybrid compound might be able to reduce the surface tension between the two partners of the blend allowing the preparation of new materials.

Almost simultaneously to the research on the use of cyclophosphazenes functionalized with 2-oxazoline groups as chain-extenders and polymer compatibilizers described above, we started investigations on the use as polymer modifiers of new cyclophosphazenes substituted with epoxide moieties. These products were prepared by reacting hexachlorocyclophosphazene and 2',2"-dichloro-4,4,6,6-bis[spiro(2',2''-dioxy-1',1''-biphenyl)]cyclotriphosphazene with eugenol (2-methoxy-4-allylphenol) according to literature[103], and successive peroxidation of the allyl residues to oxirane rings[104] by the action of *m*-chloroperbenzoic acid. The general structures of the resulting compounds are reported below, in which 2 (***CP-2-EPOX***)

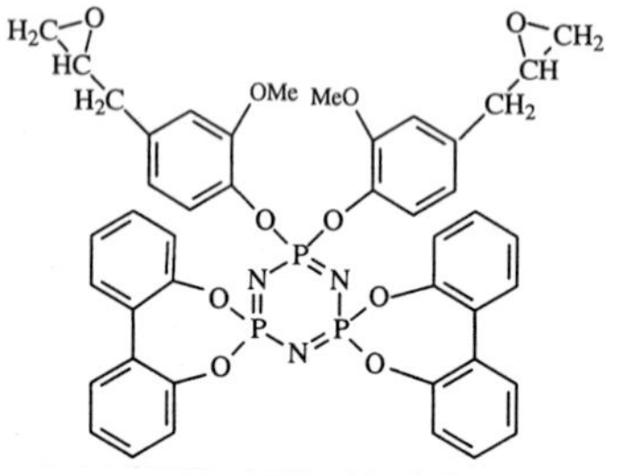

FORMULA 16

and 6 (CP-6-EPOX)

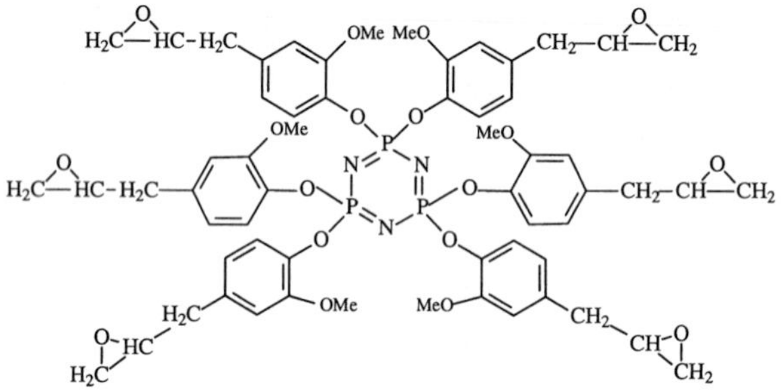

FORMULA 17

epoxide units are attached to the cyclophosphazene substrate.

These functions are known to be very reactive toward both -NH_2 and -COOH terminal groups of polyamides[105], and in this way they can act as a chain extenders in combination, for instance, with polyamide-6. The reaction between ***CP-2-EPOX*** and/or ***CP-6-EPOX*** with polyamide-6 was carried out in a Haake Minilab at 240°C, sometimes drying both polyamide-6 and the cyclophosphazene additives at 100°C prior to use to minimize the influence of hydrolytic phenomena in polyamide-6 during reactive blending. The resulting reactions between cyclophosphazenes and polyamide-6 are describe for ***CP-2-EPOX*** in the following reaction Scheme 6:

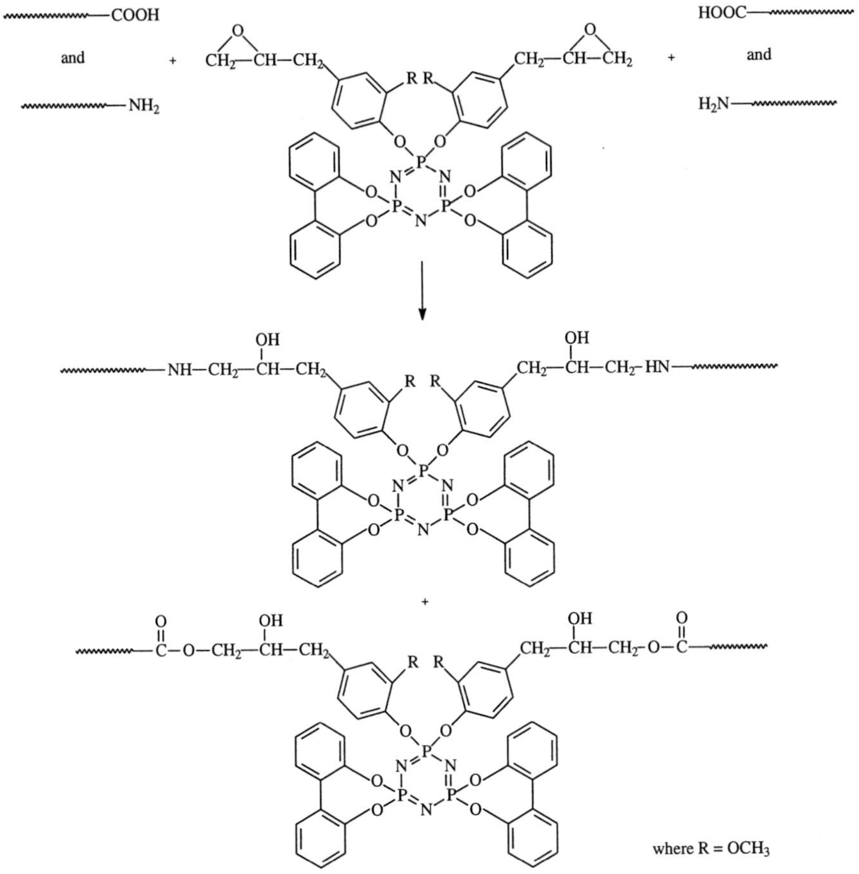

where R = OCH_3

SCHEME 6

in which different polyamide chains are linked together through a cyclophosphazene molecule.

The action of ***CP-2-EPOX*** was found to be more effective than that of ***CP-6-EPOX***, possibly due to phenomena of thermal instability which are operative for ***CP-6-EPOX*** already during the preliminary drying step of the cyclophosphazenes, resulting in the thermal degradation of the additive. These facts have been evidenced by DSC and TGA measurements on the cyclophosphazenes exploited.

In spite of these facts it could demonstrated that blending in the internal mixer described above polyamide-6 with very low amounts of ***CP-2-EPOX*** (0.2% w/w) introduces strong variations with an increase in both the torque and in the melt pressure drop during blending polyamide-6 in the presence of the cyclophosphazene with respect to the same process carried out only with the starting polymer.

The same holds true for rheological measurements in the melt, as reported in Figure 4 below for complex viscosity:

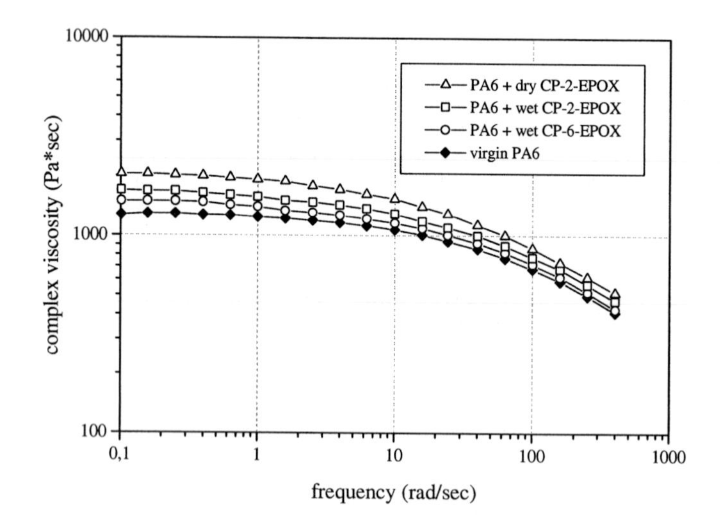

FIGURE 4. Complex viscosity at 240°C of Polyamide-6 samples treated with wet ***CP-6-EPOX*** and wet and dry ***CP-2-EPOX.***

and in Figure 5 for the shear storage modulus of polyamide-6, where it is evident that both these parameters increase very much passing from virgin polyamide-6, polyamide-6 treated with wet ***CP-6-EPOX***, polyamide-6 combined with wet ***CP-2-EPOX***, to reach the maximum value when the polyamide is mixed with dry ***CP-2-EPOX***.

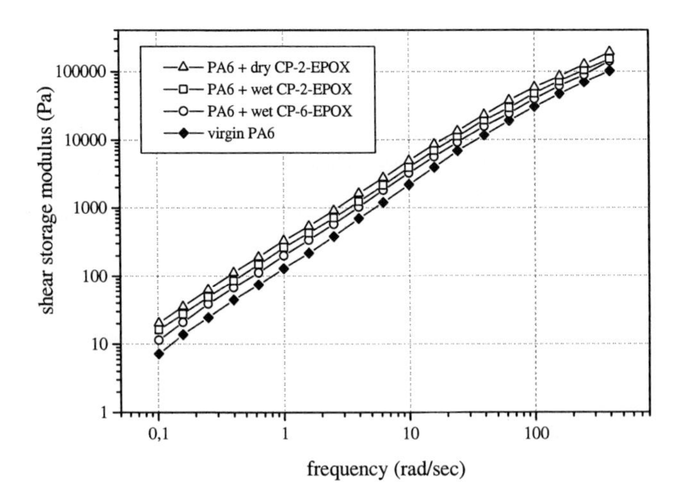

FIGURE 5. Storage and loss modulus at 240°C of Polyamide-6 samples treated with wet ***CP-6-EPOX*** and with wet and dry ***CP-2-EPOX***

Contrarily to what happens to mechanical and rheological properties, which are deeply influenced even using very low amounts of epoxide-containing cyclophosphazenes during the mixing processes, the thermal properties of the same materials seem not to be appreciably modified by the presence of cyclophosphazenes in the blend. This could be demonstrated by DSC and TGA, that showed no substantial modifications in the T_g, T_m and ΔH_m of virgin polyamide-6, compared with the value of the same parameters of the modified polymer. Moreover, during pyrolysis analysis, only ε-caprolactame production could be observed in both treated and untreated samples.

Conclusions

In this paper we highlighted literature on the utilization of cyclophosphazenes as polymer modifiers, and presented results on our activity in this field. The importance of experimental parameters such as the synthetic versatility of cyclophosphazenes, the ease of their chemical modification and/or functionalization, and the possibility of controlling the reactivity of the phosphazene trimers by carefully inserting a well defined number of chemical functions, are critically evaluated. The original work in this field carried out by K. Inoue[61] on the utilization of cyclophosphazenes containing polystyrene and polyamide chains as blend compatibilizers for poly(phenylene oxide) (PPO) and polyamide (Polyamide-6), and by H.R. Allcock[69-71,73-75] on the utilization of azide-containing trimers as crosslinkers for polyolefins, opened the route to successive research on the synthesis of cyclophosphazenes bearing new reactive functions, such as 2-oxazoline[76-81,99] and epoxide[82] groups, and on their utilization as chain extenders for polyesters (e.g. PET) and polyamides (e.g. Nylon 6), respectively, or to promote the compatibilization processes between immiscible polycarbonate (PC) and polyamides. Future developments of this research deal with the preparation of novel cyclophosphazene derivatives bearing substituents containing new reactive groups and/or with the synthesis of cyclophosphazenes containing mixed chemical functionalities.

[1] Krishnamurthy, S. S.; Sau, A. C. *Adv. Inorg. Chem. Radiochem.* **1978**, *21*, 41.
[2] Allen, C. W. In *The Chemistry of Inorganic Homo- and Hetero-Cycles*; Haiduc, I., Sowerby, D. B., Eds.; Academic Press: London, 1987, Vol. 2, p 501.
[3] Chandrasekhar, V.; Krishnan, V. *Adv. Inorg. Chem.* **2002**, *53*, 159.
[4] De Jaeger, R.; Gleria, M. *Prog. Polym. Sci.* **1998**, *23*, 179.
[5] Allcock, H. R.; Allen, R. W.; O'Brien, J. P. *J. Am. Chem. Soc.* **1977**, *99*, 3984.
[6] Winter, H.; Van de Grampel, J. C. *Rec. Trav. Chim. Pays-Bas* **1984**, *103*, 241.
[7] Krishnamurthy, S. S. *Phosphorus, Sulfur, and Silicon* **1989**, *41*, 375.
[8] Paddock, N. L.; Ranganathan, T. N.; Wingfield, J. N. *J. Chem. Soc. Dalton Trans.* **1972**, 1578.
[9] Calhoun, H. P.; Trotter, J. *J. Chem. Soc. Dalton Trans.* **1974**, 382.
[10] Calhoun, H. P.; Paddock, N. L.; Wingfield, J. N. *Can. J. Chem.* **1975**, *53*, 1765.
[11] Paddock, N. L.; Ranganathan, T. N.; Rettig, S. J.; Sharma, R. D.; Trotter, J. *Can. J. Chem.* **1981**, *59*, 2429.
[12] Gallicano, K. D.; Oakley, R. T.; Paddock, N. L.; Rettig, S. J.; Trotter, J. *Can. J. Chem.* **1977**, *55*, 304.
[13] Allcock, H. R. *Acc. Chem. Res.* **1979**, *12*, 351.
[14] Gleria, M. *Rec. Adv. in Macromolecules* **2000**, *1*, 103.
[15] *Phosphazenes: A Worldwide Insight*; De Jaeger, R.; Gleria, M., Eds.; NOVA Science Publishers: Hauppauge, N.Y., USA, 2002.
[16] Dez, I.; Levalois-Mitjaville, J.; Grützmacher, H.; Gramlich, V.; De Jaeger, R. *Eur. J. Inorg. Chem.* **1999**, 1673.
[17] Allcock, H. R.; Kellam, E. C.; Hofmann, M. A. *Macromolecules* **2001**, *34*, 5140.
[18] Labarre, J. F. *Top. Curr. Sci.* **1982**, *102*, 1.
[19] Labarre, J. F. *Top. Curr. Sci.* **1985**, *129*, 173.
[20] Van de Grampel, J. C. *Coord. Chem. Rev.* **1992**, *112*, 247.
[21] Haiduc, I. *The Chemistry of the Inorganic Ring Systems*; Wiley-Interscience: London, 1970; Vol. 2, Chapt. 5, p 623.
[22] Conesa, A. P.; Albagnac, G.; Brun, G. *C.R. Acad. Agric. Fr.* **1973**, *59*, 1457.
[23] Conesa, A. P. *C.R. Acad. Agric. Fr.* **1974**, *60*, 1353.
[24] Barel, D.; Black, C. A. *Agron. J.* **1979**, *71*, 15.
[25] Fantin, G.; Medici, A.; Fogagnolo, M.; Gleria, M.; Minto, F. Italian Patent, 1,265,090 (1996), assigned to Consiglio Nazionale delle Ricerche.
[26] Gleria, M.; Minto, F.; Facchin, G.; Bertani, R. Italian Patent, 1,270,949 (1997), assigned to Consiglio Nazionale delle Ricerche.
[27] Pond, D. M.; Wang, R. S. H. U.S., 3,936,418 (1976), Chem. Abstr. 84, 151551r (1976), assigned to Eastman

Kodak Co.
[28] Gleria, M.; Paolucci, G.; Minto, F.; Lora, S. *Chem. Ind. (Milan)* **1982**, *64*, 479.
[29] Wiezer, H. Eur. Pat. Appl. EP, 64,752 (1982), Chem. Abstr. 98, 108403w (1983), assigned to Hoechst Aktiengesellschaft.
[30] Wiezer, H. US Patent, 4,451,400 (1984), Chem. Abstr. 98, 108403w (1983), assigned to Hoechst Aktiengeselleschaft.
[31] Gleria, M.; Minto, F.; Bortolus, P.; Lora, S. Italian Patent, 1,176,618 (1987), assigned to Consiglio Nazionale delle Ricerche.
[32] Bortolus, P.; Busulini, L.; Lora, S.; Minto, F.; Pezzin, G. Italian Patent, 1,196,213 (1988), assigned to Consiglio Nazionale delle Ricerche.
[33] Inoue, K.; Takahata, H.; Tanigaki, T. *J. Appl. Polym. Sci.* **1993**, *50*, 1857.
[34] Inoue, K. Jpn. Kokai Tokkyo Koho JP, 06 135,977 [94 135,977] (1994), Chem. Abstr. 121, 179870a (1994), assigned to Nippon Shoe.
[35] Wang, R. H. S.; Irick, G. US Patent, 4,080,361 (1978), Chem. Abstr. 89, 111121x (1978), assigned to Eastman Kodak.
[36] Goins, D. E.; Li, H. M. U.S. US, 5,105,001 (1992), Chem. Abstr. 117, 61662y (1992), assigned to Ethyl Corp.
[37] Allen, C. W. *J. Fire Sci.* **1993**, *11*, 320.
[38] Sozzani, P.; Comotti, A.; Simonutti, R. In *Crystall Engineering: From Molecules and Crystals to Materials*; Braga, D., al., e., Eds.; Kluwer Academic Publisher, The Netherlands, 1999, p 443.
[39] Allcock, H. R.; Primrose, A. P.; Silverberg, E. N.; Visscher, K. B.; Rheingold, A. L.; Guzei, I. A.; Parvez, M. *Chem. Mater.* **2000**, *12*, 2530.
[40] Sozzani, P.; Comotti, A.; Simonutti, R.; Meersmann, T.; Logan, J. W.; Pines, A. *Angew. Chem. Int. Ed. Engl.* **2000**, *39*, 2695.
[41] Singler, R. E.; Bierberich, M. J., In *Synthetic Lubrificants and High Performance Functional Fluids*; Shubkin, R. L., Ed.; Marcel Dekker: New York, USA, 1993; Chap. 10; p 215.
[42] Singler, R. E.; Gomba, F. J. In *Synthetic Lubricants and High-Performance Functional Fluids*; Rudnick, L. R., Shubkin, R. L., Eds.; Marcel Dekker, Inc.: New York, USA, 1999; Chapt. 13, Vol. 77, p 297.
[43] Inoue, K.; Itaya, T.; Azuma, N. *Supramol. Sci.* **1998**, *5*, 163.
[44] Itaya, T.; Inoue, K. *Bull. Chem. Soc. Jpn* **2000**, *73*, 2615.
[45] Itaya, T.; Inoue, K. *Bull. Chem. Soc. Jpn* **2000**, *73*, 2829.
[46] Inoue, K.; Itaya, T. *Bull. Chem. Soc. Jpn* **2001**, *74*, 1381.
[47] Inoue, K.; Inoue, Y. *Polym. Bull.* **2001**, *47*, 239.
[48] Itaya, T.; Azuma, N.; Inoue, K. *Bull. Chem. Soc. Jpn* **2002**, *75*, 2275.
[49] Chandrasekhar, V. *Adv. Polym. Sci.* **1998**, *135*, 139.
[50] Inoue, K. *Prog. Polym. Sci.* **2000**, *25*, 453.
[51] Maraval, V.; Laurent, R.; Merino, S.; Caminade, A. M.; Majoral, J. P. *Eur. J. Org. Chem.* **2000**, 3555.
[52] Brauge, L.; Caminade, A. M.; Majoral, J. P.; Slomkowski, S.; Wolszczak, M. *Macromolecules* **2001**, *34*, 5599.
[53] Chang, J. Y.; Ji, H. J.; Han, M. J.; Rhee, S. B.; Cheong, S.; Yoon, M. *Macromolecules* **1994**, *27*, 1376.
[54] Mitjaville, J.; Caminade, A. M.; Majoral, J. P. *Tetrahedron Lett.* **1994**, *35*, 6865.
[55] Hochart, F.; Mouveaux, C.; Levalois-Mitjaville, J.; De Jaeger, R. *Tetrahedron Lett.* **1998**, *39*, 6171.
[56] Mouveaux, C.; Levalois-Mitjaville, J.; De Jaeger, R. *Phosphorus Res. Bull.* **1999**, *10*, 702.
[57] Inoue, K.; Kaneyuki, S.; Tanigaki, T. *J. Polym. Sci., Polym. Chem. Ed.* **1992**, *30*, 145.
[58] Kumar, D.; Fohlen, G. M.; Parker, J. A. *Macromolecules* **1983**, *16*, 1250.
[59] Allcock, H. R. *Phosphorus-Nitrogen Compounds. Cyclic, Linear, and High Polymeric Systems*; Academic Press: New York, USA, 1972.
[60] Carriedo, G. A.; Fernandez-Catuxo, L.; Garcia Alonso, F. J.; Gomez-Elipe, P.; Gonzalez, P. A. *Macromolecules* **1996**, *29*, 5320.
[61] Miyata, K.; Watanabe, Y.; Itaya, T.; Tanigaki, T.; Inoue, K. *Macromolecules* **1996**, *29*, 3694.
[62] Grundmann, C.; Rätz, R. *Z. Naturforsch.* **1955**, *10b*, 116.
[63] Sharts, C. M.; Bilbo, A. J.; Gentry, D. R. *Inorg. Chem.* **1996**, *5*, 2140.
[64] Sharts, C. M. US Patent, 3,347,876 (1967), Chem. Abstr. 67, 116942w (1967), assigned to Secretary of Navy, United States of America.
[65] Roesky, H. W.; Banek, M. *Z. Naturforsch.* **1979**, *34B*, 752.
[66] Müller, U.; Schmock, F. *Z. Naturforsch.* **1980**, *35B*, 1529.
[67] Willson, M.; Sanchez, M.; Labarre, J. F. *Inorg. Chim. Acta* **1987**, *136*, 53.

[68] Dave, P. R.; Forohar, F.; Axenrod, T.; Bedford, C. D.; Chaykovsky, M.; Rho, M. K.; Gilardi, R.; George, C. *Phosphorus, Sulfur, and Silicon* **1994**, *90*, 175.
[69] McInstosh, M. B.; Hartle, T. J.; Allcock, H. R. *J. Am. Chem. Soc.* **1999**, *121*, 884.
[70] Allcock, H. R.; McIntosh, M. B.; Hartle, T. J. *Inorg. Chem.* **1999**, *38*, 5535.
[71] Hartle, T. J.; McInstosh, M. B.; Allcock, H. R. *ACS Polym. Prep.* **1999**, *40(2)*, 908.
[72] Silvis, H. G.; McIntosh, M. B. U.S. US, 6,291,618 (2001), Chem. Abstr. 135, 242709 (2001), assigned to The Dow Chemical Co.
[73] Hartle, T. J.; Sunderland, N. J.; McIntosh, M. B.; Allcock, H. R. *Macromolecules* **2000**, *33*, 4307.
[74] Allcock, H. R.; Hartle, T. J.; McIntosh, M. B.; Sunderland, N. J.; Prange, R.; Taylor, J. P. PTC Int. Appl. WO, 00 71,589 (2000), Chem. Abstr. 134, 17876 (2001), assigned to The Penn State Research Foundation.
[75] Allcock, H. R.; Hartle, T. J.; McIntosh, M. B.; Sunderland, N. J.; Prange, R.; Taylor, J. P. U.S. US, 6,339,166 (2002), Chem. Abstr. 134, 17876 (2001), assigned to The Penn State Research Foundation.
[76] Gleria, M.; Minto, F.; Po', R.; Cardi, N.; Fiocca, L.; Spera, S. *Macromol. Chem. Phys.* **1998**, *199*, 2477.
[77] Gleria, M.; Minto, F.; Galeazzi, A.; Po', R.; Cardi, N.; Fiocca, L.; Spera, S. *Phosphorus, Sulfur, and Silicon* **1999**, *144-146*, 201.
[78] Gleria, M.; Minto, F.; Bertani, R.; Tiso, B.; Po', R.; Fiocca, L.; Lucchelli, E.; Giannotta, G.; Cardi, N. *Phosphorus Res. Bull.* **1999**, *10*, 730.
[79] Pò, R.; Fiocca, L.; Giannotta, G.; Lucchelli, E.; Cardi, N.; Minto, F.; Fambri, L.; Gleria, M. *Phosphorus, Sulfur and Silicon Relat. Elem.* **2001**, *168*, 269.
[80] Gleria, M.; Minto, F.; Tiso, B.; Bertani, R.; Tondello, E.; Pò, R.; Fiocca, L.; Lucchelli, E.; Giannotta, G.; Cardi, N. *Designed Monomers and Polymers* **2001**, *4*, 219.
[81] Bertani, R.; Fambri, L.; Fiocca, L.; Giannotta, G.; Gleria, M.; Po', R.; Scalabrin, S.; Tondello, E.; Venzo, A. *J. Inorg. Organomet. Polym.*, submitted.
[82] Bertani, R.; Boscolo-Boscoletto, A.; Dintcheva, N.; Ghedini, E.; Gleria, M.; La Mantia, F.; Pace, G.; Pannocchia, P.; Sassi, A.; Scaffaro, R.; Venzo, A., *Des. Monom. Polym.*,submitted.
[83] Aoi, K.; Okada, M. *Prog. Polym. Sci.* **1996**, *21*, 151.
[84] Meyers, A. I.; Temple, D. L. *J. Am. Chem. Soc.* **1970**, *92*, 6644.
[85] Lach, C.; Müller, P.; Frey, H.; Mülhaupt, R. *Macromol. Rapid Commun.* **1997**, *18*, 253.
[86] Lach, C.; Hanselmann, R.; Frey, H.; Mülhaupt, R. *Macromol. Rapid Commun.* **1998**, *19*, 461.
[87] Loontjens, T.; Belt, W.; Stanssens, D.; Weerts, P. *Macromol. Chem., Macromol. Symp.* **1993**, *75*, 211.
[88] Loontjens, T.; Belt, W.; Stanssens, D.; Weerts, P. *Polym. Bull.* **1993**, *30*, 13.
[89] Baker, W. E.; Saleem, M. *Polymer* **1987**, *28*, 2057.
[90] Vainio, T.; Hu, G. H.; Lambla, M.; Seppälä *J. Appl. Polym. Sci.* **1996**, *61*, 843.
[91] Vocke, C.; Anttila, U.; Heino, M.; Hietaoja, P.; Seppälä, J. *J. Appl. Polym. Sci.* **1998**, *70*, 1923.
[92] Frump, J. A. *Chem. Rev.* **1971**, *71*, 483.
[93] Kobayashi, S. *Prog. Polym. Sci.* **1990**, *15*, 751.
[94] Gant, T. G.; Meyers, A. I. *Tetrahedron* **1994**, *50*, 2297.
[95] Culberston, B. M. *Prog. Polym. Sci.* **2002**, *27*, 579.
[96] Chang, J. Y.; Park, P. J.; Han, M. J. *Macromolecules* **2000**, *33*, 321.
[97] Chang, J. Y.; Park, P. J.; Han, M. J.; Chang, T. *Phosphorus, Sulfur and Silicon Relat. Elem.* **1999**, *144-146*, 197.
[98] Kobayashi, S.; Mizutani, T.; Saegusa, T. *Makromol. Chem.* **1984**, *185*, 441.
[99] Giannotta, G.; Po', R.; Fiocca, L.; Cardi, N.; Lucchelli, E.; Braglia, R.; Fambri, L.; Pegoretti, A.; Minto, F.; Gleria, M. *Phosphorus, Sulfur and Silicon Relat. Elem.* **2001**, *169*, 263.
[100] *Reactive Extrusion. Principles and Practice*; Xanthos, M., Ed.; HANSER Publ.: Munich, 1992.
[101] Seo, K. S.; Cloyd, J. D. *J. Appl. Polym. Sci.* **1991**, *42*, 845.
[102] Giannotta, G.; Po', R.; Cardi, N.; Tampellini, E.; Occhiello, E.; Garbassi, F.; Nicolais, L. *Polym. Eng. Sci* **1994**, *34*, 1219.
[103] Lukacs, A. Eur. Pat. Appl. EP, 313,863 (1989), Chem. Abstr. 111, 154585s (1989), assigned to Hercules Inc.
[104] Fantin, G.; Medici, A.; Fogagnolo, M.; Pedrini, P.; Gleria, M.; Bertani, R.; Facchin, G. *Eur. Polym. J.* **1993**, *29*, 1571.
[105] McAdams, L. V.; Gannon, J. A. In *Encyclopedia of Polymer Science and Engineering*; Mark, H. F., Bikales, N. M., Overberger, C. G., Menges, G., Eds.; Wiley: New York, USA, 1986, Vol. 6, p 322.

New Functional Inorganic Polymers Containing Phosphorus

Derek P. Gates, Chi-Wing Tsang, Vincent A. Wright, Mandy Yam*

Department of Chemistry, University of British Columbia, 2036 Main Mall, Vancouver, British Columbia, Canada, V6T 1Z1
E-mail: dgates@chem.ubc.ca

Summary: The C=C bond plays numerous roles in polymer science. This moiety is used as a precursor to polymers by addition polymerization and has been incorporated into π-conjugated polymers. The addition polymerization reaction has been extended to P=C bonds and the first example of a poly(methylenephosphine) has been prepared. The new macromolecule is of moderate molecular weight (ca. 10^4 g/mol) and the oxidized polymers are air-stable. Poly(p-phenylenephosphaalkene), the first π-conjugated polymer containing P=C bonds in the backbone, has been prepared. The UV/Vis spectrum of this polymer shows a red shift in λ_{max} when compared with molecular model systems.

Keywords: addition polymerization; conjugated polymers; inorganic polymers; step-growth polymerization

Introduction

The development of synthetic methodologies to prepare new polymeric materials with novel structures and properties is a challenging frontier in chemistry.[1] Most known polymers contain backbones composed of combinations of carbon, nitrogen and oxygen, and their properties are tailored by structural modification of the side-group or main-chain architecture. The incorporation of inorganic elements into macromolecules is known to lead to materials possessing unique properties not obtainable by modification of known organic macromolecules. In principle, it should be possible to prepare a wide range of polymers with interesting structures and properties using various combinations of main group elements and/or transition metals. A limitation which has slowed the development of inorganic polymers has been the lack of suitable synthetic methods for their preparation.

The carbon-carbon double bond in alkenes is one of the most versatile functional groups in chemistry. The numerous transformations of this functionality form the basis for many important

© 2003 WILEY-VCH Verlag GmbH & KGaA, Weinheim CCC 1022-1360/00/$ 17,50+.50/0

industrial and academic pursuits. In polymer science, molecules containing C=C bonds are essential precursors to many commodity materials through addition polymerization processes. For example, macromolecules such as polyethylene, polypropylene, polystyrene and poly(acrylates) are prepared using this route. Recently, the incorporation of the C=C functionality into π-conjugated polymers has received significant attention due to the remarkable electronic properties exhibited by these materials.[2] For example, doped polyacetylene shows metallic conductivity and poly(p-phenylenevinylene) is an electroluminescent material.

Despite the remarkable diversity exhibited by the C=C bond in polymer synthesis, there is a virtual absence of research into using heteroatom containing (p-p)π bonds in polymer chemistry. We have embarked on a research program directed towards: (i) extending addition polymerization to other E=E' bonds (where E and E' are p-block elements); and (ii) incorporating E=E' bonds into π-conjugated macromolecules. Herein, a review of our recent work in these areas will be provided.

Addition Polymerization

The addition polymerization of olefins is the most important process for the preparation of commodity polymers. The driving force for the addition polymerization reaction is the thermodynamic preference of two sigma (σ) bonds in the polymer over a sigma (σ) plus a pi (π) bond in the alkene. This renders the process highly exothermic. For styrene, the enthalpy of polymerization (ΔH_p) is –73 kJ/mol.[3] The stability of a two σ bonds over a σ bond plus a π bond is not unique to olefins. Most element-element combinations of the heavier p-block elements (n>2) have a thermodynamic preference to form compounds with a fully sigma bonded structure rather than one containing (p-p)π bonds. Thus, by analogy to olefins, the polymerization of compounds possessing E=E' bonds should also be thermodynamically favourable.

The key feature of alkenes that makes them suitable for polymerization is that, in addition to their thermodynamic instability with respect to polyolefin, there is a large activation barrier to this reaction. This kinetic stability allows olefins to be isolated, purified and later treated with an initiator for polymerization to obtain high molecular weight polymers. Unfortunately, double bonds containing elements of the second and subsequent rows lack kinetic stability and it was long believed that heavier elements could not form stable compounds with multiple bonds involving pπ-pπ overlap.[4] For example, the structure of phosphorus is P_4 rather than P_2 and silicon dioxide (SiO_2) is polymeric whereas its congener CO_2 is molecular.

In the late seventies and early eighties methods were developed to kinetically stabilize (p-p)π bonds involving the heavier p-block elements. The employment of sterically bulky groups imposes a large activation barrier to oligomerization and facilitated the preparation of numerous compounds with (p-p)π bonds (see Figure 1). In particular, early breakthroughs in this area involved the isolation of compounds with P=N,[5] P=C,[6, 7] P≡C,[8] Si=C,[9] Si=Si[10] and P=P[11] bonds. Today, the development of multiply bonded systems containing the heavier main group elements remains an active area of investigation.[12] Like olefins, these systems are kinetically stable but thermodynamically unstable with respect to polymerization. Therefore, by employing the right combination of bulky substituents, the compounds in Figure 1 are potential monomers for the synthesis of inorganic polymers using addition polymerization.

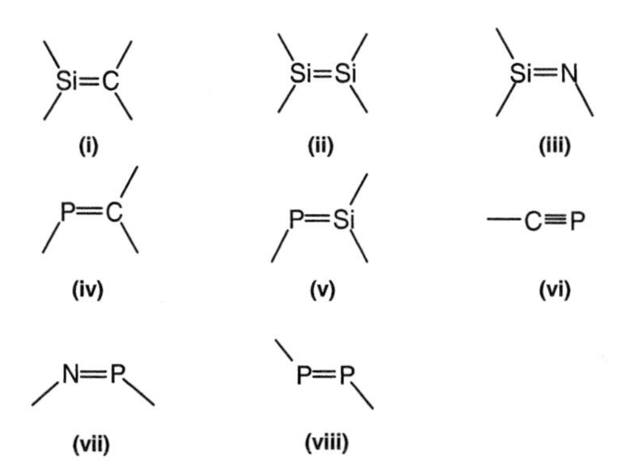

Figure 1. Examples of known systems which possess stable multiple bonds involving the heavier elements of Groups 14 and 15.

Addition Polymerization of Phosphaalkenes

As a starting point to the investigation of addition polymerization for inorganic multiple bonds, we chose to investigate the polymerization of the well-established phosphaalkenes (or methylenephosphines).[13] We have been exploring the reactions of kinetically stable phosphaalkenes with potential initiators for polymerization. Our initial studies focused on the hindered phosphaalkene (**1**) which possesses the very bulky supermesityl (Mes*=2,4,6-tri-tert-butylphenyl) substituent on phosphorus and hydrogen substituents on carbon. Compound **1** was chosen because large substituents would only be found on every second atom in a polymer which may hinder cycloaddition reactions and favour chain formation. Phosphaalkene **1** was conveniently prepared from supermesitylphosphine and CH_2Cl_2 in the presence of KOH following the literature procedure.[14]

$$\text{Mes*PH}_2 + \text{CH}_2\text{Cl}_2 \xrightarrow{\text{KOH}} \text{Mes*P=CH}_2 \; (\mathbf{1})$$

The stoichiometric reaction of **1** with Lewis and protic acids was explored as a means to understand the possible initiation step in a cationic polymerization reaction.[15] Two possible modes of coordination were considered. Based on electronegativity arguments it is possible that electrophiles would add to the double bond at carbon generating a phosphenium ion **2**. This would represent the desired initiation step in a polymerization reaction. Alternatively, due to the close energy of the (p-p)π orbital and the lone pair on phosphorus, the electophilic reagent may simply form a coordination adduct **3**. There is precedence for this mode of coordination from the reaction of the phosphaalkene ^{t}BuP=C(^{t}Bu)H with $AlCl_3$.[16]

$$\text{Mes*}\overset{\oplus}{\text{P}}\text{–CH}_2\text{–}\overset{\ominus}{\text{GaCl}_3} \; (\mathbf{2}) \qquad \text{Cl}_3\overset{\ominus}{\text{Ga}}\text{–}\overset{\oplus}{\text{P}}(\text{Mes*})\text{=CH}_2 \; (\mathbf{3})$$

Interestingly, when **1** was treated with the Lewis acid $GaCl_3$, neither the phosphenium ion **2** nor the adduct **3** were observed.[15] An X-ray crystallographic analysis of the product **4** revealed that an intramolecular C-H activation of one of the *ortho*-^{t}Bu groups of Mes* had occurred. The organometallic compound **4** may be envisaged as a coordinated ylide. Mechanistic studies of this reaction and a related reaction of **1** with triflic acid showed that the mechanism involved initial formation of the adduct **3** followed by rearrangement to **2** and subsequent rapid intramolecular oxidative addition of the C-H bond to the divalent phosphenium centre giving **4**.

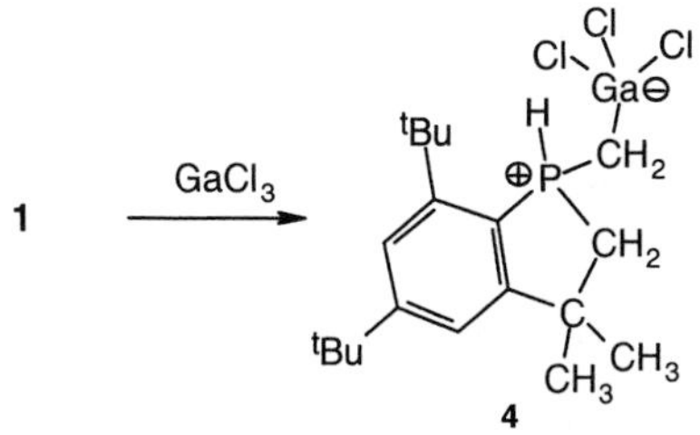

Due to the problem of C-H activation when employing the supermesityl substituent, other stable phosphaalkenes with minimal steric protection were sought which would not show the same propensity for intramolecular C-H activation. The phosphaalkene **5**, which contains the smaller mesityl substituent at phosphorus, attracted our attention. This compound was first prepared in 1978 from the base-induced dehydrochlorination of $MesP(Cl)C(H)Ph_2$.[7] We followed another procedure which involves the reaction of bis(trimethylsilyl)mesitylphosphine with benzophenone in the presence of a catalytic quantity of NaOH.[17] Compound **5** was vacuum distilled and the residue from distillation contained a gummy material which solidified upon cooling. The ^{31}P, ^{1}H and ^{13}C NMR spectra of the residue were consistent with a polymeric material containing trivalent phosphines. Poly(methylenephosphine) **6** was purified by precipitation with hexanes to remove molecular impurities.[18] The molecular weight of this new polymer was estimated to be about 14,000 g/mol using GPC vs. polystyrene.

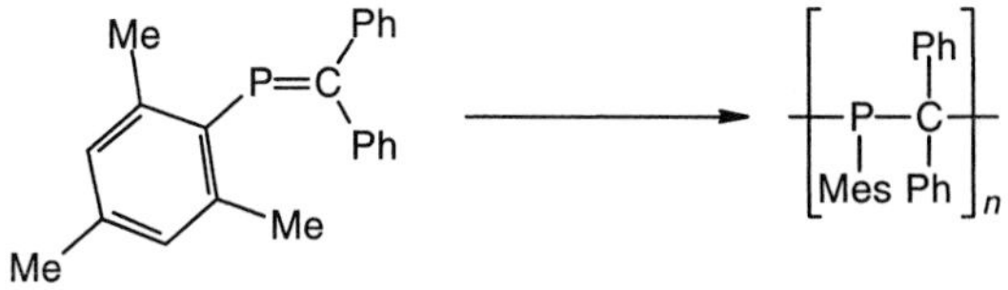

Polymer **6** has proven to be easily functionalized.[18] An air stable phosphine oxide polymer **7** was prepared rapidly and quantitatively from **6** using hydrogen peroxide. This new material shows no weight loss until 320 °C when analyzed using TGA. The phosphine sulfide polymer **8** was prepared from the reaction of **6** with elemental sulfur. The new polymers **7** and **8** had molecular weights comparable to that of **6**.

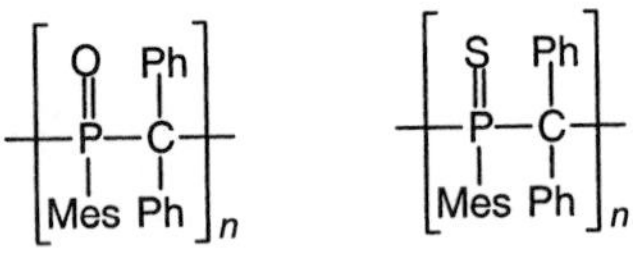

In experiments with potential initiators, we found that the polymerization of **5** can be initiated at 150 °C using anionic initiators such as MeLi or BuLi.[18] The poly(methylenephosphine) **6** was isolated in good yields and exibited spectroscopic properties identical to those for the polymer obtained by thermolysis. Reasonable molecular weights (ca. 5,000 – 10,000 g/mol) were obtained for several experiments using methyllithium as an initiator.

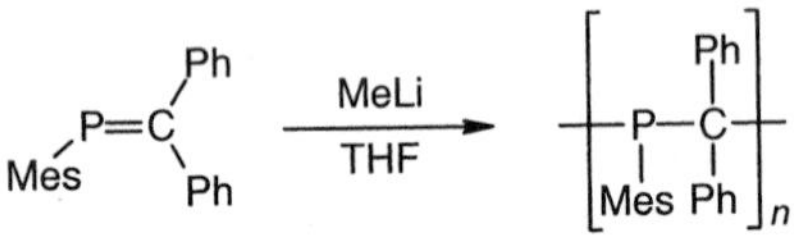

Incorporation of P=C Bonds into π-Conjugated Polymers

Polymers containing π-conjugated backbones are currently attracting significant attention due to their interesting properties. We have been investigating the possible incorporation of inorganic multiple bonds into π-conjugated polymers by the replacement of the C=C bonds in poly(p-phenylenevinylene) with P=C bonds. The synthetic strategy employed involves the condensation of a bis(trimethylsilyl)arylphosphine with an acid chloride followed by [1,3]-silatropic rearrangement as the key P=C bond forming step.[6, 19, 20] The thermal condensation reaction of the silylated 1,4-diphosphinobenzene **9** and the bis(acid chloride) **10** gave poly(p-phenylenephosphaalkene) **11** which was purified by precipitation with hexanes.[21] A molecular weight of ca. 6,300 g/mol was estimated for one sample of the polymer **11** by using end-group analysis. The ratio of *Z*-isomer (phenylenes are *trans*) to *E*-isomer (phenylenes are *cis*) for several samples of polymer ranged from 1.05-1.14.

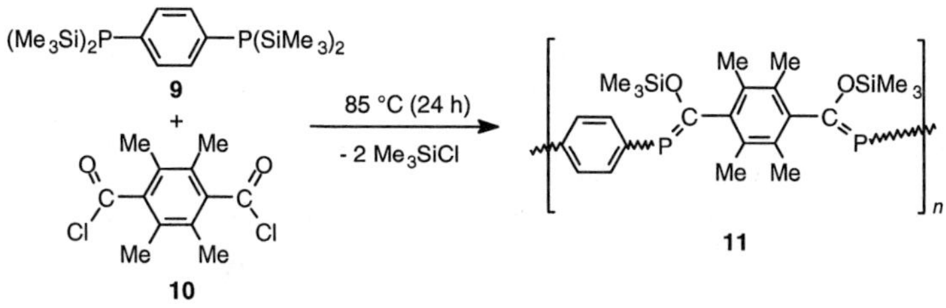

Two molecular models for the polymer were prepared using the route described for the synthesis of **11**. Compounds **12** and **13** were fully characterized using NMR spectroscopy, mass spectrometry and microanalysis.[21] The *Z*/*E* ratios were 0.85 and 1.27, respectively. The electronic structure of the polymer and the two model compounds was investigated using UV/Vis spectroscopy. Absorbances were observed for **12** (λ_{max} = 310 nm) and **13** (λ_{max} = 314 nm), whilst the polymer **11** which had a molecular weight of 6,300 g/mol showed a significant red-shift in its absorbance maximum (λ_{max} = 334 nm).[21] This red-shift suggests some degree of π-conjugation through the phenylene and P=C moieties in the polymer.

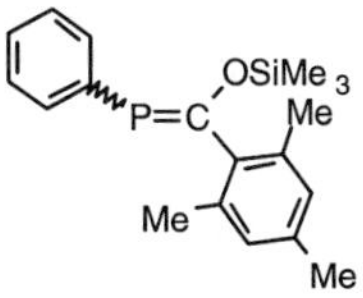

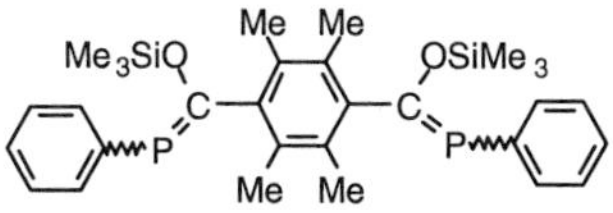

Summary

In closing, we have shown that the addition polymerization reaction, a general method for the polymerization of alkenes, can be extended to P=C bonds. A new air-stable class of macromolecule containing phosphines in the main chain has been prepared from a phosphaalkene. The first π-conjugated polymer containing alternating phenylene and phosphaalkene moieties has been prepared and is an inorganic analogue of poly(p-phenylenevinylene).

[1] See, for example: R. D. Archer, "*Inorganic and Organometallic Polymers*", Wiley-VCH, New York, 2001; I. Manners, *Angew. Chem. Int. Ed. Engl.* **1996**, *35*, 1602; H. R. Allcock, *Adv. Mater.* **1994**, *6*, 106; J. E. Mark, H. R. Allcock, R. West, "*Inorganic Polymers*", Prentice Hall, New Jersey, 1992.

[2] For reviews, see: "*Handbook of Conducting Polymers*", T. A. Skotheim, R. L. Elsenbaumer, J. R. Reynolds, Eds., Dekker, New York, 2 Ed., 1998; A. Kraft, A. C. Grimsdale, A. B. Holmes, *Angew. Chem. Int. Ed. Engl.* **1998**, *37*, 402; W. J. Feast, J. Tsibouklis, K. L. Pouwer, L. Groenendaal, E. W. Meijer, *Polymer*, **1996**, *37*, 5017.

[3] G. Odian, "*Principles of Polymerization*", Wiley & Sons, New York, **1991**.

[4] This is sometimes termed "the double bond rule" and it still holds true for simple systems. For recent discussions of this rule and its history, see: P. Jutzi, *Angew. Chem. Int. Ed.* **2000**, *39*, 3797; P. Power, *Chem. Rev.* **1999**, *99*, 3463.

[5] E. Niecke, W. Flick, *Angew. Chem. Int. Ed. Engl.* **1973**, *12*, 585.

[6] G. Becker, *Z. Anorg. Allg. Chem.* **1976**, *423*, 242.

[7] C. Klebach, R. Lourens, F. Bickelhaupt, *J. Am. Chem. Soc* **1978**, *100*, 4886.

[8] G. Becker, G. Gresser, W. Uhl, *Z. Naturforsch.* **1981**, *36b*, 16.

[9] A. G. Brook, F. Abdesaken, B. Gutekunst, G. Gutekunst, R. K. Kallury, *J. Chem. Soc. Chem. Commun.* **1981**, 191.

[10] R. West, M. J. Fink, *Science* **1981**, *214*, 1343.

[11] M. Yoshifuji, I. Shima, N. Inamoto, *J. Am. Chem. Soc.* **1981**, *103*, 4587.

[12] For reviews, see: P. P. Power, *Chem. Rev.* **1999**, *99*, 3463; P. Jutzi, *Angew. Chem. Int. Ed.* **2000**, *39,* 3797; P. P. Power, *J. Chem. Soc., Dalton Trans.* **1998**, 2939; M. Yoshifuji, *J. Chem. Soc., Dalton Trans.* **1998**, 3343; M. Driess, H. Grützmacher, *Angew. Chem. Int. Ed. Engl.* **1996**, *35*, 828; N. C. Norman, *Polyhedron*, **1993**, *12*, 2431; E. Niecke, D. Gudat, *Angew. Chem. Int. Ed. Engl.* **1991**, *30*, 217; M. Regitz, *Chem. Rev.* **1990**, *90*, 191; R. West, *Angew. Chem. Int. Ed. Engl.* **1987**, *26*, 1201; A. H. Cowley, *Polyhedron*, **1984**, *3*, 389.

[13] For reviews, see: K. B. Dillon, F. Mathey, J. F. Nixon, "*Phosphorus: The Carbon Copy*", Wiley, New York, 1998; L. Weber, *Eur. J. Inorg. Chem.* **2000**, 2425; A. C. Gaumont, J. M. Denis, *Chem. Rev.* **1994**, *94*, 1413; F. Mathey, *Acc. Chem. Res.* **1992**, *25*, 90; R. Appel, in "*Multiple Bonds and Low Coordination in Phosphorus Chemistry*", M. Regitz, O. J. Scherer, Eds. Thieme, Stuttgart, 1990.

[14] R. Appel, C. Casser, M. Immenkeppel, F. Knoch, *Angew. Chem. Int. Ed. Engl.* **1984**, *23*, 895.

[15] C.-W. Tsang, C. A. Rohrick, T. S. Saini, B. O. Patrick, D. P. Gates, *Organometallics* **2002**, *21*, 1008.

[16] E. Niecke, E. Symalla, *Chimia* **1985**, *39*, 320.

[17] O. Mundt, G. Becker, W. Uhl, *Z. Anorg. Allg. Chem.* **1986**, *540/541*, 319.

[18] C.-W. Tsang, M. Yam, D. P. Gates, *J. Am Chem. Soc.* **2003**, *125*, 1480.

[19] G. Becker, *Z. Anorg. Allg. Chem.* **1977**, *430*, 66.

[20] G. Becker, O. Mundt, *Z. Anorg. Allg. Chem.* **1978**, *443*, 53.

[21] V. A. Wright, D. P. Gates, *Angew. Chem. Int. Ed.* **2002**, *41*, 2389.

Synthesis and Properties of New Phosphorus-Functionalized Bithiophene Materials

Thomas Baumgartner

Institut für Anorganische und Analytische Chemie, Johannes Gutenberg-Universität, Duesbergweg 10-14, 55099 Mainz, Germany
E-mail: baumga@mail.uni-mainz.de

Summary: The synthesis and reactivity of the novel dithieno[3,2-*b*:2',3'-*d*]-phosphole system and its potential use in polymeric sensory materials was investigated. Due to the nucleophilic nature of the phosphorus atom, these materials were found to be easily tunable. Classical reactions at the phosphorus center were performed to modify the electronic structure of the system and the corresponding changes were detected by UV/Vis and fluorescence spectroscopy. Depending on the oxidation state of the central phosphorus atom or its substitution pattern, the dithieno[3,2-*b*:2',3'-*d*]-phospholes show varying wavelengths for absorption and emission allowing to distinguish between different compounds by means of optical spectroscopy.

Keywords: fluorescence; phospholes; polymeric sensors; thiophenes

Introduction

The incorporation of thiophene moieties into oligomeric or macromolecular systems has recently become a topic of significant interest due to their highly intriguing properties.[1] A variety of such conjugated polymers exhibit a striking ability; ultrafast energy transfers that leads to a metal-like conductivity. The resulting materials therefore show great potential for applications in electronic devices such as organic light-emitting diodes (OLEDs), photovoltaic cells, flat panel displays or polymeric sensors.[2] However, the introduction of the corresponding electronic properties is often connected with synthetic difficulties that would significantly increase the production costs of the resulting devices. This could keep them from industrial applications. Since the optoelectronic properties of, for example, the classic poly(phenylenevinylidene) (PPV) system is limited to certain emissions, doping is very important in order to increase the potential for applications.[3] A great deal of attention has therefore been focused on the tuning of the

© 2003 WILEY-VCH Verlag GmbH & KGaA, Weinheim CCC 1022-1360/00/$ 17,50+.50/0

optoelectronic structure of the polymeric systems in order to modify their electronic nature (i.e. band gap) in such a way that it suits the targeted application.[4]

As part of our research on the development of processable, electronically active, macromolecular inorganic materials we have focussed on the novel dithieno[3,2-*b*:2',3'-*d*]phosphole moiety. This system should allow selective tuning of the electronic properties of the materials by functionalization of the central phosphorus atom. Like almost no other element, phosphorus is particularly suited to act as a bridging element in the dithieno system due to its nucleophilic nature. Its ability to react with oxidizing agents such as oxygen or sulfur, its Lewis basicity which allows reactions with Lewis acids such as BH_3 in addition to the potential for coordination to transition metals offers a unique variety of synthetically facile possibilities to modify the electronic properties of the dithienophosphole materials. In this paper we report on our initial studies in this area and describe model compounds for prospective polymeric materials, their chemical modification, and their electronic properties.

Results and Discussion

We have targeted the investigation of the novel dithieno[3,2-*b*:2',3'-*d*]phosphole system (**1**) since it combines two favorable properties; excellent π-conjugation between the two thiophene subunits and 'guided' electronic doping through the phosphorus center. It should be mentioned in this context that the number of dithieno systems is very limited. The only known examples involve carbon, nitrogen, silicon and sulfur as bridging elements (**2**).[5] The use of phosphorus has only been reported in compound **3**[6]; σ^3-λ^3 phosphorus-based systems (**1**) on the other hand have been completely unknown to date.

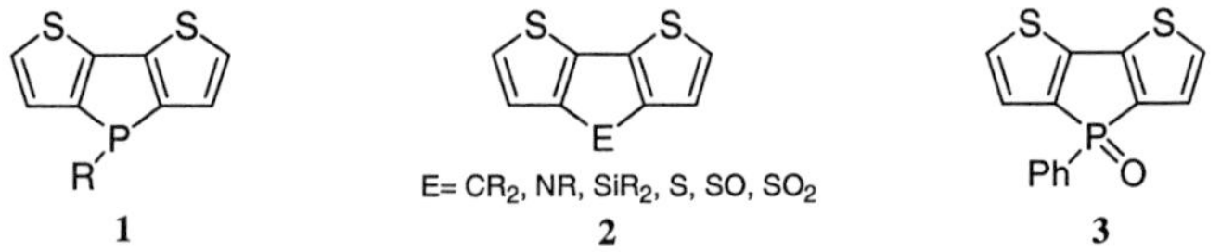

S S P R
1

S S E
E= CR_2, NR, SiR_2, S, SO, SO_2
2

S S P Ph O
3

The advantageous electronic features of the phosphorus center in phospholes have recently been pointed out by Réau and coworkers who incorporated the phosphole moiety into extended thiophene containing π-conjugated systems.[7] In the context of their work they were able to show that the reduced aromatic character (compared to furan, pyrrole or thiophene) leads to intriguing electronic properties. Due to the pyramidalization of the phosphorus center, an efficient orbital interaction with the conjugated π-system is reduced. As a result, the lone pair at the phosphorus atom only functions as a dopant for the phosphole π-system. The strength of this doping effect depends on the geometry of the phosphorus center. The effect can easily be switched from *n*-type (electron donor) to *p*-type (electron acceptor) by formal oxidation from P(III) to P(V). Réau's phenyl, pyridine or thiophene-substituted materials **4** have proven to be useful synthons for the synthesis of polymeric π-conjugated systems whose electronic properties can be easily tuned over a wide range by performing simple chemical modifications.[7] However, in their case the rings are connected through solely a single bond, which allows for twisting between the central ring and the adjacent units that can reduce the π-conjugation.

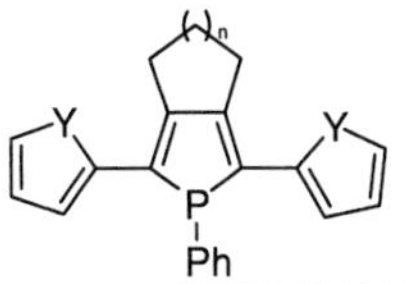

n = 3,4; Y = CH=CH; CH=N, S

4 **5**

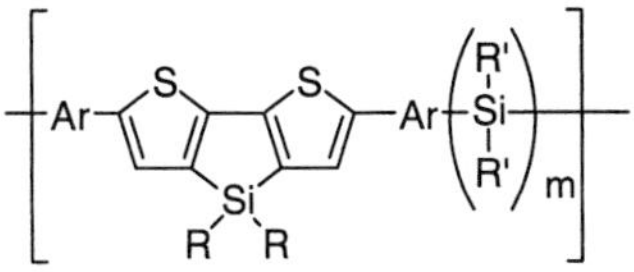

The incorporation of the phosphole moiety into a rigidified tricyclic dithieno system on the other hand should lead to a significantly higher degree of π-conjugation since the annulation of aromatic rings has been found to be a powerful approach to tune the band gap of conjugated polymers.[8] This is further supported by a recent theoretical investigation which has shown that thiophene-based, fused tricyclic polymers show a much more favorable band gap than the related polythiophenes.[9] In this context, Ohshita and coworkers have shown that the incorporation of silicon into the rigid dithieno system (**2**, R = SiR_2) reduces the HOMO-LUMO gap significantly due to the extended conjugated π-system and allows for application as hole transport materials (**5**) in electro luminescent (EL) devices.[10] Taking into account the result of the above-mentioned studies it seemed necessary for us to incorporate an electron-acceptor into the system as well to obtain the targeted HOMO-LUMO separation. We decided on silyl groups, since they can be introduced easily and exhibit suitable acceptor properties by 'negative hyperconjugation'.[11]

Applying Ohshita's strategy for dithienosiloles[10] to our system, we were able to synthesize the novel dithieno[3,2-*b*:2',3'-*d*]phospholes with the desired silyl functionality. Starting from the tetrabromo-dithiophene **6**, the trimethylsilyl groups were introduced first by reaction with *n*-butyllithium followed by addition of Me_3SiCl. Subsequent lithiation of **7** with *n*-butyllithium followed by quenching with $RPCl_2$ (R= Ph, 4-*t*Bu-C_6H_4, *t*Bu) cleanly afforded the first dithieno[3,2-*b*:2',3'-*d*]phospholes **8**:

Br S Br S Br Br **6** — 1) nBuLi 2) Me_3SiCl → Me_3Si S Br S $SiMe_3$ Br **7** — 1) nBuLi 2) $RPCl_2$ → Me_3Si S S $SiMe_3$ P R **8**

Compounds **8** exhibit ^{31}P NMR shifts (**8a**: R = Ph; δ = -25.0 ppm; **8b**: R = 4-*t*Bu-C_6H_4; δ = -25.9 ppm; **8c**: R = *t*Bu; δ = 2.8 ppm) which are significantly high field-shifted from related phospholes, e.g. those reported by Réau et al. ($\delta^{31}P$ = 11 – 45 ppm)[7]. This indicates a higher electron density at the phosphorus atom likely due to the increased degree of π-conjugation in **8**. To our satisfaction, compounds **8** showed a strong blue fluorescence, supporting the desired optical properties (*vide infra*).

In order to alter the electronic structure of the dithieno[3,2-*b*:2',3'-*d*]phosphole system, we performed classical reactions that utilized the nucleophilic nature of the σ^3-λ^3 phosphorus center. Reaction of **8a** with bis(trimethylsilyl)peroxide or sulfur resulted in the clean formation of the phosphine oxide **9** and the sulfide **10**, respectively. The oxidized species **9** exhibits a ^{31}P NMR shift at δ = 17.2 ppm which shows the expected low field shift for phosphine oxides in comparison to **8**. However, **9** is still high field-shifted from related systems ($\delta\,^{31}P \approx 45$ ppm)[7]. The same is true for the sulfurized species **10** ($\delta\,^{31}P$ = 23.6 ppm; c.f. $\delta\,^{31}P \approx 53$ ppm [7]).

In order to prove its ability to act as a ligand for transition metals, dithienophosphole **8a** was reacted with $Pd(COD)Cl_2$, $W(CO)_5(THF)$ and $Fe_2(CO)_9$. All three reactions were performed at room temperature and afforded the corresponding complexes **11-13** in good yields. In contrast to the inorganic compounds **8-10**, whose solutions were almost colorless, the solutions of these organometallic complexes were significantly colored (**11**: dark orange, **12**: red, **13**: yellow) indicating electronic interactions of the ligands with the metal centers. The ^{31}P NMR data of the metal complexes again showed the expected low field shift but were, as before, relatively high field-shifted in comparison to related known complexes (**11**: $\delta\ ^{31}P = 1.5$ ppm[12]; **12**: $\delta\ ^{31}P = -7.9$ ppm,[7,13] $^{1}J(P,W) = 228$ Hz; **13**: $\delta\ ^{31}P = 49.9$ ppm[14]).

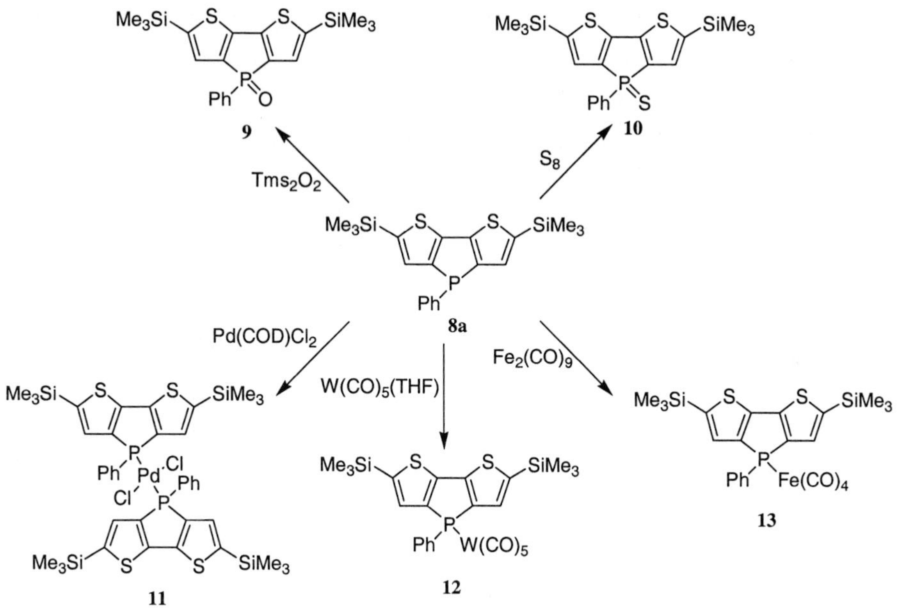

Optical Properties

As previously mentioned, the dithienophospholes **8** exhibit a strong blue fluorescence. This feature opens up a great potential for applications since the detection of fluorescence is rapidly becoming a method of choice in materials science and optoelectronics in addition to medical or environmental studies.[15] Due to its highly sensitive, selective and safe nature it allows for example the observation of living processes or electronic transitions in progress. In materials the presence of luminescence can reflect the delocalization and polarization of the electronic structure.[16] Therefore, changing this electronic structure should have an effect on the luminescence properties. This can be easily detected for the case of fluorescence phenomena, and would allow for the application of dithienophospholes as sensory materials once a library for different derivatives is created.

For this reason we investigated the UV/Vis and fluorescence properties of **8a**, **9**, **10** and **11**, four representative systems involving either a σ^3-λ^3 phosphorus center, one of two different σ^4-λ^5 phosphorus centers, or a transition metal-substituted phosphorus center. The UV/Vis spectra in CH_2Cl_2 show different absorptions for the four compounds, which are partially dependant on the oxidation state of the phosphorus center. As a result **8a** and **11** show similar absorptions at ca. 344 nm with an additional shoulder at 425 nm for **11** presumably indicating the optical features of the transition metal center. Likewise, the oxidized σ^4-λ^5 compounds **9** and **10** show similar absorptions to each other at ca. 373 nm. These initial results indicate that the oxidation state of the phosphorus atom in related systems could ultimately be detected by UV/Vis analysis.

More detailed information was obtained from the fluorescence spectroscopic data. The fluorescence spectra for **8a**, **9**, **10** and **11** (Figures 1–4) in CH_2Cl_2 not only revealed different *wavelengths* for absorption and emission but correspondingly different *intensities* for the fluorescence which may prove to be helpful for the detection of chemically closely related compounds.

As expected, the four compounds show different absorption/emission wavelengths reflecting the electronic nature of the different systems. The values for the σ^3-λ^3 phosphorus-containing system **8a** appear at lower wavelengths [x (excitation) = 344 nm; m (emission) = 422 nm] which is coherent with the electron-donating nature of the P-lone pair. Oxidation of the phosphorus atom causes switching from electron donor to electron acceptor and hence to a energetically more favorable system (**9/10**: x = 374 nm, m = 460 nm). The values for the palladium complex **11** (x = 384 nm, m = 470 nm) can be explained by electronic interactions between the dithienophosphole ligand and the transition metal center.

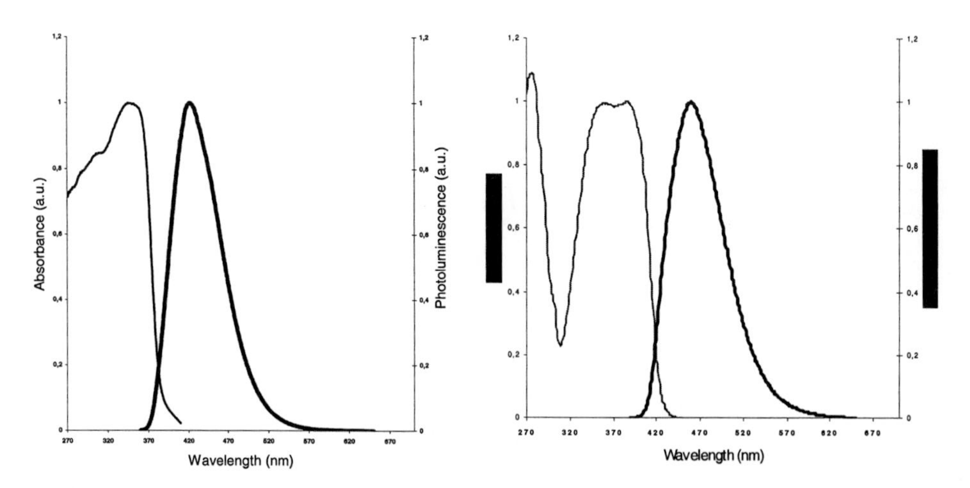

Figure 1 (left): Absorption (left) and photoluminescence (right) spectrum of a $5x10^{-5}$ M solution of **8a** in CH_2Cl_2.

Figure 2 (right): Absorption (left) and photoluminescence (right) spectrum of a $1x10^{-4}$ M solution of **9** in CH_2Cl_2.

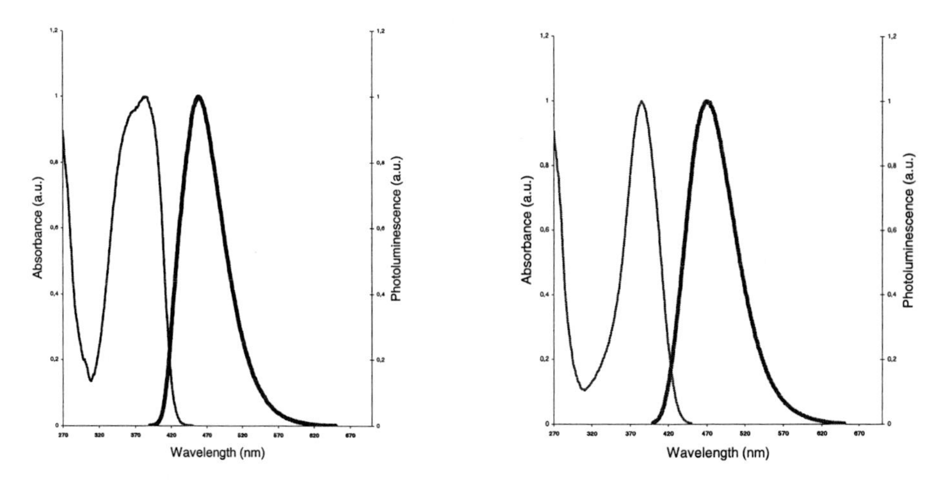

Figure 3 (left): Absorption (left) and photoluminescence (right) spectrum of a $1x10^{-4}$ M solution of **10** in CH_2Cl_2.

Figure 4 (right): Absorption (left) and photoluminescence (right) spectrum of a $3x10^{-5}$ M solution of **11** in CH_2Cl_2.

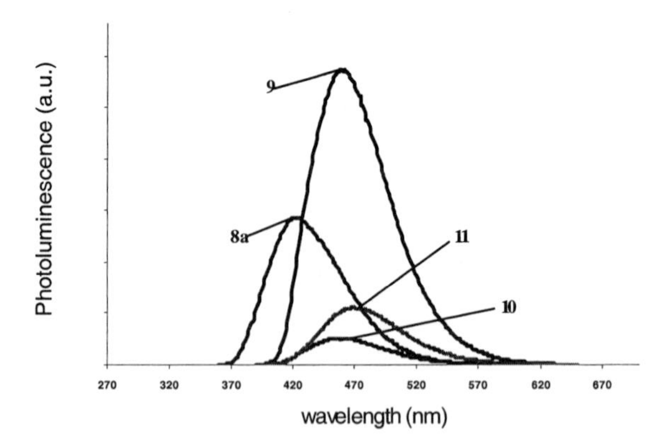

Figure 5: Photoluminescence spectra of **8a** (c = $5x10^{-5}$M), **9** (c = $1x10^{-4}$M), **10** (c = $1x10^{-4}$M) and **11** (c = $3x10^{-5}$M) - relative intensities.

Due to the similar nature of oxygen and sulfur, compounds **9** and **10** exhibit similar absorption/emission wavelengths. However, it is possible to distinguish between them, since they show significantly different fluorescence intensities (ratio approx. 5:1 for **9**:**10**) suggesting that the fluorescence phenomenon is quenched significantly by the sulfur substituent. The relative intensities (with concentrations) of **8a** – **12** are shown in Figure 5. Taking the intensity for compound **10** as a standard, the intensity for **11** is almost three times higher whereas the intensity for compounds **8a** and **9** shows the strongest fluorescence that amounts to approximately five times the intensity of **10**.

Conclusion

Our initial investigations on the novel dithieno[3,2-*b*:2',3'-*d*]-phosphole system resulted in the synthesis of a number of differently-substituted phosphole centers whose electronic properties were found to be easily tunable via simple chemical modifications at the phosphorus center. An additional tuning of this system should be possible by introduction of different exocyclic substituents at phosphorus atom, using the same synthetic strategy. The novel $\sigma^3\lambda^3$ phosphorus-based dithienophospholes, their $\sigma^4\lambda^5$ derivatives, and the corresponding organometallic complexes showed an intriguing feature - fluorescence - that opened up the detection of different species by means of optical spectroscopy. Depending on the electronic nature of the phosphorus center different wavelengths for absorption/emission and intensities were found. These results support the potential use of the novel dithieno[3,2-*b*:2',3'-*d*]phospholes as sensory materials. In the case of the strongly fluorescent compounds **8** and **9** another application as blue emitting materials seems to be possible as well.

Work in Progress

In order to achieve better processability for the use of dithienophospholes as sensory materials it would be advantageous to incorporate this moiety into macromolecular systems. Since recent investigations have shown the presence of silyl substituents is necessary for strong fluorescence[17], the incorporation of silyl centers into the polymeric system would also be highly desirable. A possible route to this goal could be realized by the following method:

HMe2Si S S SiMe2H P Ph **8d** — cat., n → Me2 Si S S Me2 Si H H P Ph n n **14**

The possibility to perform hydrosilylation reactions with the SiH-functionalized compound **8d** is currently under investigation and will be presented elsewhere.

Acknowledgments

The author would like to thank the german *Fonds der Chemischen Industrie* and Prof. Dr. J. Okuda for financial support of this work. We also wish to acknowledge U. Zmij for obtaining the UV/Vis spectra and H.J. Menges (MPI-P Mainz) for assistance with the fluorescence spectroscopy.

[1] "*Handbook of Oligo- and Polythiophenes*" (D. Fichou, Ed.), Wiley-VCH, Weinheim **1998**.
[2] see e.g.: a) *Conjugated Conducting Polymers*, (H. Kiess, Ed.), Springer-Verlag, New York **1992**, Vol.102. b) G. Tourillion, Polythiophene and its Derivatives in *Handbook of Conducting Polymers*, (T. A. Skotheim, Ed.), Marcel Dekker, New York **1986**.
[3] a) N. Hall, *Chem. Comm.* **2003**, 1. b) R. H. Friend, R. W. Gymer, A. B. Holmes, J. H. Burroughes, R. N. Marks, C. Taliani, D. D. C. Bradley, D. A. Dos Santos, J. L. Brèdas, M. Löglund, W. R. Salaneck, *Nature* **1999**, *397*, 121.
[4] see e.g.: a) F. Wudl, M. Kobayashi, A. J. Heeger, *J. Org. Chem.* **1984**, *49*, 3382. b) E. E. Havinga, W. Hoeve, H. Wynberg, *Synth. Met.* **1993**, *55-57*, 299.
[5] a) T. Benincori, V. Consomi, P. Grammatica, T. Pilati,, S. Rizzo, F. Pannicolo, R. Todeschini, G. Zotti, *Chem. Mater.* **2001**, *13*, 1665. b) D. D. Kenning, K. Ogawa, S. D. Rothstein, S. C. Rasmussen, *Polym. Mat. Sci. Eng.* **2002**, *86*, 59. c) J. Ohshita, M. Nodono, T. Watanabe, Y. Ueno, A. Kunai, Y. Harima, K. Yamashita, M. Ishikawa, *J. Organomet. Chem.* **1998**, *553*, 487. d) G. Barbarella, L. Favaretto, G. Sotgiu, L. Antolini, G. Gigli, R. Cingolati, A. Bongini, *Chem. Mater.* **2001**, *13*, 4112.
[6] J.-P. Lampin, F. Mathey, *J. Organomet. Chem.* **1974**, *71*, 239.
[7] a) C. Hay, M. Hissler, C. Fischmeister, J. Rault-Berthelot, L. Toupet, L. Nyulászi, R. Réau, *Chem. Eur. J.* **2001**, *7*, 4222. b) C. Hay, C. Fischmeister, M. Hissler, L. Toupet, R. Réau, *Angew. Chem.* **2000**, *112*, 1882; *Angew. Chem. Int. Ed.* **2000**, *39*, 1812.
[8] J. Roncali, *Chem. Rev.* **1997**, *97*, 173. b) M. Pomerantz in *Handbook of Conducting Polymers*, 2nd ed., (T. A. Skotheim, R. L. Elsenbaumer, J. R. Reynolds, Eds.), Marcel Dekker, New York **1998**, pp. 277-309.
[9] S. Y. Hong, J. M. Song, *J. Chem. Phys.* **1997**, *107*, 10607.
[10] a) J. Ohshita et al., *Organometallics* **1999**, *18*, 1453. b) J.Ohshita, T. Sumida, A. Kunai, A. Adachi, K. Sakamaki, K. Okita, *Macromolecules* **2000**, *33*, 8890. c) J. Ohshita et al., *J. Organomet. Chem.* **2002**, *642*, 137.
[11] A. Reed, P. v. R. Schleyer, *J. Am. Chem. Soc.* **1990**, *112*, 1434.
[12] a) J. J. MacDougall, J. H. Nelson, F. Mathey, J. J. Mayerle, *Inorg. Chem.* **1980**, *19*, 709. b) M. Ogasawara, K. Yoshida, T. Hayashi, *Organometallics* **2001**, *20*, 1014. c) J. Hydrio, M. Gouygou, F. Dallemer, G. G. A. Balavoine, J.-C. Daran, *J. Organomet. Chem.* **2002**, *643-644*, 19.
[13] S. Affandi, J. H. Nelson, N. W. Allcock, O. W. Howarth, E. C. Alyea, G. M. Sheldrick, *Organometallics* **1988**, *7*, 1724.
[14] J. B. M. Wit, G. T. v. Eijkel, F. J. J. de Kanter, M. Schakel, A. W. Ehlers, M. Lutz, A. L. Spek, K. Lammertsma, *Angew. Chem.* **1999**, *111*, 2716; *Angew. Chem. Int. Ed.* **1999**, *38*, 2596.
[15] G. Barbarella, *Chem. Eur. J.* **2002**, *8*, 5073.
[16] D. T. McQuade, A. E. Pullen, T. M. Swager, *Chem. Rev.* **2000**, *100*, 2537.
[17] T. Baumgartner , unpublished results.

Silole-Containing Linear and Hyperbranched Polymers: Synthesis, Thermal Stability, Light Emission, Nano-Dimensional Aggregation, and Optical Power Limiting

Jacky Wing Yip Lam,[1] *Junwu Chen,*[1] *Charles Chi Wang Law,*[1] *Han Peng,*[1] *Zhiliang Xie,*[1,2] *Kevin Ka Leung Cheuk,*[1] *Hoi Sing Kwok,*[2] *Ben Zhong Tang**[1,2]

[1]Department of Chemistry, Institute of Nano Science and Technology, and Open Laboratory of Chirotechnology of the Institute of Molecular Technology for Drug Discovery and Synthesis, Hong Kong University of Science & Technology, Clear Water Bay, Kowloon, Hong Kong, China
[2]Center for Display Research, Hong Kong University of Science & Technology, Clear Water Bay, Kowloon, Hong Kong, China

Summary: Linear polyacetylenes and hyperbranched polyphenylenes carrying 1,2,3,4,5-pentaphenylsilolyl (PS) pendants are designed and synthesized. Homopolymerization of HC≡CPS, $HC{\equiv}C(CH_2)_9OPS$, and $C_6H_5C{\equiv}C(CH_2)_9OPS$ and (co)polycyclotrimerization of $(HC{\equiv}C)_2PS$ with 1-octyne are effected by $NbCl_5$–, WCl_6–, $MoCl_5$–, and $TaCl_5$–based catalysts. High molecular weight linear (**1**–**3**) and hyperbranched polymers (**6**) are obtained in high yields (M_w up to ~70×10^3 and yield up to 85%). All the polymers are thermally stable, losing little weight when heated to 350 °C. Whereas all the polymers emit faint light when molecularly dissolved, polymers **2**, **3**, and **6** become emissive when aggregated in poor solvents or when cooled to low temperatures. Restricted intramolecular rotations of the phenyl rings upon the axes of the single bonds linked to the silole cores may be responsible for the aggregation- or cooling-induced emission. A multilayer electroluminescent device using **3** as an active layer emits a blue light of 496 nm with maximum brightness, current efficiency, and external quantum yield of 1118 cd/m^2, 1.45 cd/A, and 0.55%, respectively. Polymers **6** are non-linear optically active and strongly attenuate the optical power of intense laser pulses, whose optical limiting performances are superior to that of C_{60}, a best-known optical limiter.

Keywords: aggregation- and cooling-induced emission; hyperbranched polyphenylenes; light-emitting diode; optical limiting; organometallic polymers; polyacetylenes; polycyclotrimerization; siloles

Introduction

Siloles or silacyclopentadienes are a group of organometallic molecules that possess novel electronic and optical properties.[1],[2] We have recently observed a novel phenomenon of

© 2003 WILEY-VCH Verlag GmbH & KGaA, Weinheim CCC 1022-1360/00/$ 17,50+.50/0

aggregation-induced emission (AIE) in this group of molecules: the siloles are practically nonluminescent when molecularly dissolved but become emissive when aggregated in poor solvents or fabricated into thin films.[3],[4] The photoluminescence (PL) quantum yield (Φ_{PL}) of the silole aggregates can differ from that of its molecularly dissolved species by two orders of magnitude (>300). Utilizing this AIE property, we have fabricated silole-based light-emitting diodes (LEDs), which exhibit outstanding electroluminescence (EL) performance. The highest external quantum efficiency is 8%,[5],[6] approaching the limit of the possible.[7],[8] Siloles are thus a group of excellent organometallic molecules for LED applications. Low molecular weight compounds, however, have to be fabricated into thin films by relatively expensive techniques. One way to overcome this processing disadvantage is to make high molecular weight polymers, which can be readily processed from their solutions into thin solid films over large areas by simple spin coating or doctor blade techniques. In this paper, we report our work on synthesizing silole polymers via transition metal-catalyzed metathesis and cyclotrimerization polymerizations of acetylenes and present the novel properties of the polymers originated from their unique molecular structures.

Results and Discussion

We prepared three acetylene-silole adducts by nucleophilic substitutions of 1-chloro-1,2,3,4,5-pentaphenylsilole with a Grignard reagent ethynylmagnesium bromide and alkynyl alcohols.[9] The polymerizations of the monomers were effected by $NbCl_5$–, WCl_6–, and $MoCl_5$–Ph_4Sn catalysts, producing polymers **1–3** (Figure 1) with high molecular weights in high yields (M_w up to ~70 × 10^3 and isolation yield up to 80%) (Table 1).

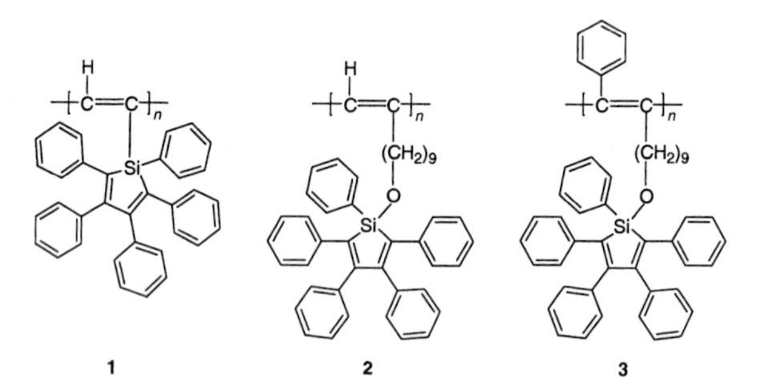

Figure 1. Linear polyacetylenes containing 1,2,3,4,5-pentaphenylsilolyl pendants.

We also prepared a diacetylene (**4**; Figure 2) by reacting 1,1-dichloro-2,3,4,5-tetraphenylsilole with ethynylmagnesium bromide.[10] Homo- and copolycyclotrimerizations of **4** with 1-octyne **5** were effected with $TaCl_5$–Ph_4Sn in toluene at room temperature, which gave completely soluble hyperbranched poly(phenylenesilolenes) **6** in high yields (Table 2). All the polymers were characterized by standard spectroscopic methods and satisfactory analysis data corresponding to their molecular structures were obtained.

Table 1. Synthesis of poly(silolylacetylenes)[a]

no.	catalyst	solvent	temp (°C)	yield (%)	M_w[b]	M_w/M_n[b]
		Poly(1,2,3,4,5-pentaphenysilolylacetylene) (1)				
1	$NbCl_5$–Ph_4Sn	toluene	60	60.0	46 400	1.7
2	$NbCl_5$–Ph_4Sn	toluene	80	78.0	68 800	1.8
		Poly{11-[(1,2,3,4,5-pentaphenylsilolyl)oxy]-1-undecyne} (2)				
3	WCl_6–Ph_4Sn	toluene	60	60.3	11 500	3.5
4	WCl_6–Ph_4Sn	dioxane	60	29.0	33 900	2.6
5	$MoCl_5$–Ph_4Sn	toluene	60	11.7	10 600	3.4
		Poly{11-[(1,2,3,4,5-pentaphenylsilolyl)oxy]-1-phenyl-1-undecyne} (3)				
6	WCl_6–Ph_4Sn	toluene	60	80.5	33 400	2.2

[a] Carried out under nitrogen for 24 h; $[M]_0 = 0.1$ M, [cat.] = [cocat.] = 10 mM.
[b] Estimated by GPC in THF on the basis of a polystyrene calibration.

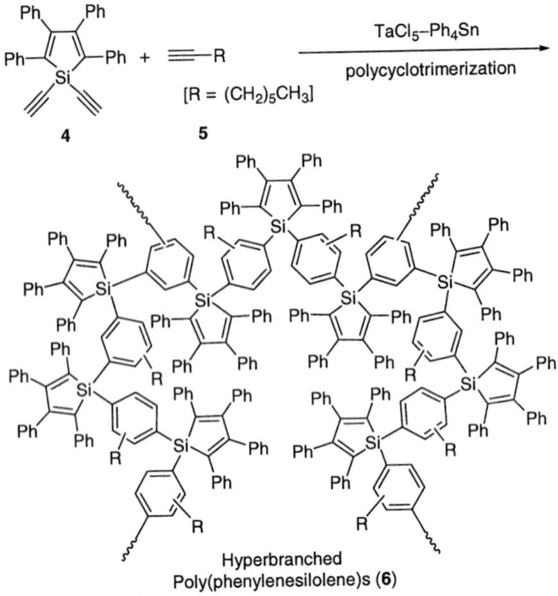

Figure 2. Synthesis of hyperbranched poly(phenylenesilolenes) by polycyclotrimerizations of alkynes catalyzed by tantalum-based initiator.

Table 2. Synthesis[a] and Properties of Hyperbranched Poly(phenylenesilolenes)

no.	feed ratio [**5**]/[**4**][b]	polymer yield (%)	M_w[c] (Da)	M_w/M_n[c]	T_d[d] (°C)	F_L[e] (mJ/cm^2)	$F_{t,m}/F_{i,m}$[f]	λ_{em}[g] (nm)	Φ_F[h]
1	0	83.0 (**6a**)	5 320	1.6	395	185	0.19	505	0.01
2	0.5	85.3 (**6b**)	5 820	1.7	378	182	0.24	493	0.01
3	1.0	67.6 (**6c**)	3 610	1.4	355	190	0.28	504	0.01
4	1.5	34.0 (**6d**)	3 530	1.4	343	1 140	0.32	499	0.01

[a] By homopolymerization of diyne **4** or copolymerizations of **4** with 1-octyne **5** catalyzed with $TaCl_5$–Ph_4Sn in toluene at room temperature under nitrogen for 24 h; [**4**] = 0.072 M, [$TaCl_5$] = [Ph_4Sn] = 20 mM.

[b] Molar ratio.

[c] Estimated by GPC in THF on the basis of a polystyrene calibration.

[d] Temperature for 5% weight loss (TGA, under nitrogen, heating rate: 20 °C/min).

[e] Optical limiting threshold (incident fluence at which the nonlinear transmittance is 50% of the linear one).

[f] Signal suppression (ratio of the saturated transmitted fluence to the maximum incident fluence).

[g] Emission maximum (in THF).

[h] Quantum yield of fluorescence (using 9,10-diphenylanthracene as standard).

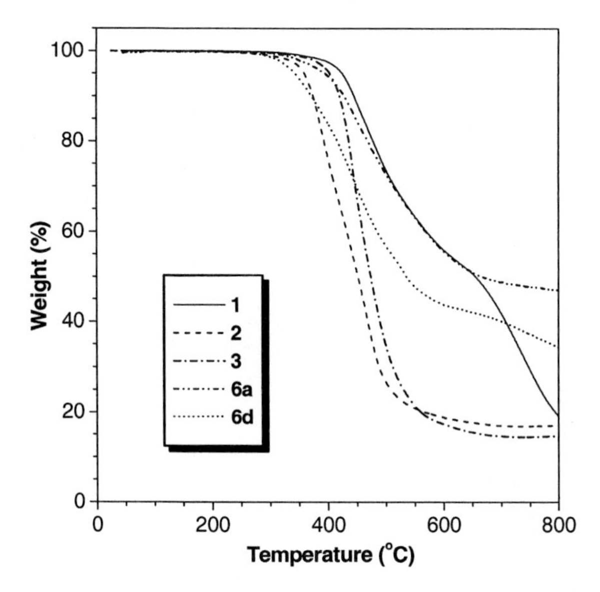

Figure 3. TGA thermograms of silole-containing linear polyacetylenes and hyperbranched polyarylenes recorded under nitrogen at a heating rate of 20 °C/min.

Figure 3 shows the TGA thermograms of the polymers. All the polymers were thermally stable and lost little of their weights at a temperature as high as ~350 °C. The degradation temperatures of polysilolyloxyacetylenes **2** and **3** were much higher than those of their structural congeners, poly(dimethylalkoxysilylacetylene)s $-\{HC{=}C[Si(CH_3)_2OC_mH_{2m+1}]\}_n-$ ($m = 2, 3$),[11] probably due to the "jacket effect"[12],[13] of the bulky silolyl pendants. Wrapping of the polyacetylene backbone in the stable silole rings may have shielded the double bonds from the harsh chemical and thermal attacks. Polymer **6a** showed higher thermal stability than that of **6d**. This is easy to understand: **6a** is an all-aromatic homopolymer, which should be thermally very stable; the incorporation of the weak aliphatic (hexyl) moiety into **6d** should increase its thermolytic susceptibility, and the copolymer should thus start to decompose at a relatively low temperature.

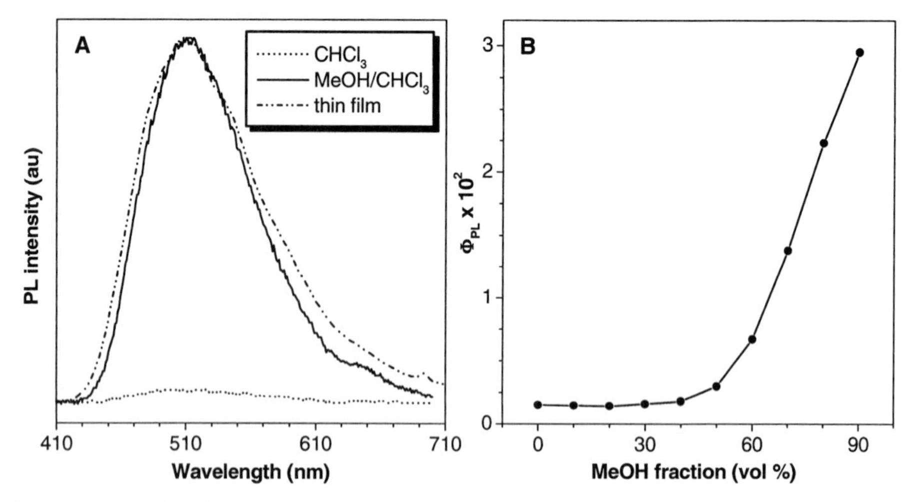

Figure 4. (A) Photoluminescence spectra of **2** in chloroform solution, methanol/chloroform mixture (9:1 by volume), and solid state (thin film); concentration of **2** in the solution and the mixture: 10 μM; excitation wavelength (nm): 400 (solution/mixture), 325 (thin film). (B) Quantum yield (Φ_{PL}) of **2** *vs.* solvent composition of methanol/chloroform mixture.

Silole's emission is characterized by its AIE feature. Do the silole polymers behave in a similar way? The answer to this question is yes or no, depending on the molecular structure of the polymer. No emission from the silole pendants was observed when the chloroform solution of **1** was photoexcited. Even when we added methanol into chloroform (while keeping the solution concentrations unchanged), hardly could the light emission of **1** be enhanced by the addition of the poor solvent. The rigid polyacetylene backbone of **1** may not allow its directly attached silole pendants to pack well in the aggregation state, thus making the polymer AIE-inactive. Thanks to the decoupling effect of the long nonanyloxy chain, polymers **2** and **3**, on the other hand, exhibited a pronounced AIE effect. As shown in Figure 4, the PL spectrum of the chloroform solution of **2** was almost a flat line with a Φ_{PL} value as low as 0.15%. The Φ_{PL} value started to swiftly increase when more than 40% of methanol was added into the mixture. When the methanol fraction was increased to 90%, Φ_{PL} rose to 2.95%, which is ~20 times higher than the solution value. Polymer **3** showed similar behavior but its Φ_{PL} in a mixture with 90% methanol was higher (9.3%), probably due to the additional contribution of its backbone emission: its poly(1-phenyl-1-alkyne) main chain is known to luminesce in the similar spectral region.[14],[15]

Similar to polymer **1**, all the hyperbranched poly(phenylenesilolenes) were AIE-inactive, possibly owing to the rigid hyperbranched polyphenylene sphere, which hampers the packing of the silole pendants.

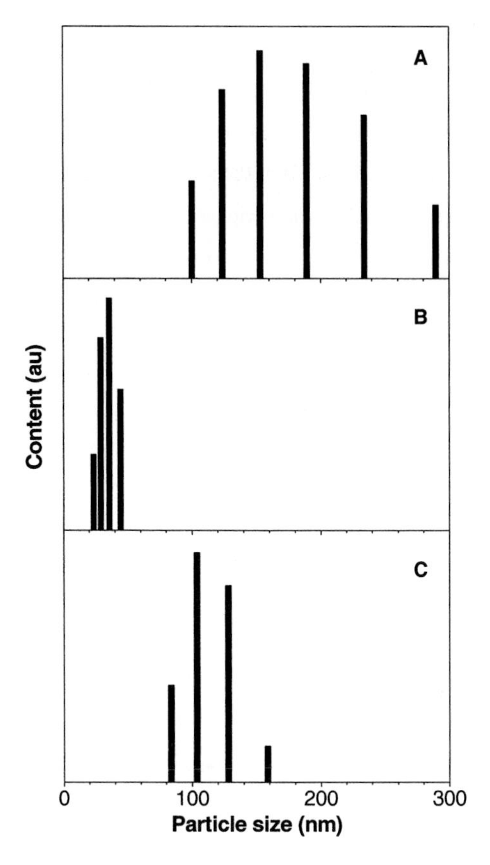

Figure 5. Particle size distributions of polymers (A) **1**, (B) **2**, and (C) **3** in methanol/ chloroform mixtures (9:1 by volume); concentration of polymers: 10 μM.

We carried out particle size analysis, which proved that the polymers did form nanoaggregates in the methanol/chloroform mixtures. Examples of the particle size histograms are shown in Figure 5. The size distribution and average size varied with polymer because of the difference in the polymer solubility in the solvent mixture. Polymer **1** is very hydrophobic and rigid and will easily associate into large nanoaggregates in the polar mixture. Polymer **2** possesses polar siloxy moieties with a better miscibility with the polar solvent and hence forms smaller aggregates under comparable conditions. Polymer **3** is a disubstituted polyacetylene with a more rigid backbone structure and its aggregates are thus understandably bigger than those of **2**.

Aggregation normally quenches light emission;[16],[17] what is the cause for this "abnormal" AIE phenomenon? To address this mechanistic question, we designed and carried out more experiments. When a dilute dioxane solution of **3** (10 μM) was cooled, the intensity of its PL spectrum was progressively increased in a nonlinear fashion (Figure 6A). When cooled from room temperature to below the mp (11.8 °C) of the solvent, the liquid solution changed to a solid "glass". The intramolecular rotations of the peripheral phenyl rings upon the axes of the single bonds linked to the central silacyclopentadiene cores may be physically restricted by the solid environmental surroundings. This restricted rotation in some sense rigidifies the chromophoric molecule as a whole, thus making the silole pendant more emissive.

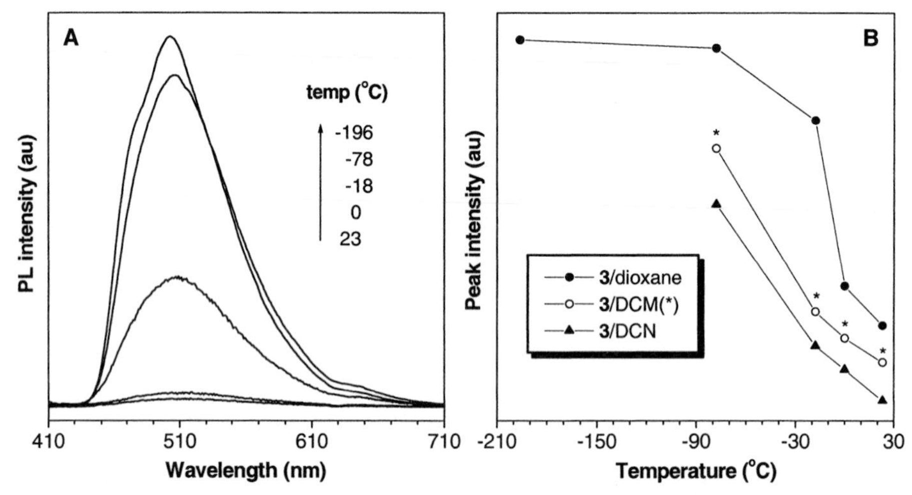

Figure 6. (A) Photoluminescence spectra of a dioxane solution of **3** at different temperatures and (B) effect of temperature on the peak intensity of the photoluminescence spectra of **3** in dioxane and DCM. All the solutions have a concentration of 10 μM except for the one marked with (*), whose concentration is 20 μM. Excitation wavelength: 407 nm.

To separate the cooling effect from the "glass" effect, we chose dichloromethane (DCM), a liquid with much lower melting point, for the PL measurement. The PL intensity of the solution increased with a decrease in temperature in a nearly "linear" fashion (Figure 6B). This enhancement in emission must be due to the restricted intramolecular rotation caused by cooling-induced conformation freezing because the melting point of the solvent (–95 °C) is lower than the lowest temperature we tested for this solution (–78 °C). We doubled the solution concentration to 20 μM and found that the emission enhancements were also roughly doubled at all the temperatures. This once again proves that the solute has indeed remained molecularly dissolved at the low temperatures because if the solute aggregates, it should evoke a clear "nonlinear" response to the concentration change. Similar to polymer **3**, polymers **2** and **6** also showed much stronger emission when their solutions were cooled.

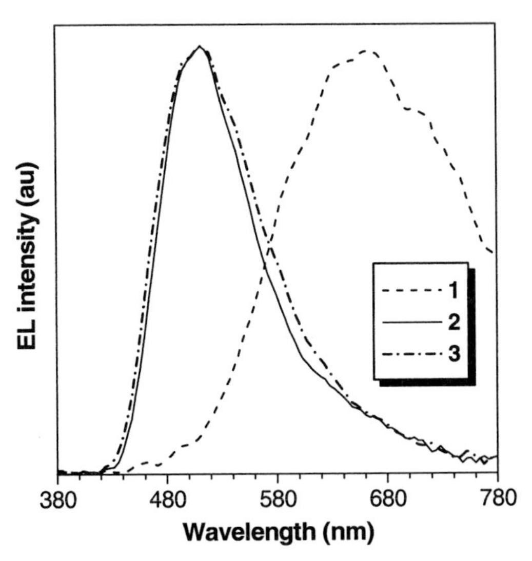

Figure 7. Electroluminescence spectra of single-layer devices of **1**, **2**, and **3** with a device configuration of ITO/polymer/LiF/Al.

Siloles are excellent materials for LED applications. How are their polymers? We checked the EL performances of polymers **1–3** using a single-layer device configuration. The EL spectrum of **1** peaked at 664 nm, while the emission maximums of **2** and **3** were at ~512 nm (Figure 7). All the EL devices exhibited similarly low current efficiencies: 0.014 (**1**), 0.013 (**2**), and 0.013 cd/A (**3**). This is not surprising because the injection and transportation of charges are generally not balanced in devices with a single-layer structure. We thus tried to modify the device configuration, using **3**, the most PL-active polymer, as the emitting material. We added poly(9-vinylcarbazole) (PVK) and tris(8-hydroxyquinolinato)aluminum (Alq_3) on the anode and cathode sides, respectively, to facilitate the charge injection and to enhance the charge transport efficiencies in the EL device. Between the PVK and Alq_3 layers, we also added a layer of bathocuproine (BCP) to prevent the holes from traveling through to reach the cathode. With these modifications, an EL device with a configuration of ITO/(**3**:PVK)(1:4)/BCP/Alq_3/LiF/Al was fabricated, which emitted a strong blue light of 496 nm with a maximum luminance of 1118 cd/m^2. The maximum current and external quantum efficiencies of the device were 1.45 cd/A and 0.55%, respectively, which are comparable to some of the best results reported by other research groups for blue-emitting LEDs.[18],[19]

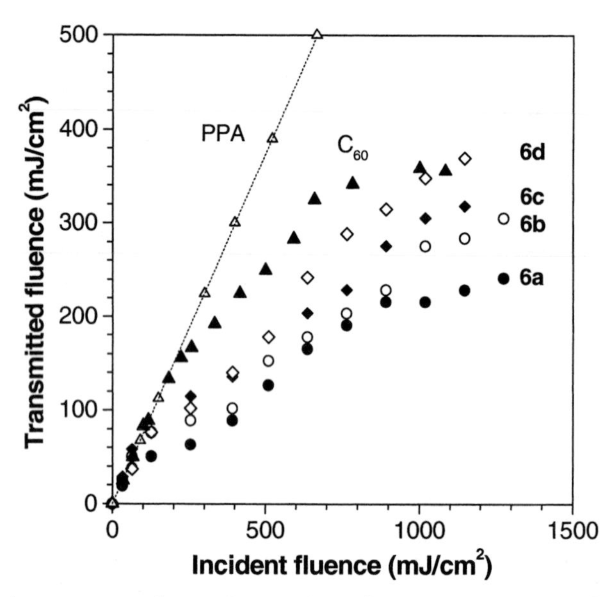

Figure 8. Optical responses to 8 ns, 10 Hz pulses of 532 nm laser light, of dichloromethane solutions (0.70 mg/mL) of hyperbranched poly(phenylenesilolene)s. Data for poly(phenylacetylene) (PPA) and C_{60} solutions are shown for comparison. Linear transmittance (%): 70 (**6a**), 80 (**6b**), 82 (**6c**), 85 (**6d**), 79 (C_{60}), 75 (PPA).

The hyperbranched polymers **6** were optically very stable. Figure 8 shows optical responses of the polymer solutions to 532 nm laser pulses, along with the data for the solutions of poly(phenylacetylene) (PPA)[20] and C_{60}.[21],[22] PPA photodegraded under the attack of harsh laser shots. The hyperbranched polyarylenes showed comparable linear transmittance in the low fluence region (T = 70–85%) but became opaque in the high fluence region. The polymers strongly attenuated the power of the intense laser pulses, whose optical limiting performances are superior to that of C_{60}, a best-known optical limiter.[21]–[23] Among the hyperbranched polymers, **6a** exhibited the best performance, which started to limit the optical power at a low threshold (185 mJ/cm^2) and suppressed the optical signals to a great extent (19%; Table 2, no. 1).

Conclusions

In this work, we succeeded in incorporating silole ring, an organometallic chromophore with novel luminescence properties, into linear polyacetylene and hyperbranched polyphenylene structures through the facile metathesis and cyclotrimerization polymerizations of alkynes catalyzed by transition-metal catalysts. All the polymerizations yielded completely soluble polymers of high thermal stability. While none of the polymers were strong luminophors when molecularly dissolved, polymers **2**, **3** and **6** became emissive when aggregated in poor solvents or when cooled to low temperatures. A multilayer EL device with a configuration of ITO/(**3**:PVK)(1:4)/BCP/Alq_3/ LiF/Al was fabricated, which emitted blue light of 496 nm and exhibited maximum luminance, current efficiency, and external quantum efficiency of 1118 cd/m^2, 1.45 cd/A, and 0.55%, respectively. The hyperbranched poly(phenylenesilolene)s strongly limited intense laser pulses, whose optical-limiting thresholds and signal-suppress power were better than those of C_{60}, a well-known optical limiter.

Acknowledgments

We thank the financial support of the Hong Kong Research Grants Council (Project Nos. HKUST 6121/01P and 6085/02P) and the University Grants Committee through an Area of Excellence Scheme (Project No.: AoE/P-10/01-1-A).

[1] C. Chuit, R. J. P. Corriu, C. Reye, J. C. Young, *Chem. Rev.* **1993**, *93*, 1371.
[2] H. Murata, G. G. Malliaras, M. Uchida, Y. Shen, Z. H. Kafafi, *Chem. Phys. Lett.* **2001**, *339*, 161.
[3] J. Luo, Z. Xie, J. W. Y. Lam, L. Cheng, H. Chen, C. Qiu, H. S. Kwok, X. Zhan, Y. Liu, D. Zhu, B. Z. Tang, *Chem. Commun.* **2001**, 1740.
[4] M. Freemantle, *Chem. Eng. News* **2001**, *79* (41), 29.
[5] B. Z. Tang, X. Zhan, G. Yu, P. P. S. Lee, Y. Liu, D. Zhu. *J. Mater. Chem.* **2001**, *11*, 2874.
[6] H. Chen, J. W. Y. Lam, J. Luo, Y. L. Ho, B. Z. Tang, D. Zhu, M. Wong, H. S. Kwok, *Appl. Phys. Lett.* **2002**, *81*, 774.
[7] Y. Cao, I. D. Parker, G. Yu, C. Zhang, A. J. Heeger, *Nature* **1999**, *397*, 414.
[8] J. S. Kim, P. K. Ho, N. C. Greenham, R. H. Friend, *J. Appl. Phys.* **2000**, *88*, 1073.
[9] J. Chen, Z. Xie, J. W. Y. Lam, C. C. W. Law, B. Z. Tang, *Macromolecules* **2003**, *36*, 1108.
[10] J. Chen, H. Peng, C. C. W. Law, Y. Dong, J. W. Y. Lam, I. D. Williams, B. Z. Tang, *Macromolecules*, submitted.
[11] M. G. Voronkov, V. B. Pukhnarevich, S. P. Sushchinskaya, V. Z. Annenkova, V. M. Annenkova, N. J. Andreeva, *J. Polym. Sci. Polym. Chem. Ed.* **1980**, *18*, 53.
[12] J. W. Y. Lam, J. Luo, Y. Dong, K. K. L. Cheuk, B. Z. Tang, *Macromolecules* **2002**, *35*, 8288.
[13] J. W. Y. Lam, Y. P. Dong, K. K. L. Cheuk, J. Luo, Z. L. Xie, H. S. Kwok, Z. S. Mo, B. Z. Tang, *Macromolecules* **2002**, *35*, 1229.
[14] J. W. Y. Lam, C. K. Law, Y. P. Dong, J. N. Wang, W. K. Ge, B. Z. Tang, *Opt. Mater.* **2002**, *21*, 321.
[15] R. G. Sun, Q. G. Zheng, X. M. Zhang, T. Masuda, T. Kobayashi, *Jpn. J. Appl. Phys.* **1999**, *38*, 2017.
[16] B.-K. An, S.-K. Kwon, S.-D. Jung, S. Y. Park, *J. Am. Chem. Soc.* **2002**, *124*, 14410.
[17] S. H. Chen, A. C. Su, Y. F. Huang, C. H, Su, G. Y, Peng, S. A. Chen, *Macromolecules* **2002**, *35*, 4229.
[18] X. Z. Jiang, S. Liu, H. Ma, A. K. Y. Jen, *Appl. Phys. Lett.* **2000**, *76*, 1813.
[19] L. C. Palilis, D. G. Lidzey, M. Redecker, D. D. C. Bradley, M. Inbasekararn, E. P. Woo, W. W. Wu, *Synth. Met.* **2001**, *121*, 1729.
[20] B. Z. Tang, H. Xu, *Macromolecules* **1999**, *32*, 2569.
[21] B. Z. Tang, S. M. Leung, H. Peng, N.-T. Yu, K. C. Su, *Macromolecules* **1997**, *30*, 2848.
[22] B. Z. Tang, H. Peng, S. M. Leung, C. F. Au, W. H. Poon, H. Chen, X. Wu, M. W. Fok, N.-T. Yu, H. Hiraoka, C. Song, J. Fu, W. Ge, K. L. G. Wong, T. Monde, F. Nemoto, K. C. Su, *Macromolecules* **1998**, *31*, 103.
[23] L. W. Tutt, A. Kost, *Nature* **1992**, *356*, 225.

Hybrid Inorganic/Organic Crosslinked Resins Containing Polyhedral Oligomeric Silsesquioxanes

Charles U. Pittman, Jr., Gui-Zhi Li, Hanli Ni*

Department of Chemistry, Mississippi State University, 9573, Mississippi State, MS 35762, USA
E-mail: cpittman@ra.mstate.edu

Summary: The incorporation of both monofunctional and multifunctional polyhedral oligomeric silsesquioxane (POSS) derivatives into crosslinked resins has been conducted as a route to synthesize hybrid organic/inorganic nanocomposites. The central cores of POSS molecules contain an inorganic cage with $(SiO_{1.5})_n$ stoichiometry where n=8,10 and 12. Each Si atom is capped with one H or R function giving an organic outer shell surrounding the nanometer-sized inorganic inner cage. By including polymerizable functions on the R groups, a hybrid organic/inorganic macromer is obtained which can be copolymerized with organic monomers to create thermoplastic or thermoset systems. We have focused on incorporating POSS derivatives into crosslinking resins of the following types: (1) dicyclopentadiene (2) epoxies (3) vinyl esters (4) styrene-DVB (5) MMA/1,4-butane dimethacrylate (6) phenolics and (7) cyanate esters. One goal has been to determine if molecular dispersion of the POSS macromers has been achieved or if various degrees of aggregation occur during crosslinked resin formation. As network formation proceeds, a kinetic race between POSS molecular incorporation into the network versus phase separation into POSS-rich regions (which then polymerize) occurs. Ultimately, we hope to determine the effects of such microstructural features on properties. Combustion of these hybrids creates a SiO_2-like surface layer that retards flame spread. Dynamic mechanical properties have been studied.

Keywords: cyanate esters methyl methacrylate resins; epoxy resins; hybrid nanocomposites; phenolic resins; poly(dicyclopentadiene); polyhedral oligomeric silsesquioxanes; styrenedivinylbenzene; vinyl esters

Introduction

The term silsesquioxane refers to all structures with the empirical formulas $RSiO_{1.5}$ where R is hydrogen or any alkyl, alkylene, aryl, arylene, or organofunctional derivative of alkyl, alkylene, aryl, or arylene groups. The silsesquioxanes include random structures, ladder structures, cage structures, and partial cage structures, as illustrated in Scheme 1.[1] The first oligomeric organosilsesquioxanes, $(CH_3SiO_{1.5})_n$, were isolated along with other volatile compounds by Scott

© 2003 WILEY-VCH Verlag GmbH & KGaA, Weinheim CCC 1022-1360/00/$ 17,50+.50/0

in 1946 through thermolysis of the polymeric products obtained from methyltrichlorosilane and dimethylchlorosilane co-hydrolysis.[2] Even though silsequioxane chemistry spans more than half a century, interest in this area continues to increase.

In 1995, Baney et al[1] reviewed the structure, preparation, properties and applications of silsesquioxanes, especially the ladder-like polysilsesquioxanes shown in Scheme 1 (structure b). These include poly(phenyl silsesquioxane) (PPSQ),[3-12] poly(methyl silsesquioxane) (PMSQ)[13-19] and poly(hydridosilsesquioxane) (PHSQ).[20,21] However, in the past few years, much more attention has been paid to the silsesquioxanes with specific cage structures [shown in Scheme 1 (structures c - f)]. These polyhedral oligomeric silsesquioxanes have been designated by the abbreviation POSS.

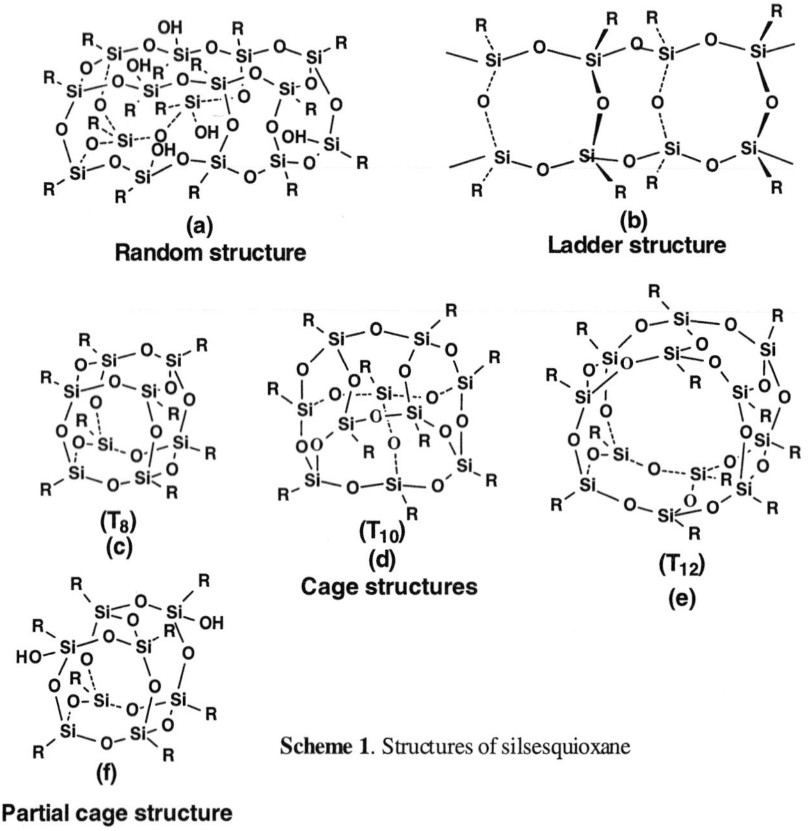

(a) Random structure

(b) Ladder structure

(T_8) (c)

(T_{10}) (d)

Cage structures

(T_{12}) (e)

(f) Partial cage structure

Scheme 1. Structures of silsesquioxane

POSS compounds embody a truly hybrid (inorganic-organic) architecture, which contains an inner inorganic framework made up of silicone and oxygen $(SiO_{1.5})_x$, that is externally covered by organic substituents. These substituents can be totally hydrocarbon in nature or they can embody a range of polar structures and functional groups. POSS nanostructured chemicals, with sizes from 1 to 3 nm in diameter, can be thought of as the smallest possible particles of silica. A variety of POSS nanostructured chemicals has been prepared which contain one or more covalently bonded reactive functionalities that are suitable for polymerization, grafting, surface bonding, or other transformations.[22,23] A group at Edwards Air Force Base, CA, has recently developed a large-scale process for POSS monomer synthesis[22,24-27] and a number of POSS reagents. As a result of this success, monomers have recently become commercially available as solids or oils from Hybrid Plastics Company (http://www.hybridplastics.com/), Fountain Valley, CA.

A selection of POSS chemicals now exist that contain various combinations of nonrective substituents and/or reactive functionalities. Thus, POSS nanostructured chemicals may be easily incorporated into common plastics via copolymerization, grafting or blending.[28] The incorporation of POSS derivatives into polymeric materials can lead to dramatic improvements in polymer properties which include, but are not limited to, increases in use temperature, oxidation resistance, surface hardening and improved mechanical properties, as well as reductions in flammability, heat evolution, and viscosity during processing. These enhancements have been shown to apply to a wide range of thermoplastics and a few thermoset systems.[27,28]

In this manuscript we report the incorporation of both monofunctional and polyfunctional POSS derivatives **1**-**7** into crosslinked organic resin systems to make a series of inorganic/organic hybrid composite materials. As POSS macromers are incorporated, a kinetic race occurs. The solubility of the POSS in the organic monomer mixture of the resin (phenolics, epoxies, vinyl esters, methacrylics, styrene-divinylbenzene, cyanate esters or dicyclopentadiene) decreases during the cure as the number of monomer molecules (hence the entropy of mixing) decreases. Therefore, phase separation of the POSS macromer may occur sometime during the cure. This would occur in competition with the POSS macromer's chemical incorporation into the resin. After such phase separation, the POSS macromer may homopolymerize (or copolymerize with smaller amounts of resin monomers) to form POSS-rich phases within the composite as the resin curing continues. This process can compete against random POSS incorporation into the developing resin network. Random incorporation leads to molecularly dispersed POSS monomer units within a homogeneous resin phase. Phase separation will be increasingly favored if the POSS monomer has a low relative reactivity. This will cause monomer drift during the cure which raises POSS concentration relative to the other monomers, favoring phase separation. Anything that increases the early incorporation of POSS will lower this phase separation.

Another type of phase separation process can also operate, even if all the POSS is polymerized randomly into the resin network. POSS macromers bonded into chain segments of the resin may preferentially self-associate with each other, forming POSS aggregates whose structures and sizes are limited by the freedom of motion permitted by the developing resin's chain segmental mobility and the POSS stoichiometry. This process is favored at low crosslink densities where segmental mobility is larger. POSS aggregation has been previously observed in uncrosslinked thermoplastics. Coughlin et al[29-32] have demonstrated that linear copolymers of ethylene with a mono-alpha-olefin-substituted POSS can form POSS nanocrystalline domains due to self-aggregation of pendant POSS moieties. The extent of such aggregation depends on the mole fraction of POSS present and the method of solidifying the polymer (cooling a melt, precipitation from solution etc.) Such aggregation in crosslinked resin matrices will certainly be dependent on (1) the buildup of crosslinked density as a function of the degree of cure, (2) the relative reactivity ratios of the monomers, (3) the solvent used (if any) and (4) other factors (solvent, processing etc.)

We have employed multifunctional POSS macromers **1** – **3** and monofunctional derivatives **4** – **7** in both condensation and addition types of resin-forming polymerization reactions. These monomers and the types of resins they have been polymerized into are listed in Scheme 2. Various resin properties have been studied and POSS aggregation has been studied in a few cases.[33-36]

1

Tetraepoxidized octa-b-styryl-POSS
M_w:1305
Vinyl ester, Phenolic, Epoxy
and Dicyclopentadiene (DCPD) resins

2

Octanorbornyl-POSS
$C_{56}H_{72}O_{12}Si_8$ M_w: 1162
DCPD resin

3

Trisilanolphenyl POSS
R = Ph
$C_{42}H_{38}O_{12}Si_7$ M_w: 931.3
Phenolic resins and
Cyanate esters

4

Styrylcyclopentyl POSS
R = Cyclopentyl
$C_{43}H_{70}O_{12}Si_8$: M_w 1003.7
Dicyclopentadiene
and Sty/DVB resins

5

Dichlorosilylisobutyl POSS
R = iso-butyl
$C_{32}H_{73}ClO_{12}Si_9$: M_w 938.1
Phenolic resins

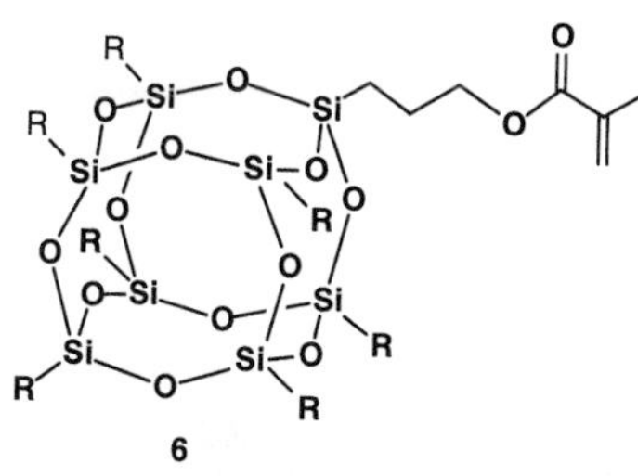

6

Methacryl Isobutyl POSS
R = isobutyl
$C_{35}H_{74}O_{14}Si_8$: M_w 943.6
For Acrylic and Vinyl Ester resins

R = Cyclopentyl

7

NorbornenylethylDisilanol
$C_{46}H_{84}O_{12}Si_8$ M_w: 1053.84
For DCPD resins, Siloxane resins
and DCPD/siloxane hybrids

Scheme 2. POSS macromers incorporated in resins in this work and the resin types

Results

Multifunctional POSS monomer incorporation

Multifunctional POSS-1 macromer (prepared for us by R. Blanski at Edwards AFB) has been cured into epoxy[34], phenolic, vinyl ester[35] and dicyclopentadiene[36] (DCPD) resins. For example, **1** in THF was blended into a resole phenolic resin (Borden SC-1008 with 35% isopropanol) and the solvents were removed *in vacuo*. This was cured via 65 °C (50 – 100 psi) 30 min., 105 °C (200 – 500 psi) 90 min., 176 °C (500 psi) 30 min. and then 250 °C post cure for 1h. The same resin was blended with vapor grown carbon fibers (VGCF) (~150nm dia., from Applied Sciences) and cured in the same manner to give a composite with 30 wt% VGCF, 3.5 wt% POSS-**1** and the 66.5 wt% resin. True curing between **1** and the phenolic resin occurs via the epoxy functions. This was confirmed by cryogrinding the resin and extensive extractions. No **1** could be extracted.

This resin (95% phenolic/5% POSS **1**) exhibited higher bending storage moduli, E', both above and below T_g, and a higher T_g than the pure phenolic resin (see Fig. 1). Even after the fibers had been added, both E' and T_g increased with POSS-**1** present.

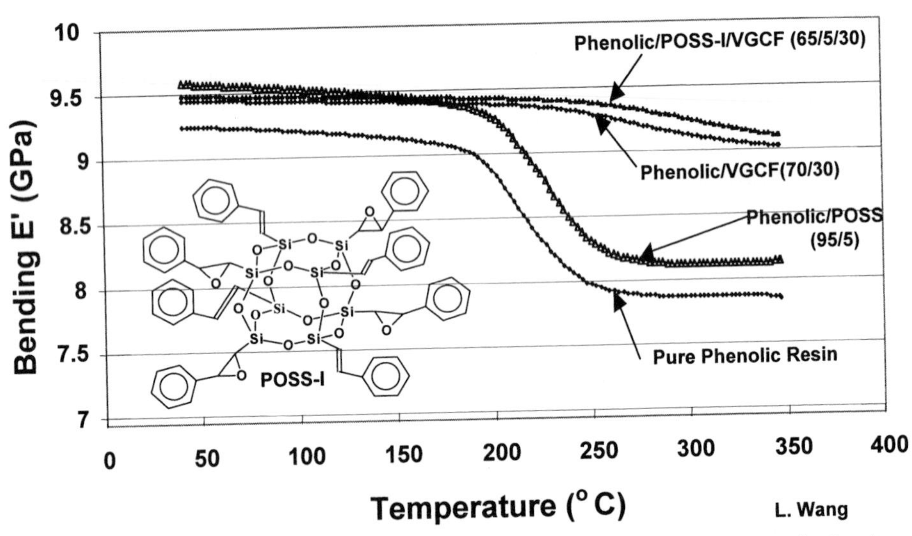

Figure 1. Bending E' (GPA) vs. temperature of different phenolic/VGCF/POSS-1 and related composites at 10 Hz.

Trisilanophenyl POSS, **3**, and dichlorosilylisobutyl-POSS, **5**, have also been cured into resole phenolic resins (Bordon SC 1008 and Hitco 134A). Monomer **3** is soluble in phenolic resins. POSS-**3** is a strongly hydrogen bonding molecule, facilitating its miscibility. This bonds **3** it into the resin during resin curing at higher temperatures. Its phenyl substituents appear to react with methylol functions to form methylene bridges. The states of aggregation of POSS-**3** in these resins is under study. POSS-**5**, in contrast to **3**, reacts very rapidly at low temperature with hydroxyl groups in the resole resin long before curing starts. This occurs due to the presence of the extremely reactive dichlorosilyl moiety. Thus, POSS-**5** is bound to resin molecules prior to cure. After curing it is likely the POSS centers are molecularly dispersed at low mole percents of **5**. This process is shown in Scheme 3 along with the curing protocol for phenolic resin with either **3** or **5**.

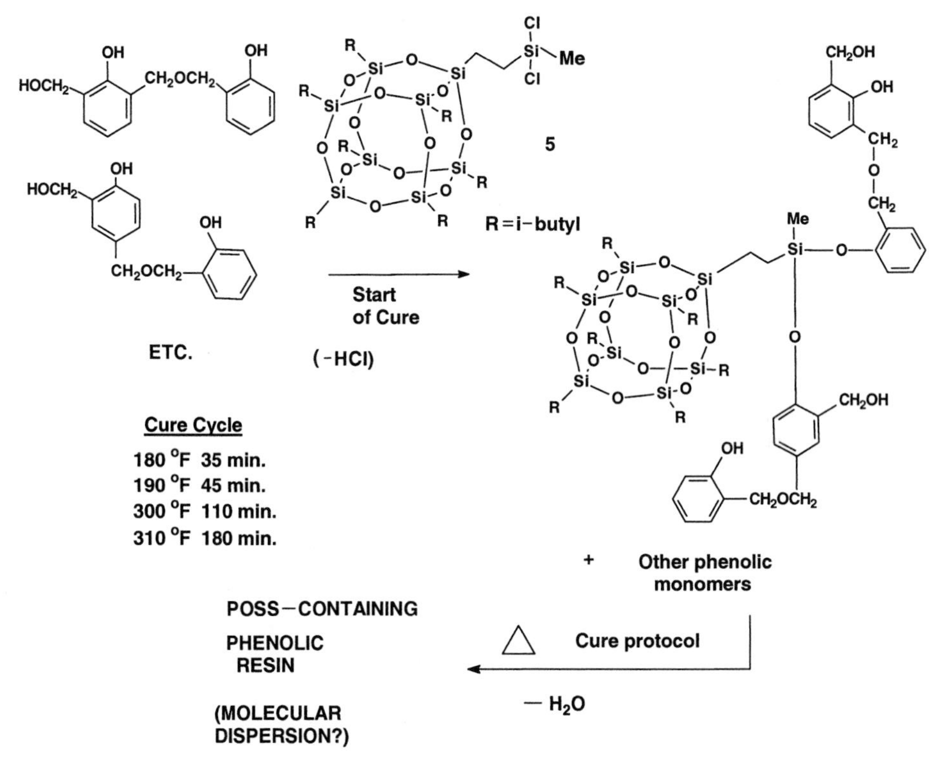

Scheme 3. Phenolic resin/POSS-**5** nanocomposite

Trisilanolphenyl POSS-**3** was dissolved in THF and blended with the liquid cyanate ester resins PT-15 and PT-30 (Lonza Inc.) Then THF was removed *in vacuo* at 80 °C. The temperature was raised to 90 °C for 3h and then to 188 °C (2h), 200 °C (5 min.) and then the temperature was raised 10 °C every 5 min. to 250 °C where it was held 2.5h. Finally, a 300 °C (30 min.) post cure was performed. Direct preblending of **3** with PT-15 resin at 120°C was also successful. In addition to thermal curing, metal acetonylacetate/nonylphenol catalyst mixtures were used. Catalysts must be added after **3** has been blended and the liquid blend is cooled, because curing occurs too rapidly to permit complete dissolution of the POSS at the higher temperatures used for this POSS mixing step. Cyanate esters cure by cyclotrimerization of isocyanate groups to form a trisubstituted triazine (Scheme 4). Therefore, no low molecular weight condensation products are driven off, unlike the curing of phenolic resins where water is extruded. For this reason, we are examining cyanate esters as thermally stable resin precurors for further pyrolysis to carbon-carbon materials. Carbonized cyanate ester resins should be denser than those made from phenolic resins. Therefore, expensive impregnation / repyrolysis / densification steps (required using phenolic resin precursors) may be unnecessary. The purpose of incorporating molecularly dispersed, or nano-sized aggregates of POSS in the cured cyanate ester resins is to use these nanocomposites to generate carbon/carbon materials with silica nanophases. These should stabilize the resulting carbon for use at higher temperatures because the silica-like nuclei should be effective char promoters and oxygen permeation barriers.

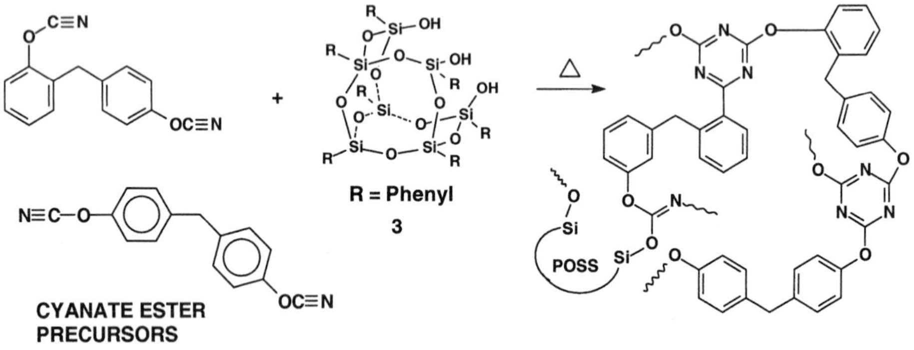

Scheme 4. Formation of cyanate ester/POSS-**3** resins

Multifunctional POSS-**1** has been cured into epoxy, vinyl ester and dicyclopentadiene resins. The four epoxy groups in **1** participate in the curing of Clearstream 9000, an aliphatic epoxy resin with its corresponding aliphatic amine curing agent (Clearstream Resins Inc.)[34] It was difficult to dissolve **1** into the resin directly. Therefore, the epoxy resin and **1** were dissolved in THF and the THF was removed to give a clear fluid which was then cured. When cured at an upper temperature of 120 °C (Figure 2) the Tg values dropped as the POSS content increased.[34] The Tg regions became wider (Figure 2) and the bending storage moduli of the resins increased at temperatures above Tg. The Tg value increments in all these systems increased when the cure temperature was raised to 150 °C. TEM studies of a 5 wt% POSS-**1**/epoxy resin appeared free of any phase separation, suggesting that molecular dispersion of the POSS had occurred.[34] However, EELS and elemental density mapping has not yet been carried out, so some degree of POSS aggregation has not yet been ruled out.

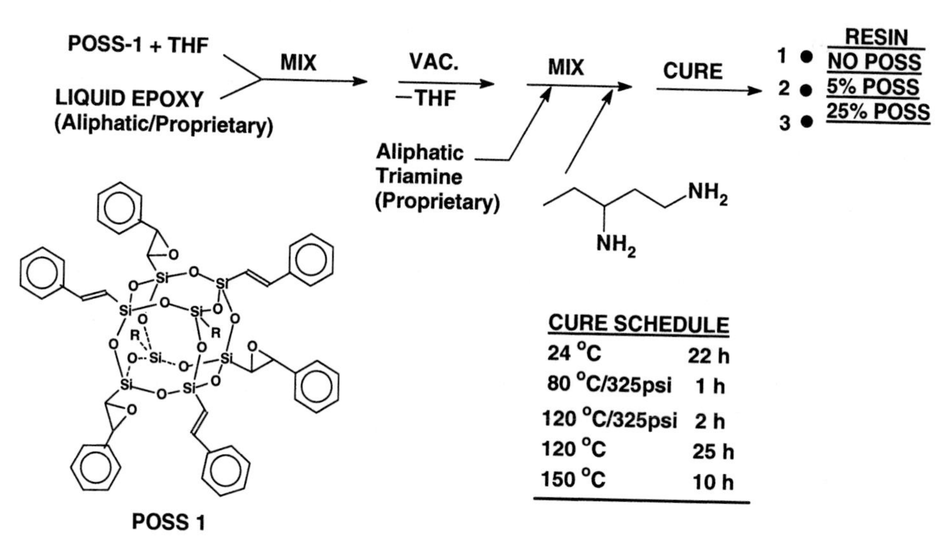

Figure 2a. Synthesis and cure of epoxy/POSS-**1** resins.

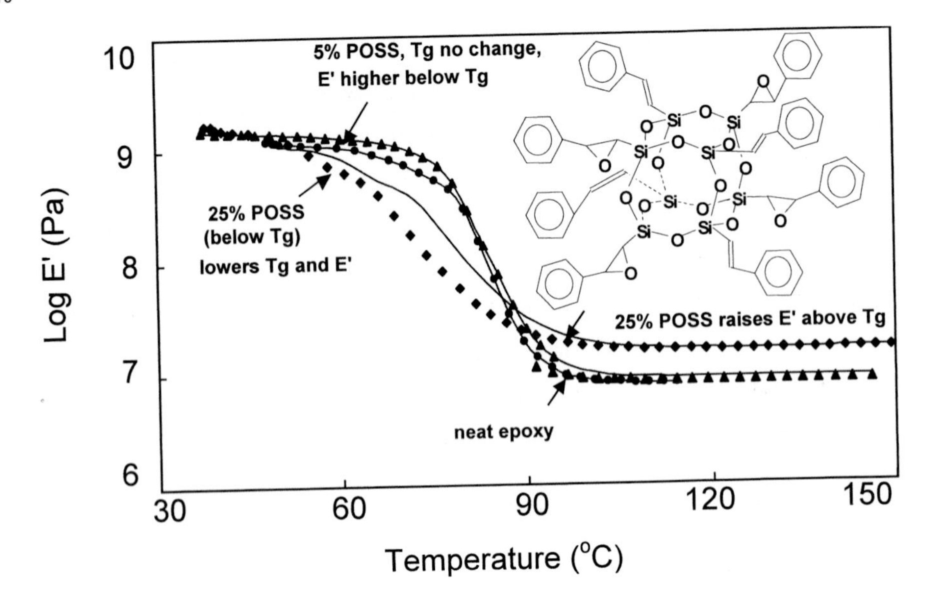

Figure 2b. Bending modulus (E') versus temperature curves at 1Hz (from DMTA) for an aliphatic epoxy resin and its epoxy/POSS 95/5 and 75/25 (wt/wt) composites after final curing at 120 °C/25h.

POSS-**1** was also dissolved in styrene and blended into the commercial vinyl ester, Derakane™ 510C-350 (from Dow Chemical Co.) This compatible liquid system was then cured at room temperature (24h), 90 °C (24h) and 150 °C (5h) using a 1% methyl ethyl ketone peroxide/0.2% cobalt naphthanate catalyst system.[35] This is shown in Scheme 5. Other vinyl ester resins were made with POSS methacrylate, **6**, but these will not be discussed here. All the resins contained 50 wt% styrene. POSS-**1** was incorporated in 5 and 10 wt% into these thermoset resins. The flexural strength dropped from 192 to 175 and 152 MPa with 5 and 10 wt% POSS added into the blends.[35] The extent of strain at failure decreased with increasing levels of POSS. DMTA measurements confirmed that the bending storage moduli of the vinyl ester/POSS-**1** resins were higher than that of vinyl ester resin without POSS at temperatures below T_g. The storage moduli increased greatly with an increase in the POSS-**1** content above T_g. The storage modulus enhancement was larger at temperatures lower than T_g (e.g. in the glassy region). Specifically, at 40 °C the bending storage moduli were 1.24, 1.96 and 1.58 GPa for the pure vinyl ester and the vinyl ester with 5 wt% and 10 wt% POSS, respectively. T_g values also increased with more POSS content.

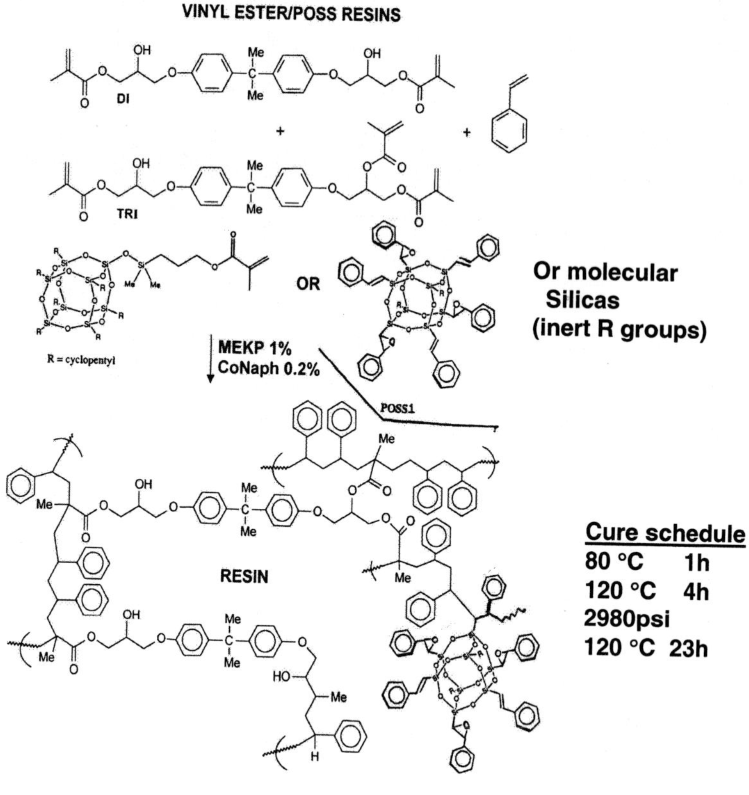

VINYL ESTER/POSS RESINS

DI

TRI

R = cyclopentyl

OR

Or molecular Silicas (inert R groups)

MEKP 1%
CoNaph 0.2%

POSS1

RESIN

Cure schedule
80 °C 1h
120 °C 4h
2980psi
120 °C 23h

Scheme 5. Vinyl ester/POSS resins

Swelling studies showed almost no uptake of THF or toluene by the vinyl ester/POSS-**1** composites after 50 days.[35] Furthermore, no POSS-**1** could be recovered on extensive extraction of fragmented samples. TEM micrographs showed no discernable phase separation of POSS at a resolution of 1-2 nm. However, EDAX and EELS studies of Si and Br (from a bromine containing label used) provided evidence of some phase separation of Si–rich regions at a small size scale in the VE/POSS-**1** (10 wt.%) sample (Figure 3).[35] POSS-rich dispersed phases with sizes of 75nm to 1 – 2nm were observed by EELS. All of the epoxy/**1** and vinyl ester/**1** (5 wt.%) composites appeared to largely exhibit molecular level dispersion of chemically-bound POSS-**1**. This illustrates the importance of early chemical bonding of a POSS derivative into the developing crosslinked resin's network structure during the cure. As curing occurs, the entropy of mixing term will contribute less and less to the thermodynamics of POSS monomer solubility. However, before phase separation of the POSS can occur due to a lowering of solubility, the chemical bonds formed between the POSS and the developing resin matrix lock POSS molecules into the resin structure. Thus, migration and phase separation don't occur, or only partial separation takes place. This stands in sharp contrast to the behavior of T_8, T_{10} and T_{12} (Scheme 1) POSS molecules where none of the corner groups have a polymerizable function (e.g. R = *i*-Bu, cyclopentyl, cyclohexyl or phenyl). These molecules regularly phase separate from (or don't dissolve in) and vinyl ester resins. We have observed dispersed POSS particle (phase) sizes from 50nm to 300μm under a variety of conditions with these types of POSS derivatives in cured resins.[36]

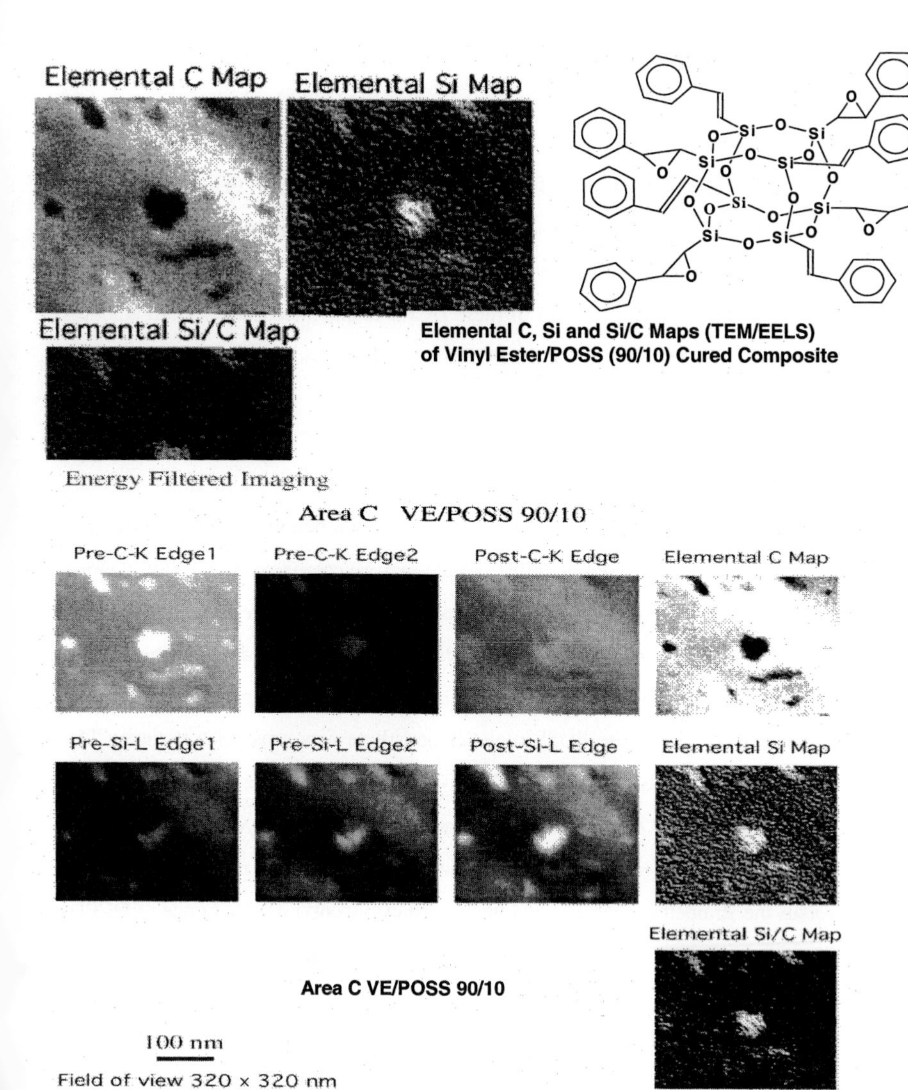

Figure 3. Nanoscale POSS aggregation in cured Derakane 510C-350 (50 wt% styrene) vinyl Ester resins, containing POSS-**1**.

POSS-**1** was dissolved into dicyclopentadiene (DCPD) and this monomer mixture was cured by ring-opening metathesis polymerization (ROMP) using the specific Grubb's catalyst: 3-methyl-2-butenylidene-bis(tricyclopropylphosphine)dichlororuthenium (Scheme 6). The metathesis proceeded on the O_3Si-corner-substituted β-phenyl olefinic moieties of POSS-**1**. Vigorous extractions of finely ground resin did not result in any recovery of **1**, confirming its chemical bonding into the resin via ROMP. Resins were made by the same curing protocol containing as much as 20 wt% POSS. The T_g and E' values of these DCPD resins increased as small amounts of POSS were incorporated but once ~10 wt% POSS-**1** had been added, no further increases occurred.

Specifically, the T_g value of the DCPD resin (182.5 °C) increased by 5.8 °C to 188.3 °C when only 1 wt% of **1** was added. The DCPD/5 wt% **1** resin exhibited a bending storage modulus at 10Hz of 1.54GPa at 40 °C and 0.032GPa at 240 °C which may be compared to those of the pure DCPD resin which were 1.45GPa and 0.027GPa, respectively. Thus, despite the effect that the volume occupied by POSS has on lowering the DCPD resin's crosslink density, the storage modulus increased because POSS-1 became chemically bound into the resin, probably by forming more than one bond per POSS. These POSS moieties constitute new crosslinks (Scheme 11). All these data confirmed that POSS-**1** actually was, on average, bound into the resin by more than one chemical bond. Thus, ROMP took place readily enough on the β-siloxy-substituted styrene function to multiply bond the cage into the resin. However, the relative rates of the ROMP reaction on **1**, versus the strained norbornene and unstrained cyclopentyl double bonds of DCPD, are unkown. Therefore, one does not know whether or not **1** is, on average, incorporated early or late into the resin. Later average ROMP of **1** would lead to an increased tendency for POSS phase separation or aggregation. TEM/EEL studies have not been done on these resins.

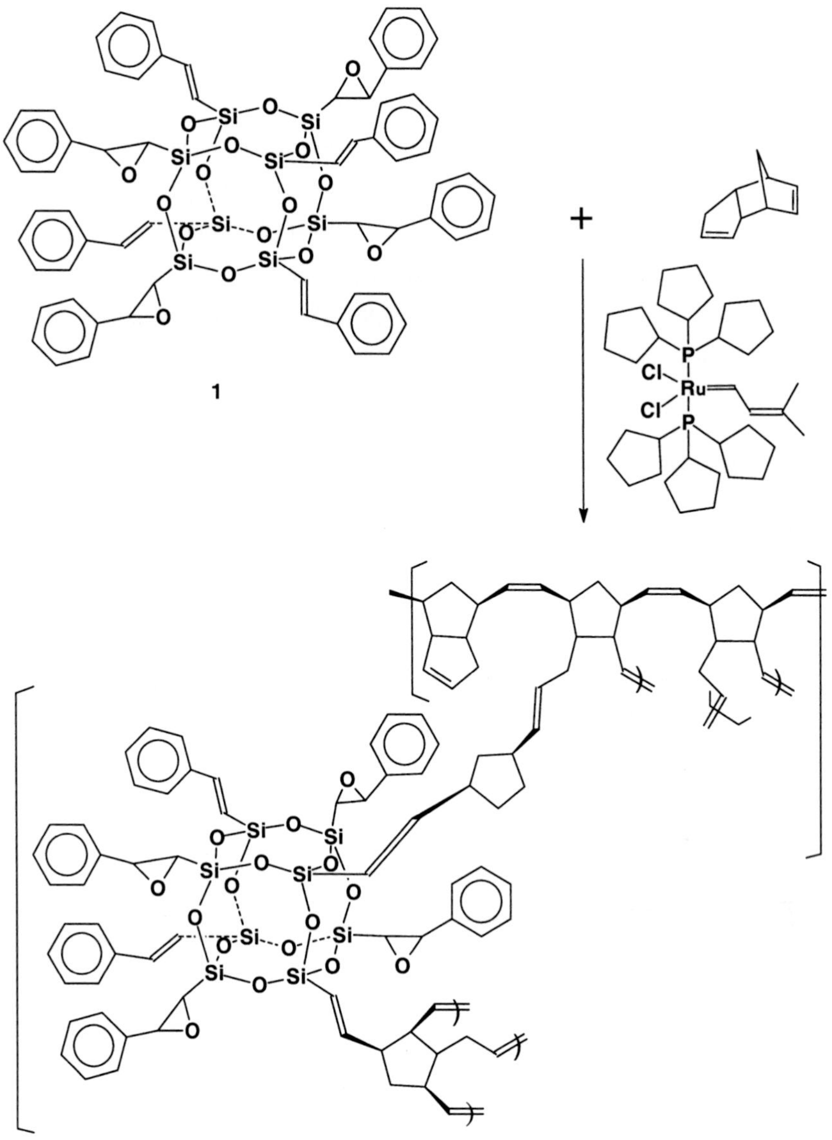

Scheme 6. Copolymerization of polyfunctional POSS-**1** with dicyclopentadiene

A second multifunctional POSS macromer, **2**, was ROMPed with DCPD using the same catalyst shown in Scheme 6 during the preparation of DCPD/POSS-**1** resins.[36] Octanorbornenyl POSS, **2**, contains eight reactive norbornene functions. Thus, it should rapidly be incorporated into the developing DCPD resin network and it should be a crosslinking center, itself. However, **2** is extremely insoluble in about 15 solvents ranging from polar to nonpolar, dipolar aprotic to aromatic etc. After high shear stirring in liquid DCPD at 40 and 80 °C, only 1/10 of 1 mole % of **2** could be dissolved! However, after ROMP (Scheme 7), the resulting DCPD resin's T_g was increased by 15 °C! Thus, this tiny amount of POSS uptake into the resin matrix enhanced significantly both the T_g and the storage modulus, E' in the rubbery region (see Figure 4).

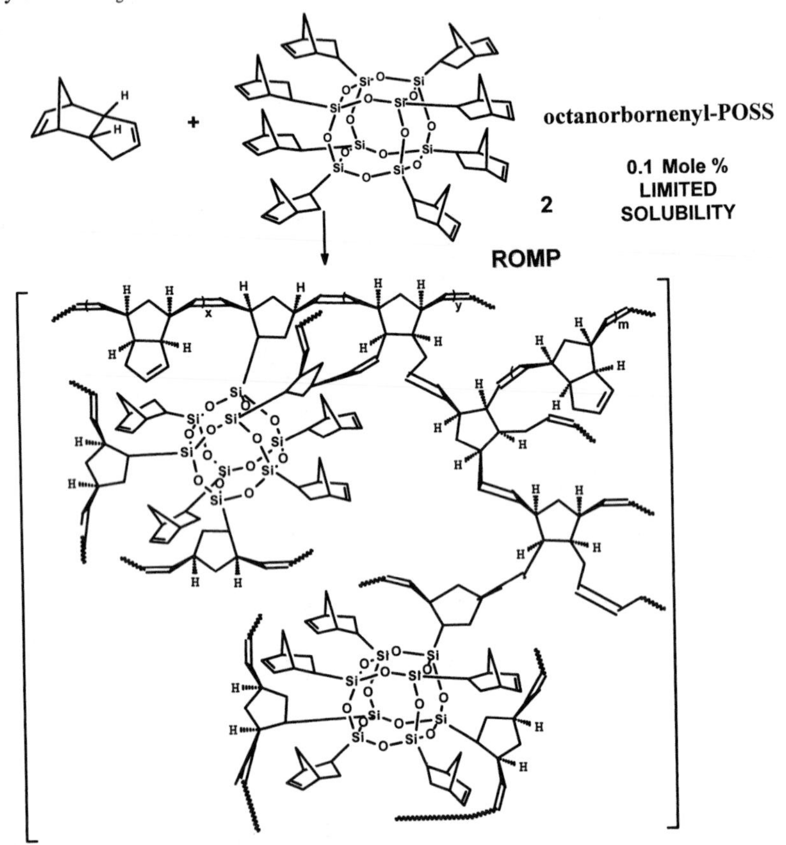

Scheme 7. Ring-opening metathesis curing of DCPD with octanorbornenyl POSS-**2**

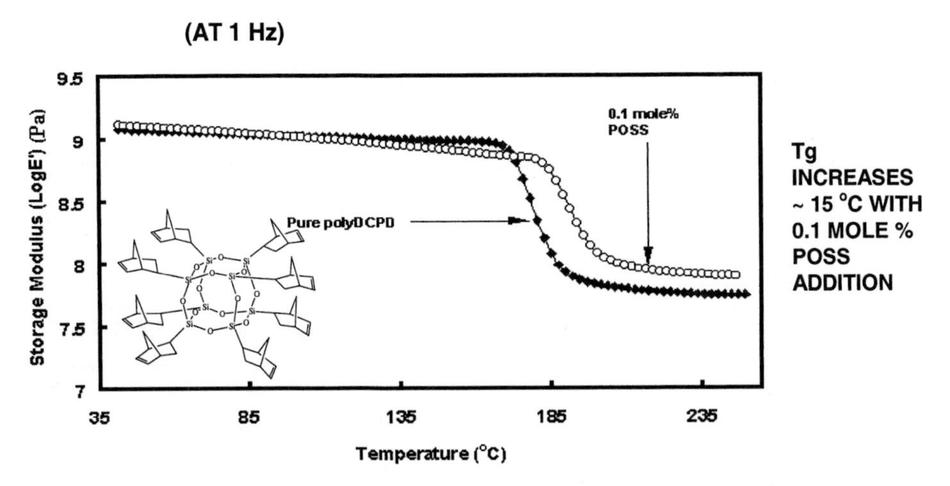

Figure 4. Bending storage moduli, E', vs. temperature for poly(DCPD) versus poly(DCPD-co-0.1 mole % octanorbornenyl POSS-**2**).

A likely explanation of the strong effect of **2** on T_g and E' involves early incorporation of the dissolved **2** into the developing matrix, since each macromer contains eight reactive functions. Furthermore, a fairly high crosslink density probably develops around each POSS cage. This may increase the net crosslink density of the overall resin, more than making up for the volume of the POSS moiety which lowers the volume average density of crosslinks, stemming from the original DCPD molecules.

Monofunctional POSS monomer incorporation

Norbornenylethyldisilanol POSS, **7**, and styrylheptacyclopentyl POSS, **4**, are both monofunctional POSS macromers. We have incorporated both of these into DCPD resins via ROMP. The partially opened cage monomer **7** was incorporated in from 0.4 to 3.0 mole % amounts (using the same ruthenium catalyst employed for **1** and **2**) with the final cure stages conducted at either 260 °C or 280 °C for 2hr. (Scheme 8). The POSS-**7** could not be extracted despite vigorous attempts to do so, confirming its chemical bonding into the matrix. However, **7** is monofunctional and cannot serve as a crosslinking site. Due to the large volume of the monomer, its incorporation will lower the cross-link density of the highly cross-linked DCPD resin. Indeed, this interpretation explains the lower T_g values of these POSS-containing resins versus that of the DCPD resin without POSS. The T_g values for resins cured to 260 °C were 182.5 °C for pure DCPD, 172.6 °C for DCPD/5 wt% **7**, and for DCPD/10 wt% **7**. When the final cure temperature was raised to 280 °C, the pure DCPD resin exhibited a higher T_g value and a lower intensity Tanδ peak temperature versus its POSS-containing counterpart. This is in accord with a lower crosslink density per unit volume for the POSS-containing resin.

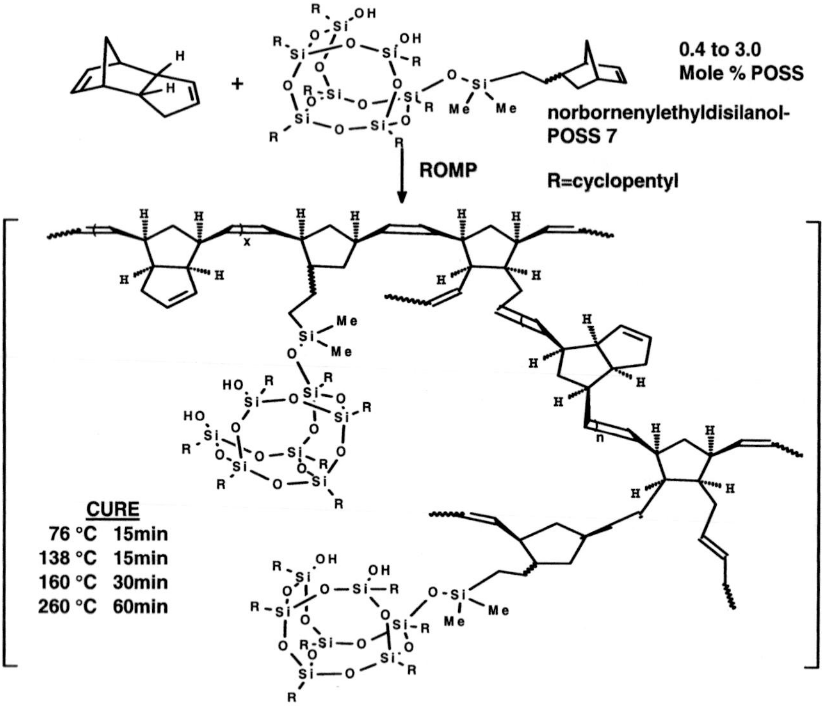

0.4 to 3.0
Mole % POSS
norbornenylethyldisilanol-
POSS 7
ROMP
R=cyclopentyl

CURE
76 °C 15min
138 °C 15min
160 °C 30min
260 °C 60min

Scheme 8. ROMP of DCPD with POSS-**7**

Styrylheptacyclopentyl POSS, **4** also was copolymerized with DCPD via ROMP using the same Grubb's catalyst mentioned earlier (Scheme 9). POSS-**4** was incorporated in 0.4, 0.8, 1.4 and 3.2 mole % amounts. Upon the addition of 0.4 mole % of **4**, the T_g increased as did the bending modulus E' both above and below T_g. However, the T_g and E' dropped after 1.4 mole % of **4** had been incorporated and these values dropped even further upon 3.2 mole % **4** addition. These results are shown in Figure 5. Apparently, monofunctional POSS-**4** addition lowers the crosslink density and increases free volume.

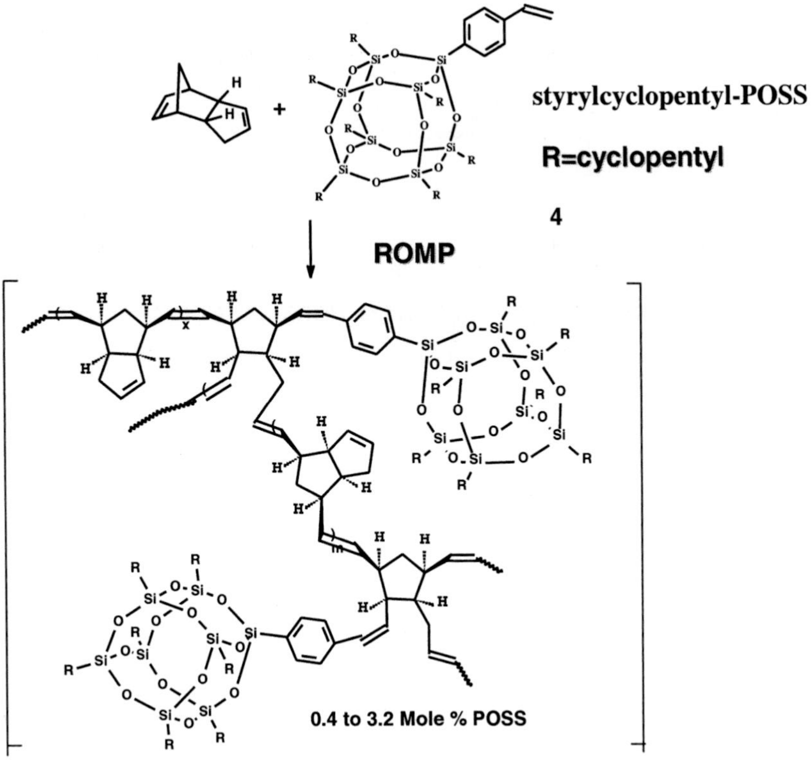

styrylcyclopentyl-POSS

R=cyclopentyl

4

ROMP

0.4 to 3.2 Mole % POSS

Scheme 9. ROMP of DCPD with styrylcyclopentyl POSS-**4**

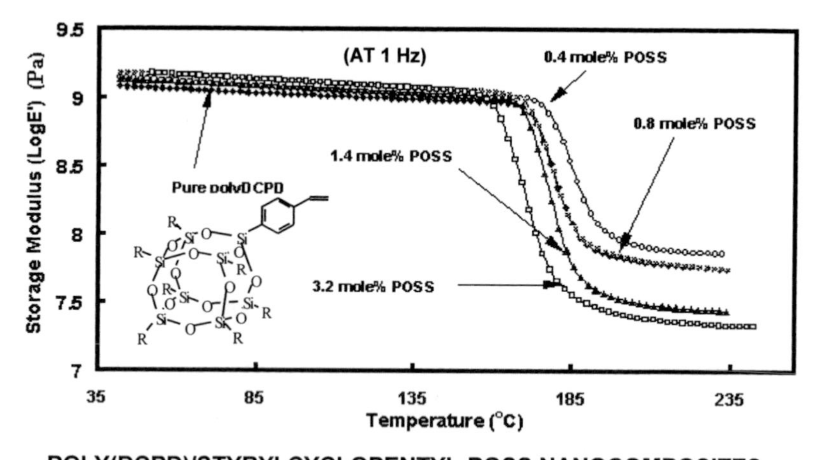

POLY(DCPD)/STYRYLCYCLOPENTYL-POSS NANOCOMPOSITES

Figure 5. A small amount of POSS: T_g increases, E' increases above and below T_g.
Adding more POSS: T_g decreases, E' decreases above T_g, E' increases below T_g.

Resin have now been synthesized using POSS-**4** with styrene and divinylbenzene (DVB).[37] A matrix of various Sty/DVB/POSS-**4** resins was made at several crosslink densities by adjusting the DVB concentrations (0.3, 1.0 and 5.0 wt%). At each crosslink density five different POSS-**4** stoichiometries were synthesized (1.0, 5.0, 10.0, 15.0 and 20 wt%) to provide a total of 15 resins. Thus, the effects high POSS-**4** contents (e.g. 15 and 20 wt%) could be studied at both low (0.3% DVB) and high (5% DVB) crosslinking densities. These effects could be compared to the effects of low POSS-**4** contents (1 and 5 wt%) at both low and high crosslink densities. This series of resins will eventually allow us to examine POSS aggregation as a function of crosslinking and POSS stoichiometry.

The values of the bending storage moduli, Tg and bending tanδ peak intensities are shown in Table 1 for the lightly crosslinked (0.3% DVB) resins as the content of POSS-**4** was increased. The E' values increased both above and below T_g. The tanδ curves broadened and their intensities decreased as the POSS-**4** content went up. However, these Tanδ intensities were higher than that for the sty/0.3 wt% DVB control resin until the POSS content exceeded 5 wt%. It would appear that lower levels of POSS incorporation induced more free volume into the resin (or lowered the crosslink density per unit volume). A similar trend was observed in the T_g values. Adding 1 wt% POSS-**4**, lowered T_g from 92.5 to 91 °C. Thereafter, the T_g was higher. Unlike the pure sty/0.3% DVB resin and lower POSS-**4** content resins, a large increase in the original T_g was observed upon reheating the 0.3% DVB resins containing 15 and 20 wt% POSS-**4**. For example, the T_g increased from 95 °C to 110 °C for the sample with 15 wt% POSS-**4** on the second heating. We suspect that conformational reorganizations of the loosely crosslinked 0.3% DVB network occurs and this internal structural change allows POSS-**4** moieties to aggregate. Thus, the second heating must overcome these forces of aggregation or clustering to activate segmental motion which had previously been activated at the original T_g.

Table 1. Properties of styrene/0.3 wt%DVB/ POSS-**4** resins as a function of POSS-**4** content.

POSS-**4** Content (wt%)	Bending Storage Moduli E' (40°C) (Gpa)	Bending Storage Moduli E' (125°C) (MPa)	T_g from Tan δ peak (°C) First Heating	T_g from Tan δ peak (°C) Second Heating	Bending Tan δ peak on first heating
0	1.496	0.375	92.5	92.6	2.35
1	1.481	0.551	91.0	90.6	2.55
5	1.525	0.546	96.5	97.3	2.48
15	1.704	0.825	95.0	110.0	2.06
20	1.705	0.733	97.0	110.0	1.91

A methacrylic resin series was synthesized which incorporated 3-methacrylylpropyl heptaisobutyl POSS-**6**. Methyl methacrylate (MMA), 1,4-butane dimethacrylate (BDMA) and POSS-**6** were cured by radical initiated polymerization using a methyl ethyl ketone peroxide/Cobalt naphthanate catayst system at 88 °C, followed by a 100 °C/24h postcure (Scheme 10). Twenty-five of these resins were made. The crosslink density was varied into five levels, where the wt% BDMA was 1%, 5%, 10%, 15% and 20%. At each of these BDMA wt% levels, four different amounts of POSS-**6** was incorporated (e.g. 0%, 5%, 10%, 15% and 20%). Then dynamic mechanical properties, swelling studies and densities were obtained.

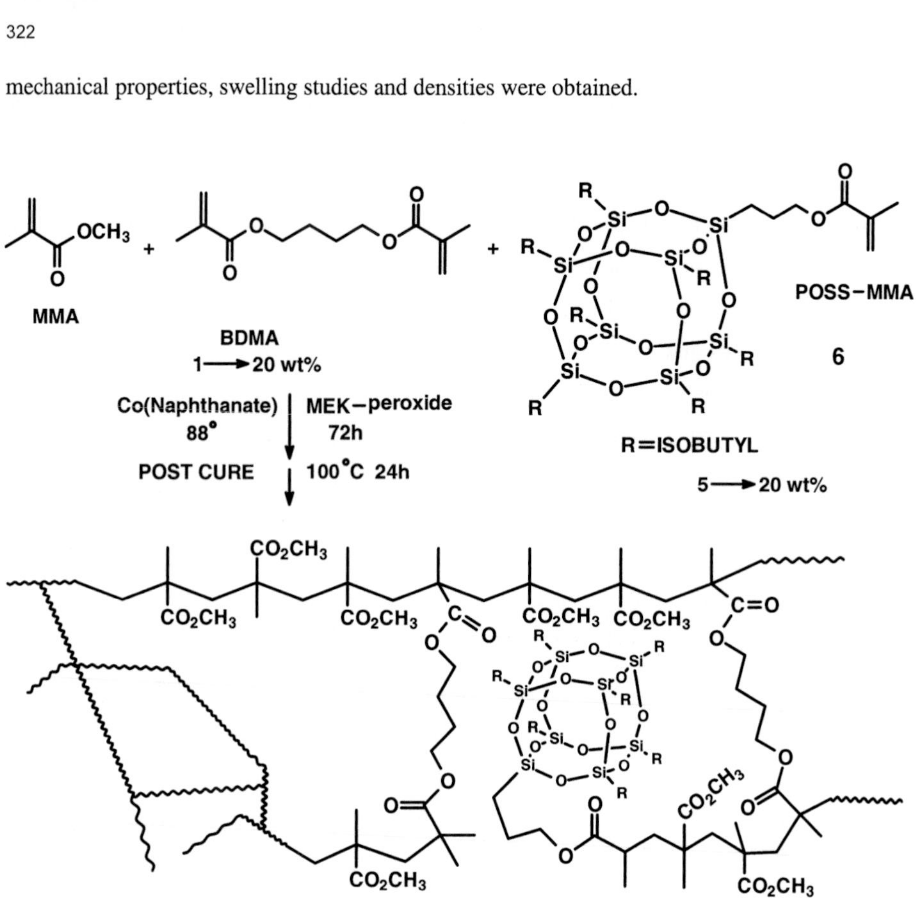

Scheme 10. Formation of MMA/BDMA/POSS-**6** resins via a radical initiated cure.

The T_g and E' (above and below T_g) values, swell ratios in THF and density measurements are summarized in Table 2 for MMA/BDMA/POSS-**6** resins with low crosslink densities (e.g. with 1 and 5 wt% BDMA). The T_g values become lower as the amount of POSS-**6** increases from 0 to 20 wt% at both 1 and 5 wt% BDMA crosslink densities. These differences are larger for the less crosslinked (1 wt% BDMA) resins. A drop also occurs in bending moduli, E', (below T_g) as the POSS-**6** content goes up. The changes in E' at 180 °C are irregular and may involve changes in POSS-**6** interactions within the resin matrices. The swelling ratios (volume in THF/dry volume) of the 1 wt% BDMA resins was always higher when POSS-**6** was present. At the higher crosslink density of 5 wt% BDMA, the swelling ratio increased going from 0 to 5 to 10 wt% POSS-**6** before dropping at 15 and 20 wt% POSS-**6**. In all cases the resin densities dropped upon increasing the amount of POSS-**6**. Therefore, it seems clear that the addition of POSS-**6** lowers crosslink density and increases the free volume in these two sets of methacrylic resins.

Table 2. Properties of MMA-co-BDMA-co-POSS-**6** resins with low crosslink densitites.

Composition wt%			Tg	E'	E'	Swell	*p*
MMA	BDMA	POSS	(°C)	(40 °C) GPa	(180 °C) MPa	Ratio (THF)	g/cm^3
99	1	0	132.2 ↓	1.573 ↓	1.599	3.14	1.188 ↓
94	1	5	126.9	1.464	0.900	3.97	1.186
89	1	10	127.6	1.416	1.251	3.69	1.183
84	1	15	126.2	1.323	1.192	3.57	1.179
79	1	20	123.7	1.136	0.920	3.66	1.176
95	5	0	133.0 ↓	1.298 ↓	2.001	2.55	1.188 ↓
90	5	5	132.2	1.585	2.141	2.66	1.185
85	5	10	130.2	1.361	2.331	3.34	1.182
80	5	15	129.4	1.117	1.551	2.11	1.180
75	5	20	129.1	1.101	2.199	2.17	1.175

Table 3 summarizes the same properties for a series of MMA/BDMA/POSS-**6** resins with high crosslink densities (10 and 20 wt% BDMA). Again, the values of T_g drop steadily with an increase in POSS-**6** at 10 wt% BDMA but at 20 wt% BDMA this trend no longer exists. As more POSS-**6** is added, the low temperature E' values and the densities steadily decrease in both sets (10 and 20 wt% BDMA) of resins. No clear trends are observed in the swelling ratios or E' (180 °C) values as the POSS-**6** content is increased in the highly crosslinked (10 and 20 wt% BDMA) resins.

Table 3. Poly(MMA-co-BDMA-co-POSS-**6**) networks with high crosslink densities.

Composition wt%			Tg (°C)	E' (40 °C) GPa	E' (180 °C) MPa	Swell Ratio (THF)	ρ g/cm^3
MMA	BDMA	POSS					
90	10	0	140.4	1.256	3.877	1.94	1.188
85	10	5	139.1	1.573	3.756	2.05	1.185
80	10	10	137.7	1.404	4.718	1.98	1.180
75	10	15	133.4	1.293	3.701	2.12	1.172
80	20	0	150.3	1.364	11.992	1.46	1.188
75	20	5	154.4	1.341	11.355	1.59	1.184
70	20	10	151.6	1.206	11.674	1.63	1.182
65	20	15	151.8	1.204	10.730	1.54	1.177
60	20	20	155.6	1.171	11.097	1.72	1.176

Conclusions

POSS-containing resin systems can exhibit a range of behaviors depending on (1) the nature of the R-groups on POSS, (2) if multiple or single polymerizable groups are present on POSS, (3) the reactivity of the polymerizable functions versus those of the other monomers, (4) the tendency of the POSS system to aggregate both before and after polymerization, (5) the POSS stoichiometry, (6) the crosslink density and many other features. The studies of crosslinked resins containing these hybrid systems is in its infancy.

Acknowledgements

This work was supported by the Air Force Office of Scientific Research, Grant No. F49620-02-1-0260 (Polymer Organic Matrix Composite Program) and by the Air Force Office of Scientific Research STTR Contract No. F49620-02-C-0086.

[1] R. H. Baney, M. Itoh, A. Sakakibara, T. Suzuki, *Chem.Rev.* **1995**, *95*, 1409.
[2] D. W. Scott, *J. Am. Chem. Soc.* **1946**, *68*, 356.
[3] J. F. Brown Jr., J. H. Vogt Jr., A. Katchman, J. W. Eustance, K. W. Kiser, K. W. Krantz, *J. Am. Chem. Soc.* **1960**, *82*, 6194.
[4] H. Adachi, E. Adachi, O. Hayashi, K. Okahashi, *Rep. Prog. Polym. Phys. Jpn.* **1985**, *28*, 261.
[5] H. Hata, S. Komasaki, *Japanese Patent* Kokai-S-59-108033, **1984**.
[6] X. Zhang, L. Shi, *Chin. J. Polym. Sci.* 1987, *5*, 197.
[7] G. Z. Li, M. L. Ye, L. H. Shi, *Chin. J. Polym. Sci.* **1994**, *12(4)*, 331.
[8] G. Z. Li, M. L. Ye, L. H. Shi, *Chin. J. Polym. Sci.* **1996**, *14(1)*, 41.
[9] G. Z. Li, T. Yamamoto, K. Nozaki, M. Hikosaka, *Polymer* **2000**, *41(8)*, 2827.
[10] G. Z. Li, T. Yamamoto, K. Nozaki, M. Hikosaka, *Macromolecular Chemistry and Physics* **2000**, *201*, 1283.
[11] J. F. Brown, Jr., *J. Polym. Sci., C* **1963**, *1*, 83.
[12] X. Zhang, L. Shi, S. Li, Y. Lin, *Polym. Degrd. Stab.* **1988**, *20*, 157.
[13] *Japanese Patent* Kokoku-S-60-17214 (1985), T. Suminoe, Y. Matsumura, O. Tomomitsu.
[14] *U.S. Patent* 4399266 (1983), Y. Matsumura, I. Nozue, O. Tomomitsu, T. Ukachi, T. Suminoe.
[15] *European Patent* 0406911A1 (1985), S. Fukuyama, Y. Yoneda, M. Miyagawa, K. Nishii, A. Matsuura.
[16] Z. Xie, Z. He, D. Dai, R. Zhang, *Chin. J. Polym. Sci.* **1989**, *7(2)*, 183.
[17] G. E. Maciel, M. J. Sullivan, D. W. Sindorf, *Macromolecules* **1981**, *14*, 1607.
[18] G. Engelhavdt, H. Jancke, E. Lippmaa, A. Samoson, *J. Organomet. Chem.* **1981**, *210*, 295.
[19] H. Adachi, E. Adachi, O. Hayashi, K. Okahashi, *Rep. Prog. Polym. Phys. Jpn.* **1986**, *29*, 257.
[20] C. L. Frye, W. T. Collins, *J. Am. Chem. Soc.* **1970**, ***92***, 5586.
[21] V. Belot, R. Corriu, D. Leclerq, P. H. Mutin, A. Vioux, *Chem. Mater.* **1991**, *3*, 127.
[22] *U.S. Patent* 5942638 (1999), J. D. Lichtenhan, J. J. Schwab, F. J. Feher, D. Soulivong.
[23] J. D. Lichtenhan, J. J. Schwab, W. A. Reinerth Sr., *Chemical Innovation*, **2001**, *1*, 3.
[24] J. D. Lichtenhan, *Comments Inorg. Chem.***1995**, *17*, 115.
[25] A. Voigt, *Orgametrallics* **1996**, *15*, 5097.
[26] F. J. Feher, K. J. Weller, *Inor. Chem.* **1991**, *30*, 880.
[27] M. W. Ellsworth, D. L. Gin, *Polymer News* **1999**, *24*, 331.
[28] T. S. Haddad, R. Stapleton, H. G. Jeon, P. T. Mather, J. D. Lichtenhan, S. Phillips, *Polym.Prepr.* **1999**, *40(1)*, 496.
[29] L. Zheng, R. J. Farris, E. B. Coughlin, *J. Polym. Sci. A. Polym. Chem.* **2001**, *39*, 2920.
[30] A. J. Waddon, L. Zheng, R. J. Farris, E. B. Coughlin, *Nanoletters* **2002**, *2(10)*, 1149.
[31] L. Zheng, R. J. Farris, E. B. Coughlin, *Macromolecules* **2001**, *34*, 8034.
[32] L. Zheng, A. J. Waddon, R. J. Farris, E. B. Coughlin, *Macromolecules* **2002**, *35*, 2375.
[33] C. U. Pittman, Jr., L. Wang, H. Ni, G. Z. Li, Air Force Office of Scientific Research Polymer Matrix Composites Contractor Review Meeting, May 11-12, Long Beach, CA, 2001.
[34] G. Z. Li, L. Wang, H. Toghiani, T. L. Daulton, K. Koyama, C. U. Pittman, Jr., *Macromolecules* **2001**, *34(25)*, 8686.
[35] G. Z. Li, L. Wang, L. T. Daulton, C. U. Pittman, Jr. , Polymer **2002**, *43(15)*, 4167.
[36] C. U. Pittman, Jr., POSS^{TM} Nanotechnology Conference, **2002**, Sept. 25-27, Huntington Beach, CA, complete papers based on this conference are available on CD (contact Hybrid Plastics by e-mail at info@hybridplastics.com. Also see H. Ni, MS Thesis **2001**, Mississippi State University.
[37] C. U. Pittman, Jr., G. Z. Li, submitted for publication.

Conformational Analysis of Siloxane-Based Enzyme-Mimic Precursors[1]

Carol A. Parish, Martel Zeldin, Sean Hilson, Jennifer Pratt*

Department of Chemistry, Hobart and William Smith Colleges, NY 14456, USA
E-mail: parish@hws.edu

Summary: The conformational flexibility of six hybrid organodisiloxane oligomers were studied using the Low Mode-Monte Carlo conformational search method with the MM2* force field and the Generalized Born/Surface Area continuum solvent model for water., These systems have enzyme-like properties as synthetic acyltransferases and contain aminopyridine groups in various states of protonation. An ensemble of low energy structures was generated and used to investigate the dependence of molecular shape and flexibility on protonation, which plays an important role in catalyst solubility and self-association. The results as measured by the number of unique conformations, end-to-end or longest intramolecular distance and radius of gyration of the conformational point cloud indicate that the number of protonated pyridines plays a significant role in the overall molecular shape. A similar study was also carried out on various POSS-substitutive organodisiloxane oligomers.

Keywords: acyltransferase; conformational analysis; 4-dialkylaminopyridine; molecular modeling; organosiloxane ladder and cage compounds; siloxane oligomers

Introduction

Oligomeric *bis*(trimethylene)aminopyridinedimethyldisiloxanes (**A**) are effective catalysts for the hydrolysis of activated esters of alkanoic acids in aqueous media.[2-4] The design of the oligomeric catalyst took into consideration (1) the supercatalytic behavior of the aminopyridine group, (2) the water solubility of the partially protonated oligomers at pH 8, and (3) the inclusion of the siloxane units in the hydrocarbon backbone to impart orientational flexibility and lipophilicity toward organic substrates. A remarkable feature of **A** is that it exhibits saturation kinetics in the hydrolysis of activated esters (e.g. *p*-nitrophenyl alkanoates) with a maximum effect at C_n=14. Moreover, the reaction kinetics conform to a Michaelis-Menten model for enzyme-like behavior with a turnover number comparable to known natural acyltransferases; e.g., for cholesterol and chymotrypsin, k_{cat}/K_M = 1.25 x 10^6 for n = 6 and 7.6 x 10^4 for n = 7,

© 2003 WILEY-VCH Verlag GmbH & KGaA, Weinheim

CCC 1022-1360/00/$ 17,50+.50/0

respectively; for **A**, $k_{cat}/K_M = 1.1 \times 10^4$ for n = 14. We have launched a computational study to gain insight into the structural conformations of **A** in polar solvents by examining some model oligomers (i.e. trimers). The atom numbering sequence and the torsional degrees of freedom for the unprotonated (**B**) and all the possible protonated (**C–G**) species that were evaluated computationally are represented in Fig. 1. We have also extended the study to other analogous catalytically active organosiloxanes that have 3D-cage structures comparable to oligomeric silsesquioxanes (**H**).

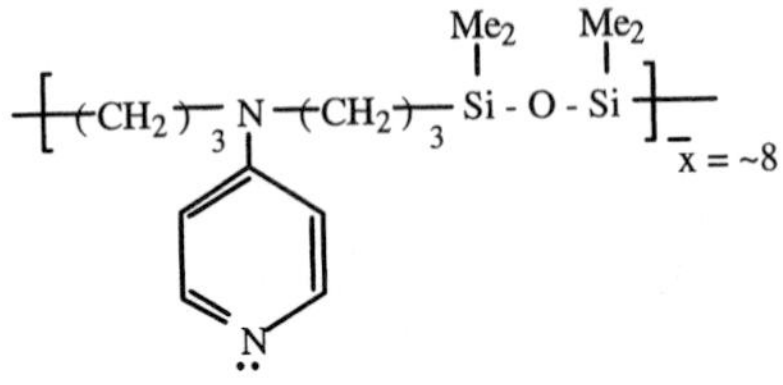

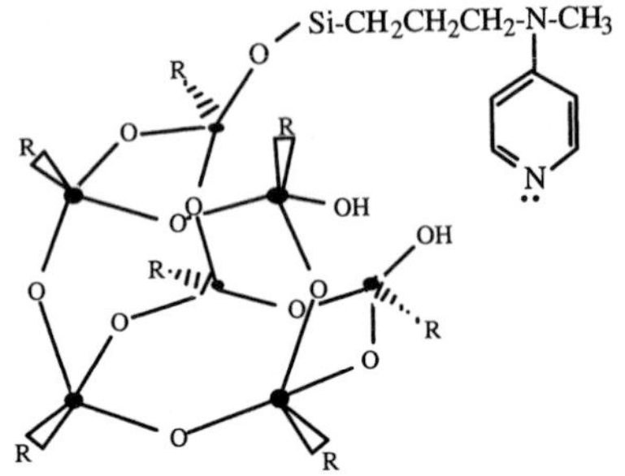

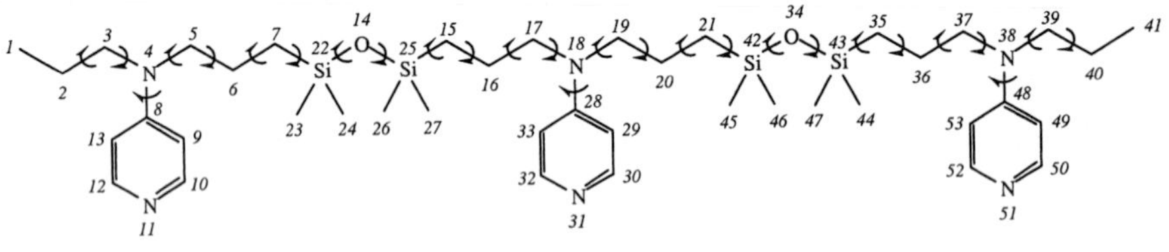

B, no H^+; **C**, H^+ on N11; **D**, H^+ on N31; **E**, H^+ on N11 & N31; **F**, H^+ on N11 & N 51; **G**, H^+ on N11, N31 & N51

Fig. 1. Model oligomers (trimers) B-G and the atom numbering sequence and torsions used in the conformational calculations.

Methods

The conformational ensembles were calculated using version 7.2 of the MacroModel[5] suite of software programs running on 800 MHz Athlon PCs under the RedHat LINUX 6.2 operating system. The MM2* force field[6] as implemented in MacroModel V7.2 was chosen since it was the most well parameterized MacroModel force field available for the molecular system under study and it contained siloxane bond lengths, angles and torsional barriers that were in good agreement with previously published results.[7] Solvent effects were included using the Generalized Born/Surface Area (GB/SA) continuum model for water.[8] The Low Mode (LM) search method[9] was used in a 1:1 combination with the Monte Carlo (MC) search method[10] to explore the potential energy surfaces of the model systems. Torsions that were varied during the searches are shown Figure 1. Ensembles generated with each conformational search method were grouped into geometrically similar families using the Xcluster program.[11]

Results and Discussion

Trimer **B** contains three aminopyridinium moieties connected through trimethylenedisiloxane units. In addition to the unprotonated **B**, the trimer can also exist as five uniquely different protonated molecules; viz. two monoprotonated (**C**, **D**), two diprotonated (**E**, **F**) and one triprotonated (**G**) species. The conformational space for each of these trimers in water was exhaustively examined. An analysis of the enthalpic and solvation energies, along with distance parameters of the lowest energy structure of each species (**B'**-**G'**) is summarized in Table 1. The size and shape of the lowest energy structures are dependent on the number of pyridinium moieties and their distribution in the oligomer. For example, **B'** has one of the more compressed low energy structures as indicated by its longest molecular dimension (13.20 Å), which is measured as the N11–N51 distance. In addition, the hydrocarbon backbone bonds in **B'** are all *trans* and the orientation about the siloxane bonds Si(22)–O(14), O(14)–Si(25), Si(42)–O(34) and O(34)–Si(43) is θ=53 (gauche$^+$), θ=–128 (2*gauche$^-$), θ=163 (*trans*) and θ=–177 (*trans*), respectively. Structure **D'**, with one protonated pyridine in the middle of the oligomer backbone, is even more compressed than **B'** with a longest molecular dimension (10.71 Å) also corresponding to a distance between atoms N11–N51. It is noteworthy that one of the hydrocarbon backbone torsions in **D'** is twisted away from the *trans* orientation producing a "kink' in the chain that allows the protonated pyridinium group to extend away from the rest of the molecule. The driving force for the structure of **D'** is a desire to extend the protonated moiety as far into the solvent medium as possible. This causes the rest of the molecule to collapse on itself leading to a very compressed structure.

Table 1. Energetic and structural comparison of the lowest energy structures (B'-G') found using the LM:MC 50:50 method.

Structure	Lowest Energy (kJ/mol)	Solvation Energy of Lowest Energy Structure (kJ/mol)	Longest Molecular Distance (Å)	End-to-End Distance (C1 – C41) (Å)
B'	98.5	-11.9	13.2 *(N11-N51)*	9.7
C'	-176.3	-302.2	15.3 *(N11-N51)*	11.2
D'	-177.3	-330.2	10.7 *(N11-N51)*	10.6
E'	-454.1	-726.5	19.5 *(N11-N31)*	9.2
F'	-470.8	-730.6	20.0 *(N11-N51)*	11.1
G'	-745.5	-1266.1	19.9 *(N11-N51)*	10.7

The driving force for the structure of **B'** appears to be an attractive intramolecular π-stacking interaction between two adjacent pyridines. The only other minimum energy structure that displays a π-stacking interaction is **C'** where the siloxane bonds sample the *trans*, gauche^{+} and gauche^{-} orientations, and all hydrocarbon backbone atoms are in the *trans* orientation but one. The one non-*trans* hydrocarbon orientation produces a kink in the backbone that allows the chain to wrap back on itself to assume less than a 4 Å interaction between the aromatic rings. Structures **E'** and **F'** with two and **G'** with three protonated sites are relatively rigid, extended structures with longest molecular distances of 19 Å or more. In these molecules, the longest molecular dimension corresponds to the distance between positively charged nitrogen atoms. This causes the chain to fold back on itself and results in a shorter end–to–end distance (viz., 9.2-11.1 Å). The lowest energy structures of **E'-G'** adopt conformations in which the positively charged pyridine nitrogens are as far apart from each other as possible, presumably driven by both intramolecular electrostatic repulsions and solvation. This is achieved by forming kinks in the hydrocarbon backbone.

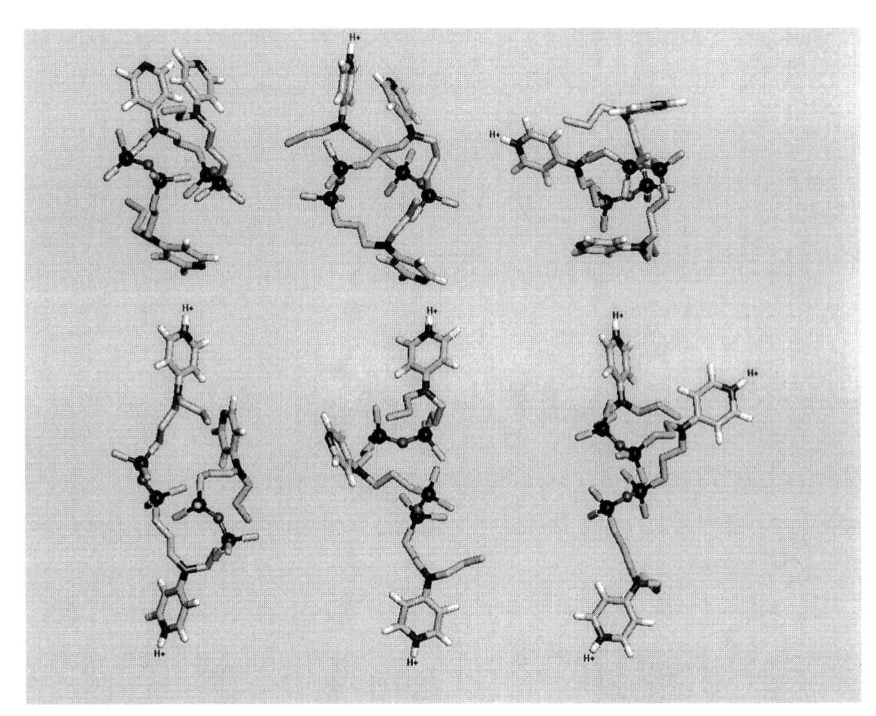

Fig. 2. Lowest energy structures, B'-G' (in sequence from upper left to lower right).

A detailed energetic comparison of isomeric pairs **C'/D'** and **E'/F'** is shown in Table 2. For example, the energies of **C'** and **D'** are similar. The stabilization of **D'** arises mainly from the more negative solvation energy (Δ=28 kJ/mol), which is determined from the polarization energy by the Generalized Born (GB) equation and the solvent accessible surface area (SASA). The SASA of **D'** is smaller than **C'** by ~28 $Å^2$ in agreement with length analysis, and indicates that **D'** is more compressed. The polarization energy for **D'** is larger than **C'** by about 26 kJ/mol, and larger than the SASA contribution. This leads to a significantly larger stabilization for **D'** by solvation. Thus, the molecule having a protonated pyridinium moiety that is sandwiched between the two-unprotonated pyridines is slightly more stable than having the pyridinium ion at a terminal position.

A comparison of the energies of the low energy structures **E'** and **F'** indicates that their isomeric stabilities differ by ~17 kJ/mol, whereas the solvation energies are comparable.

Table 2. Comparison of energetics and distances for the lowest energy isomeric structures C'-F'.

Structure	E (kJ/mol)	Solvation E (kJ/mol)	GB Polarization (kJ/mol)	SASA (Å ^{2})	Longest Distance (Å)	End–to–End Distance (Å)
C'	-176.3	-302.2	-332.7	876.4	15.3 *(N11 – N51)*	11.2
D'	-177.3	-330.2	-359.4	848.5	10.7 *(N11 – N51)*	10.6
E'	-454.1	-726.5	-759.7	954.2	19.5 *(N11 – N31)*	9.2
F'	-470.8	-730.6	-763.4	973.7	20.0 *(N11 – N51)*	11.1

The SASA and GB polarization energies are similar and lead us to conclude that the isomer having the pyridinium groups on either end is an enthalphically favored configuration.

It is informative to examine the ensembles of all low-lying structures that are found during the conformational search since the ensemble behavior is representative of the true nature of the molecule. The number of unique low energy minima for each system is shown in Table 3. After 55 000 steps, the number of new conformations found within 25 kJ/mol of the lowest energy structure represented less than a 10% increase in the number of structures on average.

Table 3. Comparison of B-G ensembles (55 000 search steps) using the LM-MC 50:50 method.

Structure	Number of Conformations Found	Weighted Average End-to-End Distance (C1 – C41) (Å)	Weighted Average N11 – N51 Distance (Å)	Radius of Gyration, R_g (Å)
	8460	9.0	10.1	3.0
C	9171	9.6	12.8	2.7
D	9198	7.1	8.6	2.3
E	7540	8.1	14.9*	2.8
F	1526	10.4	19.3	1.7
G	1401	10.4	19.4	1.9

* The N11–N31 distance is used for ensemble **E**.

An analysis of this ensemble data yields information about molecular flexibility and conformationally accessible states. Structures **B-E** are much more flexible than either **F** or **G** as evidenced by the number of unique conformations found on the respective potential energy surfaces (Table 3). As more positive charge is introduced into the molecule, the rigidity of the molecule increases; e.g., **E** has the same number of positively charged nitrogens as **F**, however, **E** has two charges on adjacent pyridines. Thus, there is an uncharged "tail" on **E**, which samples the locally available conformational space and leads to greater conformational flexibility. The introduction of positive charge at the termini restricts flexibility by anchoring the charged ends into the solvent medium.

The rotational orientation of the siloxane bonds in each molecule in each ensemble was examined. The torsional bonds favor the *trans*, gauche$^+$ and gauche$^-$ orientation. The rotational orientation of the hydrocarbon backbone atoms was also examined. Unlike the siloxane bonds the hydrocarbon torsions adopt almost exclusively *trans* conformations; in those instances where the bond was other than *trans*, a kink was introduced and readily apparent in the corresponding molecular structure. The XCluster program was used to determine if the ensembles naturally form structurally related groupings and if the groupings are the same or different for each ensemble. Clustering by atomic and backbone torsional RMS after rigid body superposition of all heavy atoms did not lead to strong clustering in any of the ensembles as evidenced by distance maps, mosaics and separation ratios. This provides further evidence of conformational flexibility.

The ensembles were examined for molecular size. The C(1)–C(41) distance was measured for each structure in each ensemble to determine the change in intramolecular distance from molecule to molecule. Ensembles for **B** and **C** contain a similar number of unique structures (8460 and 9171, respectively). Their end-to-end distance histograms are also quite similar. The distances range from 3.6 Å to 17.0 Å with a weighted average of 9.0 Å and 9.6 Å, respectively. The end–to–end distance data for **C** indicates that the structure behaves more like **B** than like **D**. The molecular ensemble for **D** displays the shortest weighted average end–to–end distance of 7.07 Å and suggests that the two unprotonated terminal pyridines are very flexible and allow the molecule to adopt a more compressed conformation. **E** displays significantly different end–to–end distances than **F**. The weighted average distance for **E** is 8.14 Å, whereas the average for **F** is 10.4 Å. **F** is clearly a more rigid system as evidenced by the significantly smaller number of unique conformations found; 7540 for **E** and 1526 for **F**. The flexibility of **E** versus **F** is also confirmed by the distribution of end–to-end distances; 3.5 Å to 14.9 Å for **E** and is 6.1 Å to 16.5 Å for **F**.

G is clearly a rigid system; i.e. 1401 unique structures with end-to-end distances ranging from 6.8 Å to 17.5 Å and a weighted average of 10.4 Å. The low energy structures for **G** are mainly extended with large end–to–end distances. It is apparent that the number of protonated pyridines controls the overall molecular shape and flexibility. To illustrate, Figure 3 superimposes the end–to–end distance data for ensembles **B** (gray tone) and **G** (black).

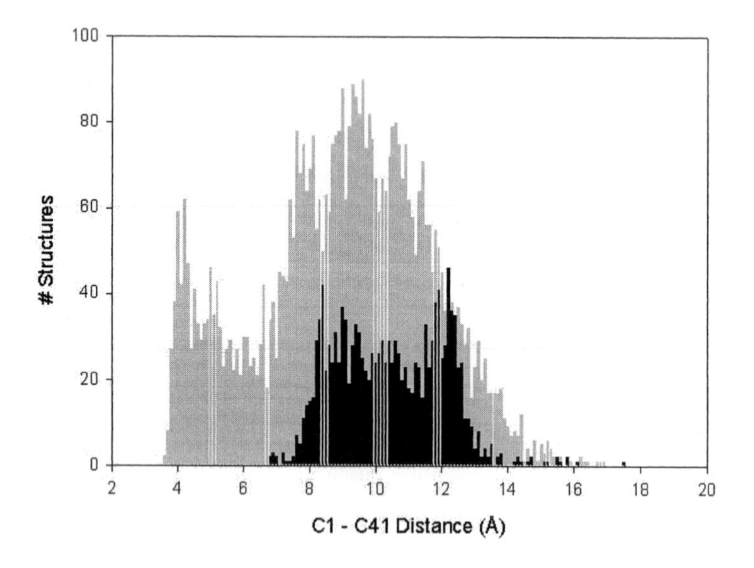

Fig. 3. Comparison of the end-to-end distance for the ensembles B (gray) and G (black).

The longest molecular distance in each lowest energy structure was also determined (Table 1). The longest molecular distance in all structures is between N11 and N51 except for **E'** where the longest distance is between N11 and N31. It is interesting to note that these lengths correspond to the space between pyridines or pyridinium nitrogens at either end of the trimeric chain except for **E'** where the distance corresponds to the length between the two charged pyridinium nitrogens, one in the middle and one at the end of the chain. The longest dimension in **E'-G'** is noticeably larger than in **B'-D'**. The difference between **E'-G'** and **B'-D'** is attributed to a repulsive interaction between positively charged pyridinium ions. If we examine the ensemble behavior of these distances, it is observed that **F'** and **G'**, with two pyridinium ions anchored at either end or three pyridinium ions, display markedly different length data than **B'-E'**. For example, the N11–N51 distances in **B'-D'** have a range between ~3 Å to ~17 Å with weighted averages of 10.1 Å, 12.8 Å and 8.6 Å, respectively. The N11–N31 distance in **E'** ranged from 6.7 Å to 19.9 Å with a weighted average distance of 14.9 Å. Yet the longest distances in **F'** and **G'** ranged from 12.3 Å to 20.9 Å and from 10.6 Å to 21.0 Å with averages of 19.3 Å and 19.4 Å, respectively.

There is no guarantee that either the end-to-end or longest molecular distances in the lowest energy structures are representative of the shape of the structures in the corresponding ensembles. Therefore, we used the XCluster program to compute the radius of gyration, R_g, of the conformational point cloud for all structures in each ensemble. The structures in each ensemble are subjected to an optimal atomic RMS superimposition and from this an ensemble centroid is determined for use in calculating the radius of gyration. Used in this sense, R_g is a practical measure of the conformational diversity of each ensemble; i.e., ensembles with smaller R_g values contain conformations that are very similar whereas ensembles with larger R_g values contain structurally more diverse conformations. This provides an approximate measure of the molecular shapes involved in each ensemble without biasing the measurement to any one intramolecular distance criterion. The overall ensemble radius of gyration for **B'-G'** is summarized in Table 3. The results indicate that the distance criteria are representative of the overall ensemble behavior.

Acknowledgment

This project was partially supported by NSF grant CHE-0116435 as part of the MERCURY supercomputer consortium, the Donors of the Petroleum Research Fund, administered by the American Chemical Society, and by a grant from the Merck Foundation administered by the AAAS.

[1] This study has been published in part: C. A. Parish, M. Zeldin and J. Pratt, *J. Inorg.Organometal. Polym.*, **12**, 31 (2002).
[2] S. Rubinsztajn; M. Zeldin; W. K. Fife, *Macromolecules* **23,** 4026 (1990).
[3] W. K. Fife; S. Rubinsztajn; M. Zeldin, *J. Am. Chem. Soc.* **113,** 8535 (1991).
[4] S. Rubinsztajn; M. Zeldin; W. K. Fife, *Macromolecules* **24,** 2682 (1991) and refs.therein.
[5] F. Mohamadi; N. G. J. Richards; W. C. Guida; R. Liskamp; M. Lipton; C. Caufield; G. Chang; T. Hendrickson; W. C. Still, *J. Comput. Chem.* **11,** 440 (1990).
[6] N. L. Allinger, *J. Am. Chem. Soc.* **99,** 8127 (1977).
[7] I. Bahar; I. Zuniga; R. Dodge; W. L. Mattice, *Macromolecules* **24,** 2986 (1991), S. Grigoras; T. H. Lane In *Advances in chemistry series 224: Silicon-based polymer science a comprehensive resource*; J. M. Zeigler, F. W. G. Fearon, Eds.; American Chemical Society: Washington, D. C., 1990; pp 127, S. Grigoras; T. H. Lane, *J. Comput. Chem.* **9,** 25 (1988).
[8] J. Weiser; P. S. Shenkin; W. C. Still, *J. Comput. Chem.* **20,** 586 (1999).
[9] I. Kolossvary; W. C. Guida, *J. Comput. Chem.* **20,** 1671 (1999).
[10] G. Chang; W. C. Guida; W. C. Still, *J. Am. Chem. Soc.* **111,** 4379 (1989).
[11] P. S. Shenkin; D. Q. Mcdonald, *J. Comput. Chem.* **15,** 899 (1994).

Lewis Acidic Organoboron Polymers

*Yang Qin, Guanglou Cheng, Kshitij Parab, Anand Sundararaman, Frieder Jäkle**

Department of Chemistry, Rutgers University-Newark, 73 Warren Street, Newark, NJ 07102, USA

Summary: We have developed a highly efficient new method for the introduction of Lewis acidic boron centers into the side chains of organic polymers. Our methodology involves three steps: (i) the controlled polymerization of a functional monomer, (ii) the exchange of the functional group for Lewis acidic boron centers, and (iii) the fine-tuning of the Lewis acidity of the individual centers through substituent exchange reactions with nucleophiles. This approach gives access to a family of new well-defined organoboron polymers including moderately Lewis acidic poly(arylboronates) and the first examples of highly Lewis acidic fluorinated triarylborane polymers.

Keywords: ATRP; free radical polymerization; inorganic polymers; Lewis acid; organoborane

Introduction

Inorganic and organometallic polymers have over the past two decades emerged as an important area of research in the field of materials chemistry. New devices have, for example, been developed based on polysiloxanes and polysilanes, new advances in medical research have been triggered by the discovery of polyphosphazenes, and more recently, transition metal containing polymers have been developed for nanoscience applications.[1] The incorporation of electron-deficient boron centers into polymer structures is particularly intriguing as it, for example, provides an opportunity to further manipulate the polymers via donor acceptor bonding.[2] Binding of nucleophiles to organoboron polymers can be exploited for the design of new supported reagents and immobilized catalysts[3] and of highly selective sensor materials[4]. Boron containing polymers also play a major role as intermediates in the synthesis of functionalized polymers with polar side-groups[5-7] and are used as polymeric electrolytes for batteries[8], sophisticated flame retardants[9], and as preceramic[10] and photoluminescent materials[11].

Despite these recent advances, a general straightforward method for the synthesis of Lewis acidic boron polymers of controlled molecular weight and well-defined architecture is currently not

© 2003 WILEY-VCH Verlag GmbH & KGaA, Weinheim CCC 1022-1360/00/$ 17,50+.50/0

known. Moreover, the boron centers in most of the reported oligomers and polymers are either electronically stabilized by π-interactions with heteroatoms such as oxygen or nitrogen and/or contain highly bulky substituents. The Lewis acidic sites are therefore often not readily available for binding of Lewis bases. Our research program is aimed at the development of well-defined soluble organoboron polymers and copolymers of controlled architecture, molecular weight and degree of functionalization, in which the substituents on boron can be readily exchanged and consequently the strength of the Lewis acid centers can be fine-tuned (Figure 1).

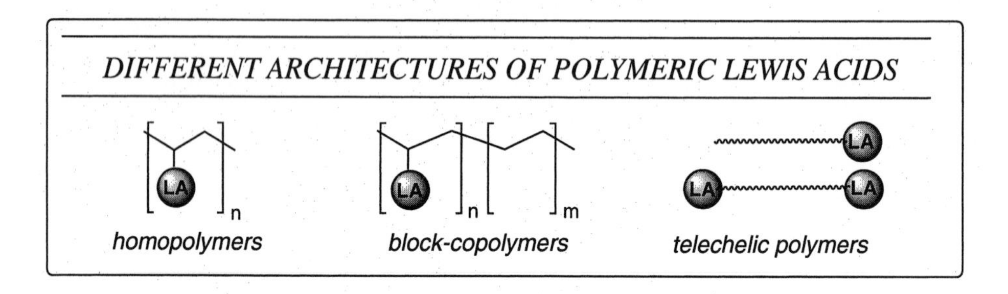

Figure 1. Schematic representation of side-chain boron-based polymeric Lewis acids.

While side-chain organoboron polymers can be prepared from organoboron monomers using a variety of polymerization techniques including standard free radical polymerization[12], metathesis polymerization[13], and Ziegler–Natta polymerization[5], the polymerization of monomers bearing strongly Lewis acidic moieties has been largely restricted to Ziegler-Natta polymerization procedures[5] due to the high reactivity of the monomers. The functionalization of organic polymers or resins in a post-polymerization modification step represents an alternative to the polymerization of borylated monomers. This method has attracted a lot of attention for the development of supported Lewis acid catalyst, in which the Lewis acid centers are attached via multi-step polymer modification reactions with typically moderate degrees of functionalization.[3] The direct borylation of polystyrene has been attempted with boranes X_2BH under forcing conditions, but occurs with low selectivity.[14] In an exciting new development, transition metal catalyzed borylation of polyolefins has been reported by Hartwig and Hillmyer and coworkers to yield boronate-functionalized polymers in one step.[7]

We have developed a new methodology for the synthesis of organoboron polymers that involves three steps: (i) the controlled (co)polymerization of a functional monomer (**S-FG**), (ii) the exchange of the functional group in polymer **PS-FG** for Lewis acidic boron centers to give the borylated polymer **PS-B**, and (iii) the fine-tuning of the Lewis acidity of the individual centers through substituent exchange reactions with nucleophiles, which gives access to a family of new well-defined polymers **PS-BR** bearing Lewis acidic boron centers (Scheme 1).[15]

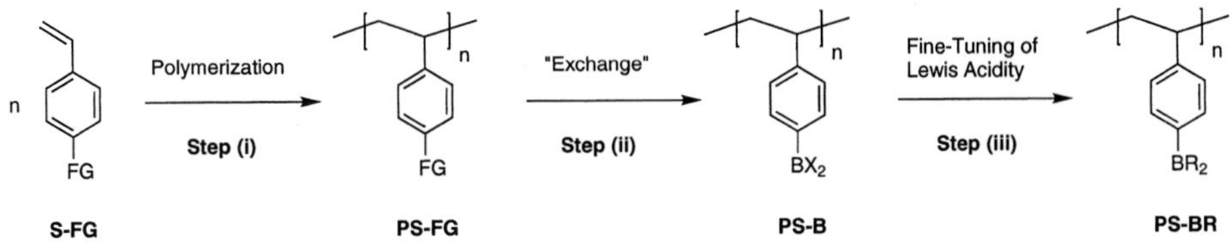

Scheme 1. General method for the synthesis of organoboron polymers of varying Lewis acidity.

Atom Transfer Radical Polymerization of 4-(Trimethylsilyl)styrene

4-Trimethylsilylstyrene was polymerized in anisole (50%) according to a typical protocol for atom transfer radical polymerization (ATRP)[16] initiated with 1-phenylethyl bromide (PEBr) and catalyzed by CuBr / pentamethyldiethylenetriamine (PMDETA) (68.5% conversion within 5 h at 110 °C). The molecular weight and polydispersity of the resulting polymer were determined by GPC analysis relative to polystyrene standards to M_w = 6,540 and PDI = 1.14.[17] Molecular weight analysis by static light scattering measurements gave a similar result (M_w = 6,810) indicating a close similarity of the MW to elution volume relationship for polystyrene and the silylated polymer **PS-Si**.

Kinetic analysis of the polymerization reaction confirmed the controlled nature of the polymerization. The linearity of $\ln[M]_o/[M]$ vs. time in the pseudo first order kinetic plot (Figure 2) indicates that the number of active species in this system remains constant throughout the reaction and the chain-termination reaction is insignificant, i.e. $K_p[P^*]$ = constant, where K_p stands for the rate constant and $[P^*]$ stands for the concentration of active propagation chains. The molecular weight of the active propagation chains increases linearly with increase in monomer conversion and the polydispersity remains narrow throughout the polymerization (*PDI*<1.2). This demonstrates the applicability of our methodology to the preparation of a wide variety of polymers of different molecular weight and architecture, including *block*-copolymers.

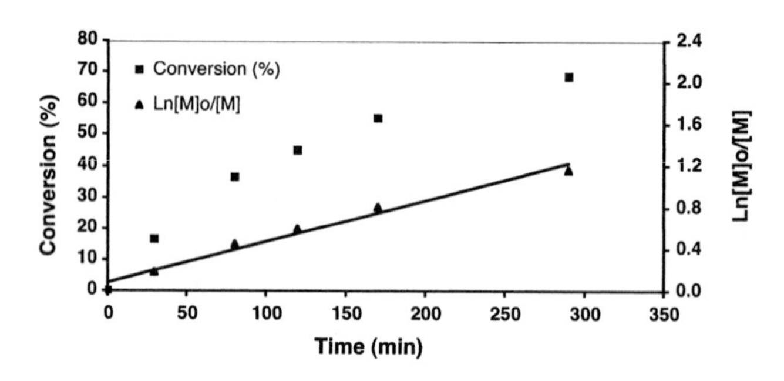

Figure 2. Kinetic data for ATRP of 4-(trimethylsilyl)styrene (50 vol.% in anisole) at 110 °C.

Borylation of Poly(4-trimethylsilyl)styrene

The silylated polymer **PS-Si** was treated with a slight excess of BBr_3, a strong Lewis acid which is known to cleave Si-C(sp^2) bonds under mild conditions with nearly quantitative yields and high selectivity[18]. Selective and quantitative cleavage of the Si-C(sp^2) bonds in **PS-Si** to form **PS-B** was confirmed by multinuclear NMR spectroscopy. Importantly, the signal due to the trimethylsilyl substituents of **PS-Si** in the ^{29}Si NMR spectrum at δ = -4.4 completely disappeared, whereas a new broad resonance developed in the ^{11}B NMR spectrum at δ = 54, in a region typical of arylboron dibromides. The presence of strongly electron-withdrawing substituents on the phenyl rings is reflected in the ^{13}C NMR spectrum (Figure 3). The signal for the carbon atoms in *ortho*-position to the boryl groups at δ = 137.9 is downfield shifted relative to that of **PS-Si** (δ = 133.8/133.5). A broad resonance at δ = 136.2 can be attributed to the carbon atom bearing the boryl substituent and two closely spaced resonances (δ = 152.3, 152.4) are observed for the *ipso*-carbon atom closest to the polymer backbone. Importantly, our data do not show any sign of isomerization reactions at the phenyl rings or of polymer degradation, but rather indicate selective and quantitative borylation of **PS-Si** by BBr_3.

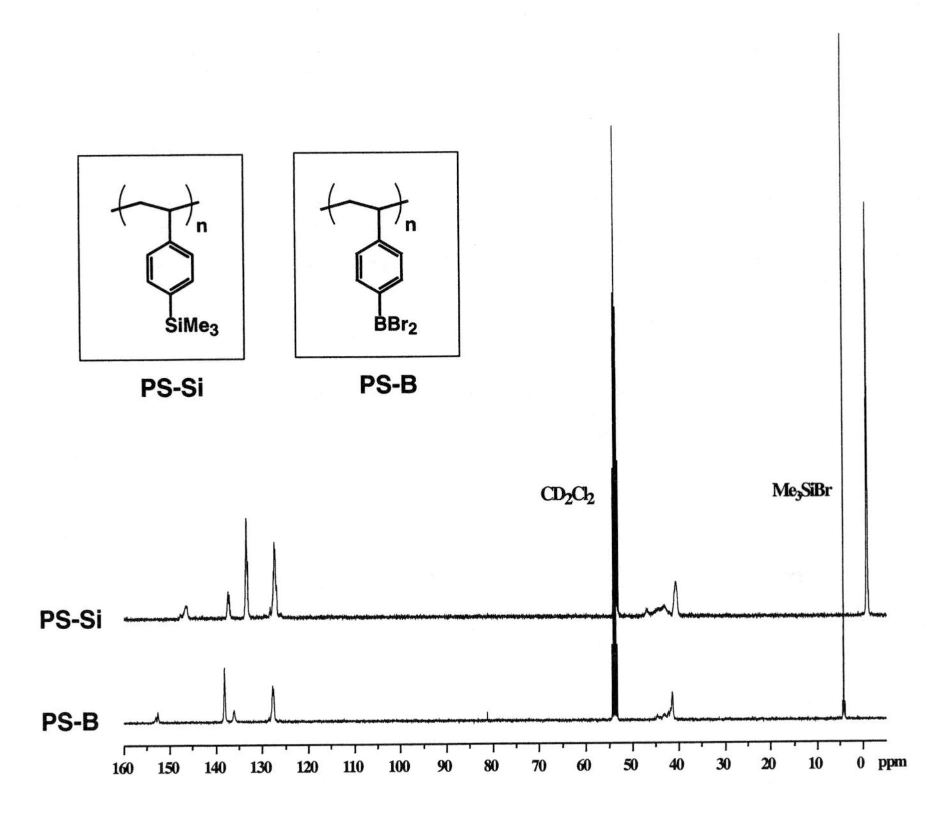

Figure 3. Comparison of the ^{13}C NMR spectra of **PS-Si** and **PS-B**.

Fine-Tuning of Lewis Acidity

The dibromoborylated polymer **PS-B** serves as a precursor to a number of other polymers with boron centers of varying Lewis acidity (Figure 4). When **PS-B** was treated with Me_3SiOEt a weakly Lewis acidic polymeric boronate, **PS-BOEt**, was obtained as a white solid in 90% isolated yield. **PS-BOEt** can readily be converted into the air-stable pinacol derivative **PS-BPin**. Polymers of high Lewis acidity such as **PS-BTh** and **PS-BAr^F** were obtained from **PS-B** with the highly selective aryl transfer reagents trimethylstannylthiophene and pentafluorophenylcopper in 82% and 74% yield, respectively.[19]

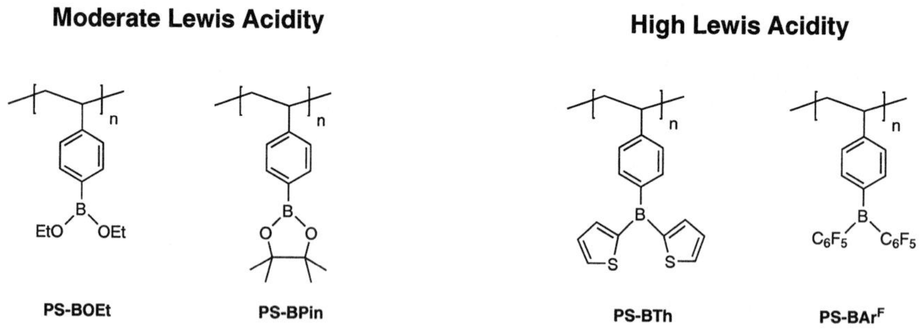

Moderate Lewis Acidity

High Lewis Acidity

PS-BOEt PS-BPin PS-BTh PS-BAr^F

Figure 4. Selected organoboron polymers obtained from reaction of **PS-B** with nucleophiles.

Molecular Weight Determination of Boron-Containing Polymers

The molecular weight of polymer **PS-BPin** was determined by GPC and light scattering analysis. Importantly, both the experimentally determined degree of polymerization and the polydispersity of **PS-BPin** are very close to those of the precursor polymer **PS-Si** (Table 1). This clearly confirms that the borylation and subsequent substituent exchange occur without significant cross-linking or cleavage of the polymer backbone.

Table 1. Comparison of light scattering data for **PS-Si** and **PS-BPin** in THF.

Polymer	*dn/dc* (ml/g)	M_w	*DP*	*PDI*	A_2 (mol·ml/g^2)	R_H (nm)
PS-Si	0.157	10,140	51	1.08	8.774e-4	2.2
PS-BPin	0.138	13,830	53	1.09	4.772e-4	2.4

Estimation of Lewis Acidity via NMR Spectroscopy

***^{11}B NMR Spectroscopy*:** Boron NMR spectroscopy provides a highly useful probe to estimate the

Lewis acidity of organoboron species. An upfield-shifted ^{11}B NMR resonance for the arylboronate polymers **PS-BOR** (R = Et, Pin) at $\delta = 25$ relative to the precursor polymer **PS-B** ($\delta = 54$) is indicative of significant π-overlap between the alkoxy substituents and the boron centers. In contrast, the large chemical shifts of $\delta = 50$ for **PS-BTh** and of $\delta = 56$ for **PS-BAr^F** confirm the expected high Lewis acidity of the triarylborane polymers.

***Formation of Crotonaldehyde Adduct (Childs' Method)*:** The relative Lewis acidity of the boron centers can also be estimated by complexation with crotonaldehyde according to Childs' method[20]. On a scale from 0 to 1.0 (BBr_3 = 1.0) a relative Lewis acidity of 0.60 was determined for **PS-BAr^F**. A pronounced Lewis acidity of **PS-BAr^F** was further confirmed by a large chemical shift difference between the *meta*- and *para*-fluorine atoms of $\Delta\delta_{m,p} = 13.0$ for the free acid (Figure 5) that significantly decreases to $\Delta\delta_{m,p} = 6.4$ in the polymeric crotonaldehyde adduct.

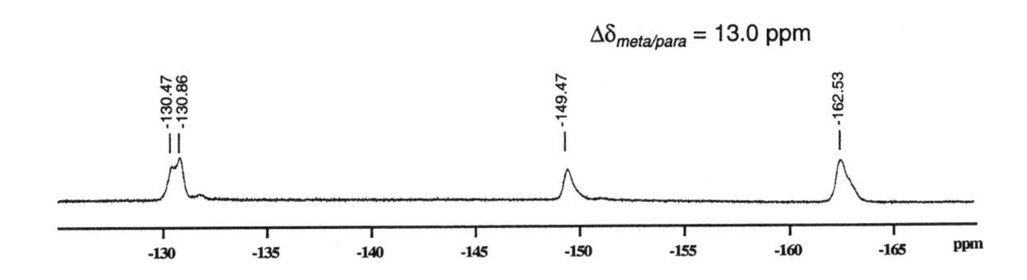

Figure 5. ^{19}F NMR spectrum of poly[4-bis(pentafluorophenyl)borylstyrene] (**PS-BAr^F**).

Conclusions

We have developed a highly efficient new method for the introduction of Lewis acidic boron centers into the side chains of organic polymers. Facile substituent exchange reactions on boron yield a new family of well-defined polymeric Lewis acids including poly(arylboronates) and the first examples of highly Lewis acidic polymeric triarylboranes. Introduction of pentafluorophenyl substituents on boron resulted in formation of **PS-BAr^F**, the first polymeric analog of highly Lewis acidic fluorinated arylboranes, which play a major role as catalyst in organic synthesis[21] and as cocatalysts in olefin polymerization[22-24]. The triarylborane polymers readily coordinate nucleophiles as exemplified in the formation of a soluble polymeric donor acceptor complex with crotonaldehyde. We are currently exploring the potential of these and of related organoboron polymers as supported catalysts and as building blocks in supramolecular polymer chemistry.

Acknowledgements

Acknowledgment is made to the donors of The Petroleum Research Fund, administered by the ACS, and the Rutgers University Research Council for financial support of this research. We also thank the National Science Foundation for GPC and LS instrumentation (NSF-MRI program).

[1] (a) Pomogailo, A. D.; Savost'yanov, V. S.; "Synthesis and polymerization of metal-containing monomers"; CRC Press: Boca Raton, 1994; (b) Interrante, L. V.; Hampden-Smith, M. J.; "Chemistry of advanced materials : an overview"; Wiley-VCH: New York, 1998; (c) Brook, M. A.; "Silicon in organic, organometallic, and polymer chemistry"; Wiley-VCH: New York, 2000; (d) Archer, R. D.; "Inorganic and organometallic polymers"; Wiley-VCH: New York, 2001; (e) Allcock, H. R.; "Chemistry and Applications of Polyphosphazenes"; Wiley-VCH: New York, 2002; (f) Manners, I.; "Synthetic Metal-Containing Polymers"; Wiley-VCH: New York, 2003.

[2] For the coordination of polymeric ligands to transition metals, see for example: (a) Fraser, C. L.; Smith, A. P. *J. Polym. Sci. A: Polym. Chem.* **2000**, *38*, 4704; (b) Lohmeijer, B. G. G.; Schubert, U. S. *Angew. Chem. Int. Ed.* **2002**, *41*, 3825.

[3] (a) De Vos, D. E.; Vankelecom, I. F. J.; Jacobs, P. A.; "Chiral Catalyst Immobilization and Recycling"; Wiley-VCH: New York, 2000; (b) Sherrington, D. C.; Kybett, A. C.; "Supported Catalysts and Their Applications"; Royal Society of Chemistry: Cambridge, 2001.

[4] See, for example: (a) Appleton, B.; Gibson, T. D. *Sensors and Actuators B* **2000**, *65*, 302; (b) Nicolas, M.; Fabre, B.; Simonet, J. *J. Electroanal. Chem.* **2001**, *509*, 73.

[5] Chung, T. C.; Janvikul, W. *J. Organomet. Chem.* **1999**, *581*, 176 and references therein.

[6] Boffa, L. S.; Novak, B. M. *Chem. Rev.* **2000**, *100*, 1479.

[7] Kondo, Y.; Garcia-Cuadrado, D.; Hartwig, J. F.; Boaen, N. K.; Wagner, N. L.; Hillmyer, M. A. *J. Am. Chem. Soc.* **2002**, *124*, 1164.

[8] See, for example: (a) Mehta, M. A.; Fujinami, T.; Inoue, T. *J. Power Sources* **1999**, *81-82*, 724; (b) Sun, X.; Angell, C. A. *Electrochimica Acta* **2001**, *46*, 1467; (c) Xiang, H.-Q.; Fang, S.-B.; Jiang, Y.-Y. *Solid State Ionics* **2002**, *148*, 35; (d) Matsumi, N.; Sugai, K.; Ohno, H. *Macromolecules* **2002**, *35*, 5731.

[9] See, for example: (a) Armitage, P.; Ebdon, J. R.; Hunt, B. J.; Jones, M. S.; Thorpe, F. G. *Polymer Degradation and Stability* **1996**, *54*, 387; (b) Gao, J.; Liu, Y.; Wang, F. *Eur. Polym. J.* **2001**, *37*, 207.

[10] See, for example: (a) Seyferth, D. *Adv. Chem. Ser.* **1995**, *245*, 131; (b) Riedel, R.; Kroke, E.; Greiner, A.; Gabriel, A. O.; Ruwisch, L.; Nicolich, J.; Kroll, P. *Chem. Mater.* **1998**, *10*, 2964; (c) Brunner, A. R.; Bujalski, D. R.; Moyer, E. S.; Su, K.; Sneddon, L. G. *Chem. Mater.* **2000**, *12*, 2770; (d) Weinmann, M.; Kamphowe, T. W.; Schuhmacher, J.; Müller, K.; Aldinger, F. *Chem. Mater.* **2000**, *12*, 2112; (e) Kho, J.-G.; Moon, K.-T.; Nouet, G.; Ruterana, P.; Kim, D.-P. *Thin Solid Films* **2001**, *389*, 78; (f) Keller, T. M. *Carbon* **2002**, *40*, 225.

[11] (a) Matsumi, N.; Naka, K.; Chujo, Y. *J. Am. Chem. Soc.* **1998**, *120*, 3112; (b) Matsumi, N.; Naka, K.; Chujo. Y. *J. Am. Chem. Soc.* **1998**, 10776.

[12] See, for example: (a) Letsinger, R. L.; Hamilton, S. B. *J. Am. Chem. Soc.* **1959**, *81*, 3009; (b) Lennarz, W. J.; Snyder, H. R. *J. Am. Chem. Soc.* **1960**, *82*, 2169; (c) Wulff, G.; Sarhan, A.; Gimpel, J.; Lohmar, E. *Chem. Ber.* **1974**, *107*, 3364; (d) Jackson, L. A.; Allen, C. W. *J. Polym. Sci., A* **1992**, *30*, 577.

[13] (a) Ramakrishnan, S.; Chung, T. C. *Macromolecules* **1989**, *22*, 3181; (b) Wolfe, P. S.; Wagener, K. B. *Macromolecules* **1999**, *32*, 7961.

[14] Paetzold, P.; Hoffmann, J. *Chem. Ber.* **1980**, *113*, 3724.

[15] Qin, Y.; Cheng, G.; Sundararaman, A.; Jäkle, F. *J. Am. Chem. Soc.* **2002**, *124*, 12672.

[16] Matyjaszewski, K.; Xia, J. H. *Chem. Rev.* **2001**, *101*, 2921.

[17] Poly(4-trimethylsilylstyrene) of higher molecular weight (M_w = 90,700) and polydispersity (*PDI* = 1.33) has been synthesized via ATRP with a different ligand system: McQuillan, B. W.; Paguio, S. *Fusion Technology* **2000**, *38*, 108; a kinetic analysis has not been reported.
[18] (a) Haubold, W.; Herdtle, J.; Gollinger, W.; Einholz, W. *J. Organomet. Chem.* **1986**, *315*, 1; (b) Kaufmann, D. *Chem. Ber.* **1987**, *120*, 853.
[19] Sundararaman, A.; Jäkle, F.; manuscript in preparation.
[20] Childs, R. F.; Mulholland, D. L.; Nixon, A. *Can. J. Chem.* **1982**, *60*, 801.
[21] Yamamoto, E. H.; "Lewis Acids in Organic Synthesis"; Wiley-VCH: New York, 2000.
[22] Chen, E. Y.-X.; Marks, T. J. *Chem. Rev.* **2000**, *100*, 1391.
[23] For an interesting related dendritic system decorated with up to 12 (perfluoroaryl)borane Lewis acid centers see: Roesler, R.; Har, B. J. N.; Piers, W. E. *Organometallics* **2002**, *21*, 4300.
[24] For the use of polymeric borates as catalyst components see, for example: (a) Roscoe, S. B.; Fréchet, J. M. J.; Walzer, J. F.; Dias, A. J. *Science* **1998**, *280*, 270. (b) Mager, M.; Becke, S.; Windisch, H.; Denninger, U. *Angew. Chem. Int. Ed.* **2001**, *40*, 1898. (c) Turner, H. W.; Patent US 5427991, 1995, Exxon Chemical Patents Inc. (d) Ono, M.; Hinokuma, S.; Miyake, S.; Inazawa, S.; Patent EP 710663, 1996, Japan Polyolefins Co., Ltd. (e) Kanamaru, M.; Okamoto, T.; Okuda, F.; Kamisawa, M.; Patent JP 2000212225, 2000, Idemitsu Petrochemical Co., Ltd.

Exploiting Catalytic Dehydrogenative Coupling in the Synthesis and Study of Polysilanes

Lisa Rosenberg

Department of Chemistry, University of Victoria, P.O. Box 3065, Victoria, BC, Canada V8W 3V6
E-mail: lisarose@uvic.ca

Summary: A reinvestigation of the catalytic activity of Wilkinson's catalyst, $(Ph_3P)_3RhCl$ (**1**), for the dehydrogenative coupling reactions of secondary silanes has pointed the way toward the synthesis of key oligosilane reagents for structure property correlation studies of polysilanes. Implications are discussed for understanding the mechanisms of coupling and redistribution reactions of silanes mediated by such late metal centres. Also described are attempts to derivatize Si-H bonds in the resulting oligosilanes, which is highly relevant to the development of improved methods for the post-polymerization functionalization of polysilanes containing Si-H bonds.

Keywords: catalysis; functionalization of polymers; oligomers; polysilanes, transition metal chemistry

Introduction

Polysilanes are receiving considerable attention for properties that may render them useful in a variety of electronic and photonic device applications. These properties are a function of σ–delocalization along the polysilicon chain, and may be correlated with *all-transoid*, or *anti*, conformations along the polymer backbone (Figure 1).[1] While the ability to tune polysilane conformations with precision and predictability will be essential for the advancement of this promising class of materials, achieving rigorous conformational control in the intrinsically flexible polysilane chains presents significant synthetic challenges. In developing methods to control and optimize conjugation in polysilanes it is desirable to assess the effects of varying steric constraints on conformational preferences at the Si–Si bonds of discrete oligomers. However, a dearth of general, straightforward routes to small molecules containing Si-Si bonds has limited our ability to prepare homologous series of oligosilanes with the requisite, systematic structural modifications.[2]

© 2003 WILEY-VCH Verlag GmbH & KGaA, Weinheim CCC 1022-1360/00/$ 17,50+.50/0

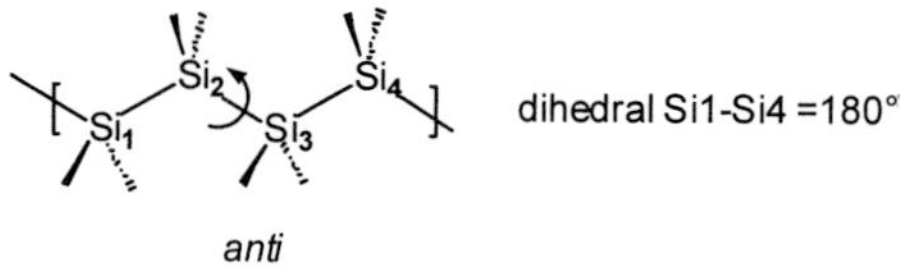

dihedral Si1-Si4 =180°

anti

Figure 1. Polysilane conformation corresponding to σ-delocalization

Recently, we have focused on the synthesis of short oligosilanes with functional groups at the terminal silicons, which will allow us to incorporate these oligomers into a range of structural environments (Figure 2). Although 1,2-dichlorotetramethyldisilane, a byproduct of the direct synthesis route to dichlorodimethylsilane, is commercially available, few other such functionalized di- or oligosilanes, with reactive Si-X bonds are commercially available or easily accessible by synthetic means.[3] Wurtz reductive coupling of dichloromonosilanes (R_2SiCl_2) is not a viable route to α,ω-dichlorooligosilanes, since these reactive species are inevitably consumed to give longer chains or cyclic compounds.[2] Although there is precedent for the synthesis of 1,2-dihydrido-substituted disilanes through reductive coupling of $R_2Si(H)Cl$, these reactions are capricious and are frequently low yielding.[4]

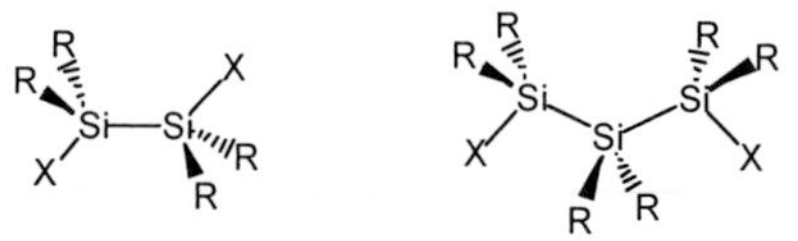

Figure 2. α,ω-Functionalized di- and trisilanes, where X = H, Cl, Br, OR', NR'_2.

Transition metal-catalyzed dehydrogenative coupling of hydrosilanes is a promising alternative to reductive coupling for the formation of Si-Si bonds (Eqn 1).[5] Since these catalysts are active principally for the coupling of primary (1°) or secondary (2°) silanes,[6] implicit in this method is the retention of reactive, terminal Si-H bonds in the catenated products.

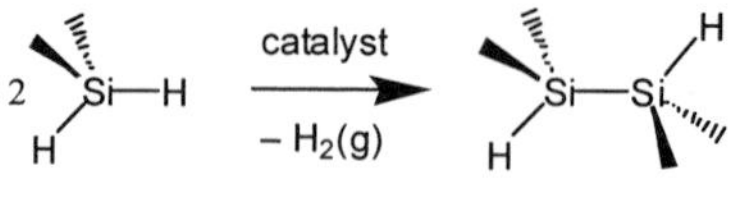

catalyst

$- H_2(g)$

(1)

Most prominent among catalysts capable of dehydrogenative coupling of silanes are group 4 metallocene systems, which can produce relatively high MW polymers (degree of polymerization as high as 70-100 monomer units) from 1° aryl silanes (Eqn 2).[7] These early metal catalysts will give short oligomers from 2° silanes with aryl substituents, but decreased activities for these reactions, relative to those of the 1° silane substrates, necessitate prolonged reaction times and high temperatures.[3b] In addition, the established group 4 catalysts exhibit little or no activity for the coupling of dialkylsilanes.[8] Although many late transition metal complexes can effect the coupling of both aryl- and alkyl-substituted 1° and 2° silanes,[5] they have received less attention in this context, since they typically exhibit competing catalytic activities for both dehydrocoupling and substituent redistribution reactions of the silane substrates.[9] Notably, however, the coupling products resulting from late metal catalysis are almost invariably limited to short chains (2-5 silicons).[10] We therefore decided to examine this class of catalysts more closely for their potential utility in the production, on a synthetically useful scale, of Si-H functionalized oligosilane reagents.

$$2n\ PhSiH_3 \xrightarrow[-\ nH_2(g)]{\text{group 4 metallocene}} H\!\left[\!-SiHPh-SiHPh-\right]_n\!H \qquad (2)$$

Coupling Reactions of 2° Silanes Catalyzed by Wilkinson's Catalyst

We have reexamined the reactions of 2° silanes in the presence of Wilkinson's catalyst, $(Ph_3P)_3RhCl$, **1**,[11] to identify optimal conditions (e.g. substrate, catalyst concentrations) to preferentially obtain the desired oligomers over redistribution products. We established a selective, high-yield route to 1,1,2,2-tetraphenyldisilane, **2**, via catalytic dehydrocoupling of diphenylsilane (Ph_2SiH_2), and, in doing so, identified general conditions for the coupling of a range of 2° silanes with little or no redistribution.[12] Our experiments demonstrated an extreme sensitivity of the chemoselectivity of this system to the rate at which the product dihydrogen is removed from reaction mixtures. Thus, production of di- and trisilanes is contingent on efficient removal of $H_2(g)$ from the reaction mixture, and thus continuously shifting a rapid, monomer-favoured equilibrium toward the desired products.[12] Because the equilibrium lies heavily toward monosilane, and because the rates of both coupling and hydrogenolysis are (apparently) extremely high, the presence of *any* residual hydrogen limits the amount of disilane produced, and, for aryl silanes (*vide infra*), allows the (slower) redistribution of substituents at silicon to occur (Scheme 1). The catalyst system is so sensitive to the local H_2 concentration that even the use of solvents in which the hydrogen has some limited solubility slows the production of coupled product relative to redistribution. Thus our dehydrocoupling reaction conditions include the use of neat substrate.

$2\ Ph_2SiH_2$

slow → $Ph_3SiH + PhSiH_3$

fast ⇌ $H\text{—}SiPh_2\text{—}SiPh_2\text{—}H + H_2(g)$

Scheme 1.

Along with mechanistic insights arising from this work (*vide infra*), we discovered that **1** has reasonably high activity for the dehydrogenative coupling of dialkylsilanes (R_2SiH_2, R = Et, *n*-Hex) to di- and trisilanes. Although the coupling turnover frequencies are slightly lower for these substrates than for Ph_2SiH_2, we observe *no* redistribution products in the reaction mixtures. This is in agreement with our experiments using $MePhSiH_2$ as substrate, in which we see facile transfer of the Si-Ph group (yielding redistribution products Ph_2MeSiH and $MeSiH_3$) relative to the Si-Me group (giving $PhMe_2SiH$ and $PhSiH_3$). It is also consistent with other reports of the relative ease of migration of groups at silicon, in the presence of transition metals.[10a,b,13] This Rh(I) catalyst system will allow the preparation of oligosilanes with a broad range of substituents, under mild conditions and at low catalyst concentrations.

Mechanistic Implications

The mechanisms of the coupling and redistribution reactions of silanes at late metal centres remain a subject of some debate.[14] Our work clearly establishes distinct rate regimes for the two reaction types. That redistribution does *not* compete with coupling at a Rh(I) catalyst, under conditions allowing kinetic control (efficient $H_2(g)$ removal), was previously unrecognized, and should now provide insight in searching for likely catalytic intermediates (or for relevant, turnover-limiting steps) in these reactions. Our enhanced understanding of dehydrocoupling and redistribution processes has positioned us to further probe mechanisms for these reactions occurring at **1**. We are now examining, through catalyst modification, such issues as the role/fate of the chloride ligand in these reactions, the possibility of mononuclear versus dinuclear active species, and the precise nature of the Si-Rh interaction in the active species. This work will be reported in future publications.

Derivatizing Si-H Bonds in Oligosilanes

Another area of focus for us is the derivatization of the Si-H bonds in our coupled oligosilanes, to yield more substitutionally labile Si-X bonds (where X = Cl, Br, OR, NR_2) via halogenation, alcoholysis, or aminolysis reactions. This expands the range of methods available to us for the incorporation of these catenated units into conformationally constrained macrostructures. These studies are also helping us to refine techniques for the post-polymerization modification of higher MW polysilanes containing Si-H bonds.[15]

We are particularly interested in accessing Si-OR or Si-NR_2 oligosilane derivatives, since, while these functional groups are readily substituted by nucleophiles such as Grignards or alkyllithiums, they should exhibit milder reactivity and decreased hydrolytic sensitivity relative to analogous halide compounds. Catalytic dehydrogenative heterocoupling of monosilanes with alcohols or amines is known to be catalyzed by **1** and other late metals (Eqn 3),[5b] although we know of no examples of alcoholysis or aminolysis of Si-H bonds in compounds containing Si-Si bonds.

$$\text{Si—H} + \text{HX} \xrightarrow[-\,H_2(g)]{\text{catalyst}} \text{Si—X} \qquad \boxed{X = OR, NR_2} \qquad (3)$$

Nevertheless, based on this well established catalytic heterodehydrogenation chemistry, we envisaged a simple one-pot procedure for the preparation of 1,2-bis(ethoxy)tetraphenyldisilane, whereby, upon completion of the coupling reaction producing 1,1,2,2-tetraphenyldisilane, we could simply add ethanol to the catalytic reaction mixture. Conversion of Si-H to Si-OEt bonds should then proceed rapidly in the presence of residual Rh catalyst. We first tested this idea by adding ethanol to 1,1,2,2-tetraphenyldisilane, in the presence of **1**. However, instead of the desired bis(alkoxy)disilane, we obtained mixtures of two monosilanes: $Ph_2Si(OEt)H$ and $Ph_2Si(OEt)_2$ (Scheme 2), in varying ratios depending on the reaction time and the extent to which ethanol was added in excess.[16] Clearly, once $H_2(g)$ is formed from the heterodehydrocoupling reaction, a monosilane-disilane equilibrium is once again accessible, in the presence of Rh catalyst. Avoiding the competition of alcoholysis and catalytic Si-Si bond cleavage is a significant challenge, which we continue to address through the screening of other classes of alcoholysis catalysts,[17] including Lewis acids.[18]

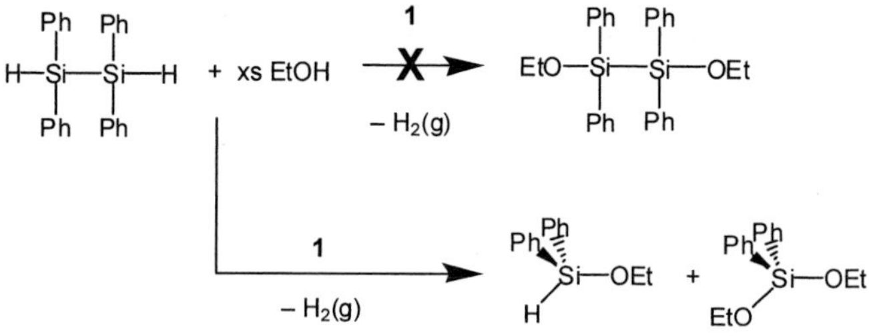

Scheme 2.

In the meantime, we have begun to use established methods for Si-H bond halogenation on our di and trisilanes. For example, we have chlorinated 1,1,2,2-tetraphenyldisilane to give 1,2-dichlorotetraphenyldisilane using chlorine gas,[3a] as confirmed by characterization of its 1,2-dimethyl derivative, obtained from the *in situ* addition of methyl Grignard (Eqn 4).[19]

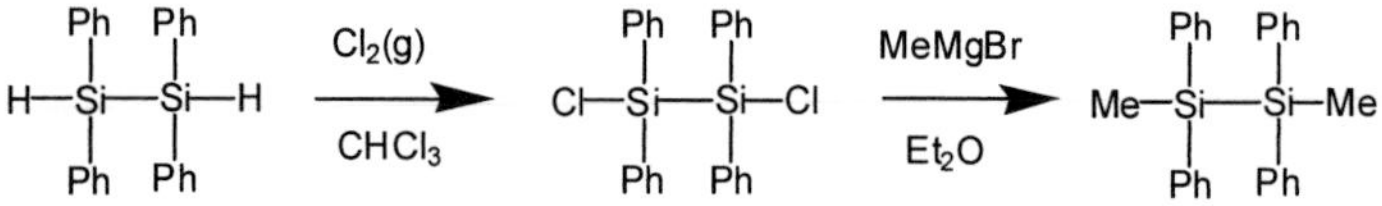

(4)

Conclusion

Our discoveries so far have pointed us toward understanding the challenges and conditions for preparing reactive oligosilane fragments via catalytic dehydrogenative coupling. They also indicate great potential for selectivity, in dehydrocoupling reactions catalyzed by **1**, through a careful balance of catalyst concentration and the method of removal of H_2(g). These studies are laying the groundwork for us to develop general routes to new classes of polysilanes with enhanced conformational preferences.

[1] R. D. Miller, J. Michl, *Chem. Rev.* **1989**, *89*, 1359.
[2] For leading references on the synthesis of small molecules containing Si-Si bonds, see M. A. Brook, "*Silicon in Organic, Organometallic, and Polymer Chemistry*" John Wiley & Sons, Inc., New York 2000, p.344.

[3] Examples of the preparation of *functionalized* oligosilanes include: (a) K. H. Pannell, J. M. Rozell, C. Hernandez, *J. Am. Chem. Soc.* **1989**, *111*, 4482; (b) J. Y. Corey, D. M. Kraichely, J. L. Huhmann, J. Braddock-Wilking, A. Lindeberg, *Organometallics* **1995**, *14*, 2704; (c) H. Sakurai, Y. Eriyama, Y. Kamiyama, Y. Nakadaira, *J. Organomet. Chem* **1984**, *264*, 229; (d) J. Zech, H. Schmidbaur, *Chem. Ber.* **1990**, *123*, 2087; (e) M. Söldner, A. Schier, H. Schmidbaur, *J. Organomet. Chem.* **1996**, *521*, 295
[4] See for example: H. J. S. Winkler, H. Gilman, *J. Org. Chem.* **1961**, *26*, 1265.
[5] (a) T. D. Tilley, *Comments Inorg. Chem.* **1990**, *10*, 37; (b) J. Y. Corey, in: "*Dehydrogenative Coupling Reactions of Hydrosilanes*", G. L. Larson, Ed., JAI Press Inc., 1991. Vol. 1, 327; (c) F. Gauvin, J. F. Harrod, H. G. Woo, *Adv. Organomet. Chem.* **1998**, *42*, 363; (d) J. A. Reichl, D. H. Berry, *Adv. Organomet. Chem.* **1999**, *43*, 197.
[6] An exception is the reported activity of a Pt catalyst in the coupling of 3° silanes to hexa-organo-substituted disilanes. M. Tanaka, T. Kobayashi, T. Hayashi, T. Sakakura, *Appl. Organomet. Chem.* **1988**, *2*, 91.
[7] (a) T. Imori, T. D. Tilley, *Polyhedron* **1994**, *13*, 2231; (b) V. K. Dioumaev, J. F. Harrod, *J. Organomet. Chem.* **1996**, *521*, 133.
[8] The activity of a titanocene-based catalyst for the coupling of $(n\text{-Pr})_2SiH_2$ has been reported, but requires the presence of cyclic olefins, which are hydrogenated in tandem with the coupling process. Complex product mixtures result, which include hydrosilation products. J. Y. Corey, X.-H. Zhu, *Organometallics* **1992**, *11*, 672.
[9] M. D. Curtis, P. S. Epstein, *Adv. Organomet. Chem.* **1981**, *19*, 213.
[10] See for example (a) M. D. Fryzuk, L. Rosenberg, S. J. Rettig, *Inorg. Chim. Acta* **1994**, *222*, 345. There are very few reports of late metal activities for the production of longer (DP>6) silicon chains. These include: (b) B. P. S. Chauhan, T. Shimizu, M. Tanaka, *Chem. Lett.* **1997**, 785; (c) F.-G. Fontaine, T. Kadkhodazadeh, D. Zargarian, *Chem. Commun.* **1998**, 1253.
[11] The reactivity of **1** and other Rh(I) phosphine complexes for dehydrocoupling of silanes, albeit invariably accompanied by redistribution reactions, was already established. (a) I. Ojima, S.-I. Inaba, T. Kogure, *J. Organomet. Chem.* **1973**, *55*, C7; (b) M. F. Lappert, R. K. Maskell, *J. Organomet. Chem.* **1984**, *264*, 217; (c) J. Y. Corey, L. S. Chang, E. R. Corey, *Organometallics* **1987**, *6*, 1595; (d) K. A. Brown-Wensley, *Organometallics* **1987**, *6*, 1590; (e) L. Rosenberg, M. D. Fryzuk, S. J. Rettig, *Organometallics* **1999**, *18*, 958.
[12] L. Rosenberg, C. W. Davis, J. Yao, *J. Am. Chem. Soc.* **2001**, *123*, 5120.
[13] P. Burger, R. G. Bergman, *J. Am. Chem. Soc.* **1993**, *115*, 10462.
[14] The debate centres around whether a metal silylene ("M=Si") intermediate is involved in (or required for) Si-Si bond formation, or whether oxidative addition and reductive elimination steps involving only M-Si single bonds can be responsible for Si-Si coupling. See Reference 5.
[15] Our preliminary attempts, using methods described in references b-d below, to halogenate the Si-H bonds in poly(phenylsilane), $H\text{-}[SiPh(H)]_n\text{-}H$, have resulted either in low or incomplete substitution (Reference 15b) or in significant polymer degradation.(References 15c,d). (a) D. J. Harrison, L. Rosenberg, unpublished results; (b) J. P. Banovetz, Y.-L. Hsiao, R. M. Waymouth, *J. Am. Chem. Soc.* **1993**, *115*, 2540; (c) Reference 3a; (d) D. M. Friesen, R. McDonald, L. Rosenberg, *Can. J. Chem.* **1999**, *77*, 1931.
[16] J. Yao, L. Rosenberg, unpublished results.
[17] E. Lukevics, M. Dzintara, *J. Organomet. Chem.* **1985**, *295*, 265.
[18] J. M. Blackwell, K. L. Foster, V. H. Beck, W. E. Piers, *J. Org. Chem.* **1999**, *64*, 4887.
[19] (a) D. J. Harrison, L. Rosenberg, unpublished results. (b) An alternative technique for this halogenation involves the use of 4 equivalents of $CuCl_2$ in the presence of catalytic amounts of CuI; A. Kunai, T. Kawakami, A. Toyoda, M. Ishikawa, *Organometallics* **1992**, *11*, 2708.